INTRODUCTION TO THE THEORY OF RELATIVITY

by

PETER GABRIEL BERGMANN

PROFESSOR OF PHYSICS
SYRACUSE UNIVERSITY

WITH A FOREWORD

BY

ALBERT EINSTEIN

DOVER PUBLICATIONS, INC.
NEW YORK

Published in Canada by General Publishing Company, Ltd., 30 Lesmill Road, Don Mills, Toronto, Ontario.
Published in the United Kingdom by Constable and Company, Ltd.

This Dover edition, first published in 1976, is a corrected and enlarged republication of the work originally published by Prentice-Hall, Inc., Englewood Cliffs, New Jersey, in 1942. The author has written, especially for the Dover edition, a new Preface; Appendix A: "Ponderomotive Theory by Surface Integrals"; and Appendix B: "Supplementary Notes."

International Standard Book Number: 0-486-63282-2
Library of Congress Catalog Card Number: 75-32903

Manufactured in the United States of America
Dover Publications, Inc.
180 Varick Street
New York, N.Y. 10014

Foreword by Albert Einstein

Although a number of technical expositions of the theory of relativity have been published, Dr. Bergmann's book seems to me to satisfy a definite need. It is primarily a textbook for students of physics and mathematics, which may be used either in the classroom or for individual study. The only prerequisites for reading the book are a familiarity with calculus and some knowledge of differential equations, classical mechanics, and electrodynamics.

This book gives an exhaustive treatment of the main features of the theory of relativity which is not only systematic and logically complete, but also presents adequately its empirical basis. The student who makes a thorough study of the book will master the mathematical methods and physical aspects of the theory of relativity and will be in a position to interpret for himself its implications. He will also be able to understand, with no particular difficulty, the literature of the field.

I believe that more time and effort might well be devoted to the systematic teaching of the theory of relativity than is usual at present at most universities. It is true that the theory of relativity, particularly the general theory, has played a rather modest role in the correlation of empirical facts so far, and it has contributed little to atomic physics and our understanding of quantum phenomena. It is quite possible, however, that some of the results of the general theory of relativity, such as the general covariance of the laws of nature and their nonlinearity, may help to overcome the difficulties encountered at present in the theory of atomic and nuclear processes. Apart from this, the theory of relativity has a special appeal because of its inner consistency and the logical simplicity of its axioms.

Much effort has gone into making this book logically and pedagogically satisfactory, and Dr. Bergmann has spent many hours with me which were devoted to this end. It is my hope that many students will enjoy the book and gain from it a better understanding of the accomplishments and problems of modern theoretical physics.

A. Einstein

The Institute for Advanced Study

Preface to the Dover Edition

THIS book was first published by Prentice-Hall, Inc., in 1942. The new Dover edition reproduces and expands the original text. In the intervening three decades entirely new aspects of the theory of relativity have been opened up, and related laboratory and astronomical investigations have led to new discoveries as well. Together with my co-workers and students I have taken part in this research. My wish to revise the book completely has had to be postponed again and again.

For this edition I have derived anew, in Appendix *A*, the laws of motion of ponderable bodies. Whereas the pioneering work of Einstein, Infeld, and Hoffmann relied on a weak-field slow-motion approximation, the new approach, which was originated by my colleague and former student J. N. Goldberg and by myself, is based on the full field equations and leads to rigorous relations between surface integrals. In Appendix *B* I have collected a number of brief notes that deal with recent progress in specific areas discussed in the original text; some include selected references to the literature. Footnotes added to the text indicate entries in Appendix *B*.

For anyone seriously concerned with the foundations of physics I consider the study of relativity indispensable. Twice in the history of physics gravitation has been crucial. When Newton related the laws of free fall to the laws of planetary motion, he established physics as an exact science. Three hundred years later Einstein revolutionized our concepts of space and time with his new theory of gravitation. These new concepts are still evolving. No doubt they will influence the development of physics for a long time to come, though their ultimate impact remains conjectural. I hope that my book will continue to be useful as an introduction to the fundamentals of relativity.

July, 1975 P. G. B.

Contents

PART I

THE SPECIAL THEORY OF RELATIVITY

PART III

UNIFIED FIELD THEORIES

Preface to the First Edition

This book presents the theory of relativity for students of physics and mathematics who have had no previous introduction to the subject and whose mathematical training does not go beyond the fields which are necessary for studying classical theoretical physics. The specialized mathematical apparatus used in the theory of relativity, tensor calculus, and Ricci calculus, is, therefore, developed in the book itself. The main emphasis of the book is on the development of the basic ideas of the theory of relativity; it is these basic ideas rather than special applications which give the theory its importance among the various branches of theoretical physics.

The material has been divided into three parts, the special theory of relativity, the general theory of relativity, and a report on unified field theories. The three parts form a unit. The author realizes that many students are interested in the theory of relativity mainly for its applications to atomic and nuclear physics. It is hoped that these readers will find in the first part, on the special theory of relativity, all the information which they require. Those readers who do not intend to go beyond the special theory of relativity may omit one section of Chapter V (p. 67) and all of Chapter VIII; these passages contain material which is needed only for the development of the general theory of relativity.

The second part treats the general theory of relativity, including the work by Einstein, Infeld, and Hoffmann on the equations of motion. The third part deals with several attempts to overcome defects in the general theory of relativity. None of these theories has been completely satisfactory. Nevertheless, the author believes that this report rounds out the discussion of the general theory of relativity by indicating possible directions of future research. However, the third part may be omitted without destroying the unity of the remainder.

The author wishes to express his appreciation for the help of Professor Einstein, who read the whole manuscript and made many valuable suggestions. Particular thanks are due to Dr. and Mrs. Fred Fender, who read the manuscript carefully and suggested many stylistic and other improvements. The figures were drawn by Dr. Fender. Margot Bergmann read the manuscript, suggested improvements, and did almost all of the technical work connected with the preparation of the manuscript. The friendly co-operation of the Editorial Department of Prentice-Hall, Inc. is gratefully acknowledged.

P. G. B.

Introduction

Almost all the laws of physics deal with the behavior of certain objects in space in the course of time. The position of a body or the location of an event can be expressed only as a location relative to some other body suitable for that purpose. For instance, in an experiment with Atwood's machine, the velocities and accelerations of the weights are referred to the machine itself, that is, ultimately to the earth. An astronomer may refer the motion of the planets to the center of gravity of the sun. All motions are described as motions relative to some reference body.

We imagine that conceptually, at least, a framework of rods which extends into space can be rigidly attached to the reference body. Using this conceptual framework as a Cartesian coördinate system in three dimensions, we characterize any location by three numbers, the coördinates of that space point. Such a conceptual framework, rigidly connected with some material body or other well-defined point, is often called a *frame of reference.*

Some bodies may be suitable as reference bodies, others may not. Even before the theory of relativity was conceived, the problem of selecting suitable frames of reference played an important part in the development of science. Galileo, the father of post-medieval physics, considered the choice of the heliocentric frame to be so important that he risked imprisonment and even death in his efforts to have the new frame of reference accepted by his contemporaries. In the last analysis, it was the choice of the reference body which was the subject of his dispute with the authorities.

Later, when Newton gave a comprehensive presentation of the physics of his time, the heliocentric frame of reference had been generally accepted. Still, Newton felt that further discussion was necessary. To show that some frames of reference were more suitable for the description of nature than others, he devised the famous pail experiment: He filled a pail with water. By twisting the rope which supported the pail, he made it rotate around its axis. As the water gradually began tc participate in the rotation, its surface changed from a plane to a paraboloid. After the water had gained the same speed of rotation as the pail, he stopped the pail. The water slowed down and eventually came to complete rest. At the same time, its surface resumed the shape of a plane.

The description given above is based on a frame of reference connected with the earth. The law governing the shape of the water's surface could be formulated thus: The surface of the water is a plane whenever the water does not rotate. It is a paraboloid when the water rotates. The state of motion of the pail has no influence on the shape of the surface.

Now let us describe the whole experiment in terms of a frame of reference rotating relatively to the earth with a constant angular velocity equal to the greatest velocity of the pail. At first, the rope, the pail, and the water "rotate" with a certain constant angular velocity with respect to our new frame of reference, and the surface of the water is a plane. Then the rope, and in turn the pail, is "stopped," and the water gradually "slows down," while its surface becomes a paraboloid. After the water has come to a "complete rest," its surface still a paraboloid, the rope, and in turn the pail, is again made to "rotate" relatively to our frame of reference (that is, stopped with respect to the earth); the water gradually begins to participate in the "rotation," while its surface flattens out. In the end, the whole apparatus is "rotating" with its former angular velocity, and the surface of the water is again a plane. With respect to this frame of reference, the law would have to be formulated like this: Only when the water "rotates" with a certain angular velocity, is its surface a plane. The deviation from a plane increases with the deviation from this particular state of motion. The state of rest produces also a paraboloid. Again the rotation of the pail is immaterial.

Newton's pail experiment brings out very clearly what is meant by "suitable" frame of reference. We can describe nature and we can formulate its laws using whatever frame of reference we choose. But there may exist a frame or frames in which the laws of nature are fundamentally simpler, that is, in which the laws of nature contain fewer elements than they would otherwise. Take the instance of Newton's rotating pail. If our description of nature were based on the frame of reference connected with the pail, many physical laws would have to contain an additional element, the angular velocity ω of the pail relative to a "more suitable" frame of reference, let us say to the earth.

The laws of motion of the planets become basically simpler when they are expressed in terms of the heliocentric frame of reference instead of the geocentric frame. That is why the description of Copernicus and

Galileo won out over that of Ptolemy, even before Kepler and Newton succeeded in formulating the underlying laws.

Once it was clearly recognized that the choice of a frame of reference determined the form of a law of nature, investigations were carried out which established the effect of this choice in a mathematical form.

Mechanics was the first branch of physics to be expressed in a complete system of mathematical laws. Among all the frames of reference conceivable, there exists a set of frames with respect to which the law of inertia takes its familiar form: In the absence of forces, the space coördinates of a mass point are linear functions of time. These frames of reference are called *inertial systems.* It was found that all of the laws of mechanics take the same form when stated in terms of any one of these inertial systems. Another frame of reference necessitates a more involved physical and mathematical description, for example, the frame of reference connected with Newton's rotating pail. The characterization of the motions of mass points not subject to forces is possible in terms of this frame of reference, but the mathematical form of the law of inertia is involved. The space coördinates are not linear functions of time.

Since the laws of mechanics take the same form in all frames of reference which are inertial systems, all inertial systems are equivalent from the point of view of mechanics. We can find out whether a given body is "accelerated" or "unaccelerated" by comparing its motion with that of some mass point which is not subject to any forces. But whether a body is "at rest" or "in uniform motion" depends entirely on the inertial system used for the description; the terms "at rest" and "in uniform motion" have no absolute meaning. The principle that all inertial systems are equivalent for the description of nature is called the *principle of relativity.*

When Maxwell developed the equations of the electromagnetic field, these equations were apparently incompatible with the principle of relativity. For, according to this theory, electromagnetic waves in empty space should propagate with a universal, constant velocity c of about 3×10^{10} cm/sec, and this, it appeared, could not be true with respect to both of two different inertial systems which were moving relatively to each other. The one frame of reference with respect to which the speed of electromagnetic radiation would be the same in all directions could be used for the definition of "absolute rest" and of "absolute motion." A number of experimenters tried hard to find this frame of reference and to determine the earth's motion with respect to it.

All these attempts, however, were unsuccessful. On the contrary, all experiments seemed to suggest that the principle of relativity applied to the laws of electrodynamics as well as to those of mechanics. H. A. Lorentz proposed a new theory, in which he accepted the existence of one privileged frame of reference, and at the same time explained why this frame could not be discovered by experimental methods. But he had to introduce a number of assumptions which could not have been checked by any conceivable experiment. To this extent his theory was not very satisfactory. Einstein finally recognized that only a revision of our fundamental ideas about space and time would resolve the impasse between theory and experiment. Once this revision had been made, the principle of relativity was extended to the whole of physics. This is now called the special theory of relativity. It establishes the fundamental equivalence of all inertial systems. It preserves fully their privileged position among all conceivable frames of reference. The so-called general theory of relativity analyzes and thereby destroys this privileged position and is able to give a new theory of gravitation.

In this book we shall first discuss the role of different frames of reference, from a classical point of view, in mechanics and to some extent in electrodynamics. Only when the student understands fully the deadlock between theoretical conclusions and experimental results in classical electrodynamics can he appreciate the necessity of revising classical physics along relativistic lines. Once the new ideas of space and time are grasped, "relativistic mechanics" and "relativistic electrodynamics" are easily understood.

The second part of this book is devoted to the general theory of relativity, while the third part discusses some recent attempts to extend the theory of gravitation to the field of electrodynamics.

PART I

The Special Theory of Relativity

CHAPTER I

Frames of Reference, Coördinate Systems, and Coördinate Transformations

We have spoken of frames of reference and have mentioned Cartesian coördinate systems. In this chapter we shall examine more closely the relationships between different frames of reference and different coördinate systems.

Coördinate transformations not involving time. As a specific instance, let us consider a frame of reference connected rigidly with the earth, that is, a geocentric frame of reference. In order to express quantitatively the location of a point relative to the earth, we introduce a coördinate system. We choose a point of origin, let us say the center of the earth, and directions for the three axes; for instance, the X-axis may go from the earth's center through the intersection of the equator and the Greenwich meridian, the Y-axis through the intersection of the equator and the 90°E-meridian, and the Z-axis through the North Pole. The location of any point is then given by three real numbers, the coördinates of that point. The motion of a point is completely described if we express the three point coördinates as functions of time. A point is at rest relatively to our frame of reference if these three functions are constant.

Without abandoning the earth as the body with which our frame of reference is rigidly connected, we can introduce another coördinate system. We may, for instance, choose as the point of origin some well-defined point on the earth's surface, let us say one of the markers of the United States Coast and Geodetic Survey; and as the direction for the X-axis the direction due East; for the Y-axis, the direction due North; and for the Z-axis, the direction straight up, away from the earth's center, the earth assumed to be a sphere.

The relationship between the two coördinate systems is completely determined if the coördinates of any given point with respect to one coördinate system are known functions of its coördinates with respect to the other coördinate system. Let us call the first coördinate system S

and the second coördinate system S', and the coördinates of a certain point P with respect to S (x, y, z) and the coördinates of the same point with respect to S' (x', y', z'). Then, x', y', and z' are connected with x, y, and z by equations of the form:

$$\left.\begin{aligned} x' &= c_{11}x + c_{12}y + c_{13}z + \mathring{x}', \\ y' &= c_{21}x + c_{22}y + c_{23}z + \mathring{y}', \\ z' &= c_{31}x + c_{32}y + c_{33}z + \mathring{z}'. \end{aligned}\right\} \tag{1.1}$$

$(\mathring{x}', \mathring{y}', \mathring{z}')$ are the coördinates of the point of origin of S with respect to S'. The constants c_{ik} are the cosines of the angles between the axes of S and S', c_{11} referring to the angle between the X- and the X'-axis, c_{12} to the angle between the Y- and the X'-axis, c_{21} to the angle between the X- and the Y'-axis, and so forth.

The transition from one coördinate system to another is called a *coördinate transformation*, and the equations connecting the point coordinates of the two coördinate systems are called *transformation equations.*

A coördinate system is necessary not only for the description of locations, but also for the representation of vectors. Let us consider some vector field, for example, an electrostatic field, in the neighborhood of the point P. The value and direction of the field strength $\mathbf{E}$ at P is completely determined when we know the components of $\mathbf{E}$ with respect to some stated coördinate system S. Let us call the components of $\mathbf{E}$ at P with respect to S, E_x, E_y, and E_z. The components of $\mathbf{E}$ at P with respect to another system, for instance S', can be computed if we know the transformation equations defining the coördinate transformation S into S'. These new components, E'_x, E'_y, and E'_z, are independent of the translation of the point of origin, that is, the constants x'_0, y'_0, and z'_0 of (1.1). E'_x is the sum of the projections of E_x, E_y, and E_z on the X'-axis, and E'_y and E'_z are determined similarly,

$$\begin{aligned} E'_x &= c_{11}E_x + c_{12}E_y + c_{13}E_z, \\ E'_y &= c_{21}E_x + c_{22}E_y + c_{23}E_z, \\ E'_z &= c_{31}E_x + c_{32}E_y + c_{33}E_z. \end{aligned}$$

A law which expresses the components of a certain quantity at a point in terms of the components of the same quantity at the same point with respect to another coördinate system is called a *transformation law.*

Coördinate transformations involving time. We have thus far considered only transformations which lead from one coördinate system to

another one rigidly connected with the same reference body, such as the earth. But coördinate transformation offers an important method for investigating the relationship between two different frames of reference which move relatively to each other. In such a case, we represent each of the two frames by one coördinate system.

Let us compare a frame of reference rigidly connected with the earth and another one connected with Newton's pail, which we assume is rotating with constant angular velocity. We can introduce two coördinate systems which enable us to describe quantitatively the location of any point with respect to either frame of reference. Let us call these two coördinate systems S (this S is not identical with the former S) and S^*, respectively, and let us choose the points of origin so that they both lie on the axis of the pail and coincide; the Z-axis and the Z^*-axis may be identical and pointing straight up. If the pail rotates with a constant angular velocity ω relative to the earth, and if at the time $t = 0$ the X-axis is parallel to the X^*-axis, the coördinate transformation equations take the form

$$\left.\begin{aligned} x^* &= \cos \omega t \cdot x + \sin \omega t \cdot y, \\ y^* &= -\sin \omega t \cdot x + \cos \omega t \cdot y, \\ z^* &= z. \end{aligned}\right\} \tag{1.2}$$

(Fig. 1)

Eqs. (1.2) have a form similar to eqs. (1.1), except that the cosines are no longer constant, but functions of time. The relative motion of the two frames of reference expresses itself in the functional dependence of the c_{ik} on time.

Eqs. (1.2) express the relationship between two frames of reference which are rotating relatively to each other. Very often we are interested in the relationship between two frames of reference which are in a state of uniform, translatory motion relative to each other. In that case, it is convenient to choose the two coördinate systems S and S^* so that their corresponding axes are parallel to each other and so that the points of origin coincide at the time $t = 0$. The transformation equations have the form:

$$\left.\begin{aligned} x^* &= x - v_x t, \\ y^* &= y - v_y t, \\ z^* &= z - v_z t, \end{aligned}\right\} \tag{1.3}$$

where v_x, v_y, and v_z are the components of the velocity of S^* relatively to S.

The form of the transformation equations (1.2) and (1.3) depends, o

course, on the relative motion of the two frames of reference, but it also depends on certain assumptions regarding the nature of time and space: We assume that it is possible to define a time t independently of any particular frame of reference, or, in other words, that it is possible to build clocks which are not affected by their state of motion. This

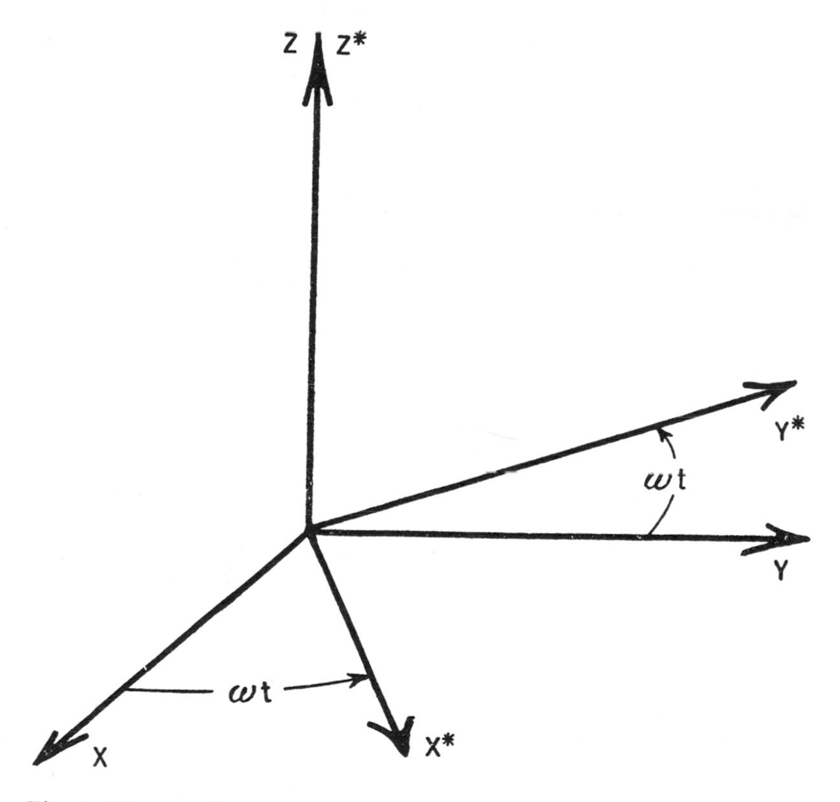

Fig. 1. The coördinate system S^* with the coördinates (x^*, y^*, z^*) rotates relatively to the coördinate system S with the coördinates (x, y, z) with the angular velocity ω.

assumption is expressed in our transformation equations by the absence of a transformation equation for t. If we wish, we can add the equation

$$t^* = t, \tag{1.4}$$

expressing the universal character of time explicitly.

The other assumption concerns length measurements. We assume that the distance between two points—they may be particles—at a given time is quite independent of any particular frame of reference;

that is, we assume that we can construct rigid measuring rods whose length is independent of their state of motion. Eqs. (1.3) show with particular clarity how this assumption is expressed by the form of the transformation equations. For the distance between two points P_1 and P_2 with the coördinates (x_1, y_1, z_1) and (x_2, y_2, z_2) is

$$s_{12} = \sqrt{(x_2 - x_1)^2 + (y_2 - y_1)^2 + (z_2 - z_1)^2}, \tag{1.5}$$

and obviously

$$\sqrt{(x_2 - x_1)^2 + (y_2 - y_1)^2 + (z_2 - z_1)^2} = \sqrt{(\overset{*}{x}_2 - \overset{*}{x}_1)^2 + (\overset{*}{y}_2 - \overset{*}{y}_1)^2 + (\overset{*}{z}_2 - \overset{*}{z}_1)^2} \tag{1.6}$$

is satisfied for any time t.

We shall have to consider these assumptions again at a later time.

CHAPTER II

Classical Mechanics

The law of inertia, inertial systems. The branch of physics which from the first was most consistently developed as an experimental science was Galilean-Newtonian mechanics. The first law to be formulated was the law of inertia: *Bodies when removed from interaction with other bodies will continue in their states of rest or straight-line uniform motion. In other words, the motion of such bodies is unaccelerated.*

To express the law of inertia in mathematical form, we designate the location of a body by its three coördinates, x, y, and z. When a body is not at rest, its coördinates are functions of time. According to the law of inertia, the second time derivatives of these three functions, the accelerations, vanish when the body is not subjected to forces, that is,

$$\ddot{x} = 0, \qquad \ddot{y} = 0, \qquad \ddot{z} = 0. \tag{2.1}$$

We use the usual notation $\ddot{x}$ for $\frac{d^2x}{dt^2}$. The first integral of eqs. (2.1) expresses the constancy of the three velocity components,

$$\dot{x} = \mathring{u}_x\,, \qquad \dot{y} = \mathring{u}_y\,, \qquad \dot{z} = \mathring{u}_z\,. \tag{2.2}$$

The equations expressing the law of inertia contain coördinates and refer, therefore, to a certain coördinate system. As long as this coördinate system is not specified, the italicized statement does not have a precise meaning. For, given any body, we can always introduce a frame of reference with respect to which it is at rest and, therefore, unaccelerated. The real assertion is, rather: *There exists a coördinate system (or coördinate systems) with respect to which all bodies not subjected to forces are unaccelerated.* Coördinate systems with this property and the frames of reference represented by them are called *inertial systems.*

Of course, not all frames of reference are inertial systems. For instance, let us start out from an inertial coördinate system S, and carry out a transformation (1.2), leading to S^*, a system rotating with a constant angular velocity ω relative to S. In order to obtain the transformation laws of eqs. (2.1) and (2.2), we differentiate the transformation equations (1.2) once and then a second time with respect to t. The resulting equations contain x, y, z, x^*, y^*, z^*, and the first and second time derivatives of these quantities.

We assumed that the coördinate system S is an inertial system. We substitute, therefore, for $\ddot{x}$, $\ddot{y}$, and $\ddot{z}$ and for $\dot{x}$, $\dot{y}$, and $\dot{z}$ the expressions (2.1) and (2.2) respectively. Thus, we obtain for the starred coördinates and their derivatives

$$\left.\begin{aligned} \dot{x}^* &= \omega y^* + \mathring{u}_x \cos \omega t + \mathring{u}_y \sin \omega t, \\ \dot{y}^* &= -\omega x^* + \mathring{u}_y \cos \omega t - \mathring{u}_x \sin \omega t, \\ \dot{z}^* &= \mathring{u}_z, \end{aligned}\right\} \tag{2.3}$$

and

$$\left.\begin{aligned} \ddot{x}^* &= \omega^2 x^* + 2\omega \dot{y}^*, \\ \ddot{y}^* &= \omega^2 y^* - 2\omega \dot{x}^*, \\ \ddot{z}^* &= 0. \end{aligned}\right\} \tag{2.4}$$

It turns out that in the coördinate system S^* the second time derivatives do not all vanish. Occasionally it is desirable to work with frames of reference in which accelerations occur which are not caused by real interactions between bodies. These accelerations, multiplied by the masses, are treated like real forces, often called "transport forces," "inertial forces," and so forth. In spite of these names, these expressions are not actual forces; they merely appear in the equations formally in the same way as forces do. In our case, the first terms, $\omega^2 x^*$, $\omega^2 y^*$, multiplied by the mass, are called "centrifugal forces," and the last terms, also multiplied by the mass, are the so-called Coriolis forces.

On the other hand, there are also types of coördinate transformations which leave the form of the law of inertia (2.1) unchanged. As a case in point, we shall consider first a transformation which involves no transition to a new frame of reference, of the type (1.1). The differentiation of eqs. (1.1) with substitution of $\dot{x}$, $\ddot{x}$, and so forth, from eqs. (2.1) and (2.2) produces the equations

$$\left.\begin{aligned} \dot{x}' &= c_{11}\mathring{u}_x + c_{12}\mathring{u}_y + c_{13}\mathring{u}_z = \mathring{u}'_x, \\ \dot{y}' &= c_{21}\mathring{u}_x + c_{22}\mathring{u}_y + c_{23}\mathring{u}_z = \mathring{u}'_y, \\ \dot{z}' &= c_{31}\mathring{u}_x + c_{32}\mathring{u}_y + c_{33}\mathring{u}_z = \mathring{u}'_z, \end{aligned}\right\} \tag{2.5}$$

and

$$\ddot{x}' = \ddot{y}' = \ddot{z}' = 0. \tag{2.6}$$

The velocity components transform just as we would expect a vector to transform, and eqs. (2.1) are reproduced in the new coördinates without change.

Another transformation preserving the law of inertia is the type (1.3). It corresponds to the transition from one frame of reference to another one which is in a state of straight-line, uniform motion relative to the first frame. Taking the second derivatives of eqs. (1.3), we obtain

$$\ddot{x}^* = \ddot{x}, \qquad \ddot{y}^* = \ddot{y}, \qquad \ddot{z}^* = \ddot{z}; \tag{2.7}$$

and if the motion of the body satisfies the law of inertia (2.1) in the coördinate system S, we have also:

$$\ddot{x}^* = \ddot{y}^* = \ddot{z}^* = 0, \tag{2.8}$$

while the first derivatives of the starred coördinates (if eqs. (2.2) apply to the unstarred coördinates) are

$$\left.\begin{aligned} \dot{x}^* &= \mathring{u}_x - v_x = \mathring{u}^*_x\,, \\ \dot{y}^* &= \mathring{u}_y - v_y = \mathring{u}^*_y\,, \\ \dot{z}^* &= \mathring{u}_z - v_z = \mathring{u}^*_z\,. \end{aligned}\right\} \tag{2.9}$$

Eq. (2.8) shows that the law of inertia holds in the new system as well as in the old one. Eqs. (2.9) express the fact that the velocity components in the new coördinate system S^* are equal to those in the old system minus the components of the relative velocity of the two coördinate systems themselves. This law is often referred to as the (classical) law of the addition of velocities.

Frames of reference and coördinate systems in which the law (2.1) is valid are inertial systems. All Cartesian coördinate systems which are at rest relative to an inertial coördinate system are themselves inertial systems. Cartesian coördinate systems belonging to a frame of reference which is in a state of straight-line, uniform motion relative to an inertial system are also inertial systems. On the other hand, when we carry out a transition to a new frame of reference which is in some state of accelerated motion relative to the first one, the corresponding coördinate transformation does not reproduce eqs. (2.1) in terms of the new coördinates. The acceleration of the new frame of reference relative to an inertial system manifests itself in apparent accelerations of bodies not subject to real forces.

Galilean transformations. If the form of a law is not changed by certain coördinate transformations, that is, if it is the same law in terms of either set of coördinates, we call that law *invariant* or *covariant* with respect to the transformations considered. The law of inertia (2.1) is covariant with respect to transformations (1.1) and (1.3), but not with respect to (1.2).

Transformations (1.1) and (1.3) are of the greatest importance for our further discussions. They are usually referred to as *Galilean transformations.* According to classical physics, any two inertial systems are connected by a Galilean transformation.

The force law and its transformation properties. We shall now discuss the transformation properties of the basic laws of classical mechanics. These laws may be formulated thus.

When bodies are subject to forces, their accelerations do not vanish, but are proportional to the forces acting on them. The ratio of force to acceleration is a constant, different for every individual body; this constant is called the mass of the body.

The total force acting on one body is the vector sum of all the forces caused by every other body of the mechanical system. In other words, the total interaction among a number of bodies is the combination of interactions of pairs. The forces which two bodies exert on each other lie in their connecting straight line and are equal except that they point in opposite directions; that is, two bodies can either attract or repel each other. The magnitude of these forces is a function of their distance only; neither velocities nor accelerations have any influence.

These laws apply to such phenomena as gravitation, electrostatics, and Van Der Waals forces, but electrodynamics is not included because the interaction between magnetic fields and electric charges produces forces whose direction is not in the connecting straight line, and which depend on the velocity of the charged body as well as on its position.

But whenever the italicized conditions are satisfied, the forces can be represented by the negative derivatives of the potential energy. The latter is the sum of the potential energies characterizing the interaction of any two bodies or "mass points,"

$$\left.\begin{aligned} V &= \sum_{i=1}^{n} \sum_{k=i+1}^{n} V_{ik}(s_{ik}), \\ s_{ik} &= \sqrt{(x_i - x_k)^2 + (y_i - y_k)^2 + (z_i - z_k)^2}. \end{aligned}\right\} \tag{2.10}$$

The indices i and k refer to the interaction between the ith and the kth mass points, and s_{ik} is the distance between them. The functions $V_{ik}(s_{ik})$ are given by the special nature of the problem, for example, Coulomb's law, Newton's law of gravitation, and so forth.

The force acting on the ith mass point is given by

$$\left.\begin{aligned} f_{i,x} &= -\frac{\partial V}{\partial x_i} = -\sum_{k=1}^{n}{}' \frac{dV_{ik}}{ds_{ik}} \frac{x_i - x_k}{s_{ik}}, \\ f_{i,y} &= -\frac{\partial V}{\partial y_i} = -\sum_{k=1}^{n}{}' \frac{dV_{ik}}{ds_{ik}} \frac{y_i - y_k}{s_{ik}}, \\ f_{i,z} &= -\frac{\partial V}{\partial z_i} = -\sum_{k=1}^{n}{}' \frac{dV_{ik}}{ds_{ik}} \frac{z_i - z_k}{s_{ik}}, \end{aligned}\right\} k \neq i. \tag{2.11}$$

The set of equations (2.11), by its form, implies that the force components due to the interaction of the ith and the kth bodies alone are equal, except for opposite signs, that is,

$$\frac{\partial V_{ik}}{\partial x_i} = -\frac{\partial V_{ik}}{\partial x_k}.$$

Therefore, the sum of all forces acting on all n mass points vanishes,

$$\sum_{i=1}^{n} f_{i,x} = \sum_{i=1}^{n} f_{i,y} = \sum_{i=1}^{n} f_{i,z} = 0. \tag{2.12}$$

The differential equations governing the motions of the bodies are

$$\left.\begin{aligned} m_i \ddot{x}_i &= f_{i,x}, \\ m_i \ddot{y}_i &= f_{i,y}, \\ m_i \ddot{z}_i &= f_{i,z}, \end{aligned}\right\} \tag{2.13}$$

where m_i is the mass of the ith body.

We are now going to show that *the system of equations determining the behavior of a mechanical system*, (2.10), (2.11), *and* (2.13) *is covariant with respect to Galilean transformations*.

Let us start with eq. (2.10). V depends on the distances s_{ik} of the various mass points from each other. How do the s_{ik} change (transform) when a coördinate transformation (1.1) or (1.3) is carried out? In order to answer that question, it must be kept in mind that the coördinates of the ith and of the kth body are to be taken at the same time; in other words, that the distance between the two bodies is itself a function of time. Of course, the coördinates of the various mass points transform independently of each other, each set (x_i, y_i, z_i) by itself, according to the transformation equations (1.1) or (1.3), respectively.

Considering these points, it is seen immediately that transformation

(1.3), corresponding to the straight-line, uniform motion, leaves the coördinate differences, for example, $x_i - x_k$, unchanged, or,

$$x_i^* - x_k^* = x_i - x_k . \tag{2.14}$$

Therefore, the s_{ik} themselves take the same form in the new coördinate system S^* which they have in S.

The transformation equations (1.1) express the relationship between two coördinate systems which are at rest relative to each other and whose axes are not parallel. Obviously, the distance between two points is expressed in the same way in either coördinate system; so that

$$\left.\begin{aligned} &\sqrt{(x_i - x_k)^2 + (y_i - y_k)^2 + (z_i - z_k)^2} \\ &\qquad = \sqrt{(x_i' - x_k')^2 + (y_i' - y_k')^2 + (z_i' - z_k')^2}, \\ &s_{ik} = s_{ik}' . \end{aligned}\right\} \tag{2.15}$$

A quantity which does not change its value (at a given point) when a coördinate transformation is carried out is called an invariant with respect to that transformation. The distance between two points is an invariant.

We have seen that the arguments of the function V, the s_{ik}, are invariant with respect to Galilean coördinate transformations. Therefore, the function V itself, the total potential energy of the mechanical system, is an invariant, too; expressed in terms of the new coördinates, it has the same form and takes the same values as in the original coordinate system. Eq. (2.10) is covariant with respect to Galilean transformations.

Let us proceed to eqs. (2.11) and again begin with transformation (1.3). The right-hand sides of eqs. (2.11) contain the derivatives of a quantity which we already know is an invariant. These derivatives with respect to the two sets of coördinates are related to each other by the equations

$$\frac{\partial V}{\partial x_i} = \frac{\partial V}{\partial x_i^*}, \qquad \frac{\partial V}{\partial y_i} = \frac{\partial V}{\partial y_i^*}, \qquad \frac{\partial V}{\partial z_i} = \frac{\partial V}{\partial z_i^*}, \tag{2.16}$$

and therefore, the right-hand side of eqs. (2.11) is invariant with respect to transformations (1.3). Whether the same holds true for the left-hand side, we shall be able to decide after discussing the transformation properties of eqs. (2.13). It is clear, however, that the equation remains valid in the new coördinate system only if both sides transform the same way. Otherwise, it is not covariant with respect to the transformation considered. We shall have to find out whether the transformation properties of the right-hand side of eqs. (2.11) are compatible with those of the left-hand side of eqs. (2.13), as both determine the transformation properties of the forces, $\mathbf{f}_i$.

Let us first transform the right-hand side of eqs. (2.11) by a transformation (1.1). Applying the rules of partial differentiation, we obtain

$$\left.\begin{aligned} \frac{\partial V}{\partial x_i} &= \frac{\partial V}{\partial x_i'}\cdot c_{11} + \frac{\partial V}{\partial y_i'}\cdot c_{21} + \frac{\partial V}{\partial z_i'}\cdot c_{31}\,, \\ \frac{\partial V}{\partial y_i} &= \frac{\partial V}{\partial x_i'}\cdot c_{12} + \frac{\partial V}{\partial y_i'}\cdot c_{22} + \frac{\partial V}{\partial z_i'}\cdot c_{32}\,, \\ \frac{\partial V}{\partial z_i} &= \frac{\partial V}{\partial x_i'}\cdot c_{13} + \frac{\partial V}{\partial y_i'}\cdot c_{23} + \frac{\partial V}{\partial z_i'}\cdot c_{33}\,. \end{aligned}\right\} \tag{2.17}$$

The $3n$ equations (2.17) can be separated into n groups of 3 equations each, these groups being identical save for the value of i. Each group transforms as the components of a vector, that is, each component in one system is equal to the sum of the projections upon it of the three components in the other system.

Whether the left-hand sides of eqs. (2.11) also have vector character must be decided after discussion of the transformation properties of eqs. (2.13).

The left-hand sides of eqs. (2.13) are products of masses and accelerations. We have already stated that in classical physics the mass is considered to be a constant of a body, independent of its state of motion and invariant with respect to coördinate transformations.

That the accelerations of a body are invariant with respect to transformation (1.3) we have already seen in eq. (2.7). Therefore, the left-hand sides of eqs. (2.13) transform with respect to (1.3) in the same way as the right-hand sides of eqs. (2.11).

Turning to transformations (1.1), we know that

$$\ddot{x}_i' = c_{11}\ddot{x}_i + c_{12}\ddot{y}_i + c_{13}\ddot{z}_i\,, \text{ and so forth,} \tag{2.18}$$

but because the c_{ab} have the significance of cosines of angles, and because the value of a cosine does not depend on the sign of the angle,

$$\cos\alpha = \cos(-\alpha),$$

it is also true that

$$\ddot{x}_i = c_{11}\ddot{x}_i' + c_{21}\ddot{y}_i' + c_{31}\ddot{z}_i'\,, \text{ and so forth.} \tag{2.18a}$$

Again, the left-hand sides of eqs. (2.13) transform in exactly the same way as the right-hand sides of (2.11), in this case as n vectors.

Eqs. (2.13) can be considered as the equations defining the forces $\mathbf{f}_i$. We conclude, therefore, that the forces themselves transform so that both eqs. (2.11) and (2.13) are covariant. *With respect to spatial, orthogonal transformations of the coördinate system, the forces are vectors, and they are invariant with respect to transformations representing a straight-line uniform motion of one system relative to the other.* These relations can be expressed in a slightly different form. By eliminating the quantities $\mathbf{f}_i$ from eqs. (2.11) and (2.13) we can combine them into new equations of the form

$$\left.\begin{aligned} \frac{\partial V}{\partial x_i} + m\ddot{x}_i &= 0, \\ \frac{\partial V}{\partial y_i} + m\ddot{y}_i &= 0, \\ \frac{\partial V}{\partial z_i} + m\ddot{z}_i &= 0. \end{aligned}\right\} \tag{2.19}$$

These equations contain the essential physical statements of eqs. (2.11) and (2.13), but do not bring out so clearly the force concept.

The result of the above consideration is that the two sides of each of equations (2.11) and (2.13) transform in the same way, and that, therefore, these equations remain valid when arbitrary Galilean transformations are carried out.

Equations which do not change at all with the transformation (that is, the terms of which are invariants) are called *invariant.* Equations which remain valid because their terms, though not invariant, transform according to identical transformation laws (such as the terms $\frac{\partial V}{\partial x_i}$ and $m\ddot{x}_i$, and so forth, in eqs. (2.19)) are called *covariant.*

The covariance of equations is the mathematical property which corresponds to the existence of a relativity principle for the physical laws expressed by those equations. In fact, the relativity principle of classical mechanics is equivalent to our result, that the laws of mechanics take the same form in all inertial systems, that is, in all those coördinate systems which can be obtained by subjecting any one inertial system to arbitrary Galilean transformations.

The other branches of mechanics, such as the treatment of continuous matter (the theory of elastic bodies and hydrodynamics) or the mechanics of rigid bodies, can be deduced from the mechanics of free mass points by introducing suitable interaction energies of the type (2.10), and by carrying out certain limiting processes. It is evident, even without a detailed treatment of these branches of mechanics, that the results obtained apply to them as well as to the laws of motion of free mass points.

CHAPTER III

The Propagation of Light

The problem confronting classical optics. During the nineteenth century, a new branch of physics was developed which could not be brought within the realm of mechanics. That branch was electrodynamics. As long as only electrostatic and magnetostatic effects were known, they could be treated within the framework of mechanics by the introduction of electrostatic and magnetostatic potentials which depended only on the distance of the electric charges or magnetic poles from each other.

The interaction of electric and magnetic fields required a different treatment. This was brought out clearly by Oersted's experiment. He found that a magnetized needle was deflected from its normal North-South direction by a current flowing through an overhead North-South wire. The sign of the deflection was reversed when the direction of the current was reversed. Obviously, the magnetic actions produced by electric currents, that is, by moving charges, depend not only on the distance but also on the velocity of these charges. Furthermore, the force does not have the direction of the connecting straight line. The concepts of Newton's mechanics are no longer applicable.

Maxwell succeeded in formulating the laws of electromagnetism by introducing the new concept of "field." As we have seen in the preceding chapter, in mechanics a system is completely described when the locations of the constituent mass points are known as functions of time. In Maxwell's theory, we encounter a certain number of "field variables," such as the components of the electric and magnetic field strengths. While the point coördinates of mechanics are defined as functions of the time coördinate alone, the field variables are defined for all values both of the time coördinate and of the three space coördinates, and are thus functions of four independent variables.[1]

---

[1] In the mechanics of continuous media, we find variables which resemble field variables: The mass density, momentum density, stress components, and so forth; but they have only statistical significance. They are the total mass of the average number of particles per unit volume, and so forth. In electrodynamics however, the field variables are assumed to be the basic physical quantities.

In Maxwell's field theory it is further assumed that the change of the field variables with time at a given space point depends only on the immediate neighborhood of the space point. A disturbance of the field at a point induces a change of the field in its immediate neighborhood, this in turn causes a change farther away, and thus the original disturbance has a tendency to spread with a finite velocity and to make itself felt eventually over a great distance. "Action at a distance" may thus be produced by the field, but always in connection with a definite lapse of time. The laws of a field theory have the form of partial differential equations containing the partial derivatives of the field variables with respect to the space coördinates and with respect to time.

The force acting upon a mass point is determined by the field in the immediate neighborhood of the mass point. Conversely, the presence of the mass point may and usually does modify the field.

Since the structure of Maxwell's theory of electromagnetism is so different from Newtonian mechanics, the validity of the relativity principle in mechanics by no means implies its extension to electrodynamics. Whether or not this principle applies to the laws of the electromagnetic field must be the subject of a new investigation.

A complete investigation of this kind would have to establish the transformation laws of the electric and magnetic field intensities with respect to Galilean transformations, and then determine whether the transformed quantities obey the same laws with respect to the new coördinates. Such investigations were carried out by various scientists, among them H. Hertz and H. A. Lorentz. But we can obtain the most important result of these investigations by much simpler considerations. Instead of treating Maxwell's field equations themselves, we shall confine ourselves at present to one of their aspects, the propagation of electromagnetic waves.

Maxwell himself recognized that electromagnetic disturbances, such as those produced by oscillating charges, propagate through space with a velocity which depends on the electric nature of the matter present in space. In the absence of matter, the velocity of propagation is independent of its direction. and equal to about 3×10^{10} cm sec^{-1}. This is equal to the known speed of light; Maxwell assumed, therefore, that light was a type of electromagnetic radiation. When Hertz was able to produce electromagnetic radiation by means of an electromagnetic apparatus, Maxwell's theory of the electromagnetic field and his electromagnetic theory of light were accepted as an integral part of our physical knowledge.

Electromagnetic radiation propagates in empty space with a uniform, constant velocity (hereafter denoted by c). This conclusion can be formulated without taking into account the involved interrelation between magnetic and electric fields. That is why it is valuable for us, for we can study the transformation properties of this law of propagation without working out the transformation laws of the field variables.

It is now possible to decide whether the laws of the electromagnetic field are covariant with respect to Galilean transformations. Let us consider a coördinate system S, with respect to which the law of the uniform speed of light holds. If we carry out a Galilean transformation of the type (1.3), corresponding to a uniform translatory motion of the new system S^* relative to S, then the speed of the same light rays cannot be equal to c in all directions in terms of the new system S^*. If the direction of a light ray is designated by the cosines of its angles α, β, and γ, measured from the three axes of S, its velocity components with respect to S are

$$\left.\begin{aligned} u_x &= c\cdot\cos\alpha, \\ u_y &= c\cdot\cos\beta, \\ u_z &= c\cdot\cos\gamma, \end{aligned}\right\} \tag{3.1}$$

with

$$\cos^2\alpha + \cos^2\beta + \cos^2\gamma = 1. \tag{3.2}$$

According to eq. (2.9), the velocity components with respect to the new system S^* are

$$\left.\begin{aligned} u_x^* &= c\cos\alpha - v_x, \\ u_y^* &= c\cos\beta - v_y, \\ u_z^* &= c\cos\gamma - v_z, \end{aligned}\right\} \tag{3.3}$$

and the speed of light depends on its direction as indicated by the equation

$$\begin{aligned} &\sqrt{u_x^{*2} + u_y^{*2} + u_z^{*2}} \\ &\qquad = \sqrt{c^2 + v^2 - 2c(v_x\cos\alpha + v_y\cos\beta + v_z\cos\gamma)}. \end{aligned} \tag{3.4}$$

It equals c only for a certain cone of directions with the vector $\mathbf{v}$ as its axis. In the direction of $\mathbf{v}$, the speed of the light with respect to S^* will be equal to $c - v$, and in the opposite direction it will be $c + v$.

It appears, thus, that the principle of relativity is incompatible with the laws of electromagnetic radiation, and therefore with the theory of electromagnetic fields. If confidence in Maxwell's equations is at all justified, there must exist one frame of reference, presumably an inertial system, with respect to which the field equations take their standard form. Any frame of reference which is in a state of motion relative to this standard frame would have to be considered as a less suitable frame, at least from the point of view of electromagnetism, even though the relative motion might be a uniform, straight-line, translatory motion. The principle of relativity, as we have formulated it in the preceding chapter, would apply only to mechanics, not to the whole of physics.

This conclusion would not have been accepted without resistance. Mechanics was generally regarded as the most trustworthy part of our physical knowledge, and the principle of relativity was held to be a fundamental feature of the whole of nature. Several attempts were made to overcome these difficulties. We shall now consider the more important ones.

The corpuscular hypothesis. One hypothesis was that the speed of light equals c with respect to a frame of reference connected with the source of radiation, just as the speed of a bullet fired from a moving train would be referred to a frame connected with the train.

This assumption is, of course, incompatible with a field theory of light, as it was proposed by Maxwell, and would be rather suggestive of a corpuscular theory of the type which Newton had believed in. But it is consistent with the principle of relativity. The law of propagation contains explicitly the velocity of a material body, the source of light. Thus, the speed of light relative to an inertial system would transform just like the velocity of a material body, and the law of propagation would be covariant with respect to Galilean transformations.

But experimental evidence spoke against this hypothesis. If the speed of light depended on its source, then double stars should give rise to peculiar phenomena. Two double stars are separated by a very small distance compared with their distance from our solar system. They also have comparatively great velocities relative to each other. We would, therefore, expect that whenever they are in such a position that one of them is rapidly moving away from us while the other's motion is directed toward us, the light emitted by them simultaneously should arrive here at very different times. Consequently, their motion around each other and together through space would appear to us completely distorted. In some cases, we should observe the same component of the double star system simultaneously at different places, and these "ghost stars" would disappear and reappear in the course of their periodic motions.

These effects would be proportional to the distance of the double star system from the earth, for the time of arrival would be equal to the distance divided by the speed of light. If v is the variation of velocity of one component of the double star, one would have

$$t = d/c, \qquad \Delta t/\Delta c \sim -\frac{d}{c^2}, \qquad \Delta t \sim -\frac{vd}{c^2}$$

(d is the distance between double star and earth, c is the speed of light, and t is the mean time for light to reach the earth). Reasonable assumptions for the order of magnitude of the quantities v, d, and c are:

$$c^2 \sim 10^{21}\ \mathrm{cm}^2\ \mathrm{sec}^{-2},$$

$$v \sim 10^6\ \mathrm{cm\ sec}^{-1},$$

$$d > 10^{18}\ \mathrm{cm};$$

therefore,

$$\Delta t > 10^3\ \mathrm{sec}.$$

As there are many double star systems for which d exceeds 10^{21} cm and which have periods less than 10^6 sec, the resulting effects could not escape observation.

However, no trace of any such effect has ever been observed. This is sufficiently conclusive to rule out further consideration of this hypothesis.

The transmitting medium as the frame of reference. Another hypothesis was that, whenever light was transmitted through a material medium, this medium was the "local" privileged frame of reference. Within the atmosphere of the earth, the speed of light should be uniform with respect to a geocentric frame of reference.

This hypothesis, too, is unsatisfactory in many respects. Let us assume, for the sake of the argument, that it is the transmitting medium and its state of motion which determine the speed of light. Suppose, now, that electromagnetic radiation goes from one medium in a certain state of motion to a second medium in a different state of motion. The speed of light would be bound to change, this change depending on the relative velocity of the two media and on the direction of the radiation (also, of course, on the difference of indices of refraction). If this experiment should be carried out with increasingly rarefied media, the interaction between matter and radiation would become less and less, as far as refraction, scattering, and so forth, are concerned, but the change of u would remain the same. In the case of infinite dilution, that is, of a vacuum, we should have a finite jump in u without apparent cause.

There is also experimental evidence bearing on this hypothesis. In order to obtain information on the influence of a moving medium on the speed of light, Fizeau carried out the following experiment. He sent a ray of light through a pipe filled with a flowing liquid, and measured the speed of light in both the negative and positive directions of flow. He determined these speeds accurately by measuring the position of interference fringes.

The experiment showed that the speed of light does depend on the velocity of the flowing liquid, but not to the extent that the velocities of the light and of the medium could simply be added. If we denote the speed of light by c, the velocity of the liquid by v, and the index of refraction by n, we should expect, according to our present assumption, that the observed speed of light is

$$u = \frac{c}{n} \pm v, \tag{3.5}$$

the sign depending on the relative directions of the light and the flow. The actual result was that the change of speed of light is, within the limits of experimental error, given by

$$u = \frac{c}{n} \pm v\left(1 - \frac{1}{n^2}\right). \tag{3.6}$$

This experimental result is consistent with the first objection. For, as the medium is increasingly rarefied, the index of refraction n approaches the value 1, the dependence of u on v becomes negligible, and, in the limiting case of infinite dilution, u becomes simply c.

Another effect which indicates that the speed of light does not depend on the motion of a rarefied medium of transmission is that of *aberration*. Fixed stars at a great distance change their relative positions in the sky in a systematic way with a period of one year. Their paths are ellipses around fixed centers, with the major axis in all cases approximately 41″ of arc. Stars near the celestial pole carry out movements that are approximately circles, while stars near the ecliptic have paths which are nearly straight lines.

Fig. 2 illustrates the way the star is seen away from its "normal" position (the center of the ellipse) at two typical points of the path of the earth around the sun.

Aberration can be explained thus (Fig. 3): As the telescope is rigidly connected with the earth, it goes through space at an approximate rate of 3×10^6 cm sec^{-1}. Therefore, when a light ray enters the telescope, let us say from straight above, the telescope must be inclined in the indicated manner, so that the lower end will have arrived straight below the former position of the upper end by the time that the light ray has arrived at the lower end. The tangent of the angle of aberration, α, must be the ratio between the distance traversed by the earth and the distance traversed by the light ray during the same time interval, or the ratio between the speed of the earth and the speed of light. This ratio is

$$\frac{v_{\text{earth}}}{c} \sim \frac{3 \times 10^6}{3 \times 10^{10}} \sim 10^{-4};$$

the angle corresponding to that tangent is 20.5″, the amount of greatest aberration from the center of the ellipse.

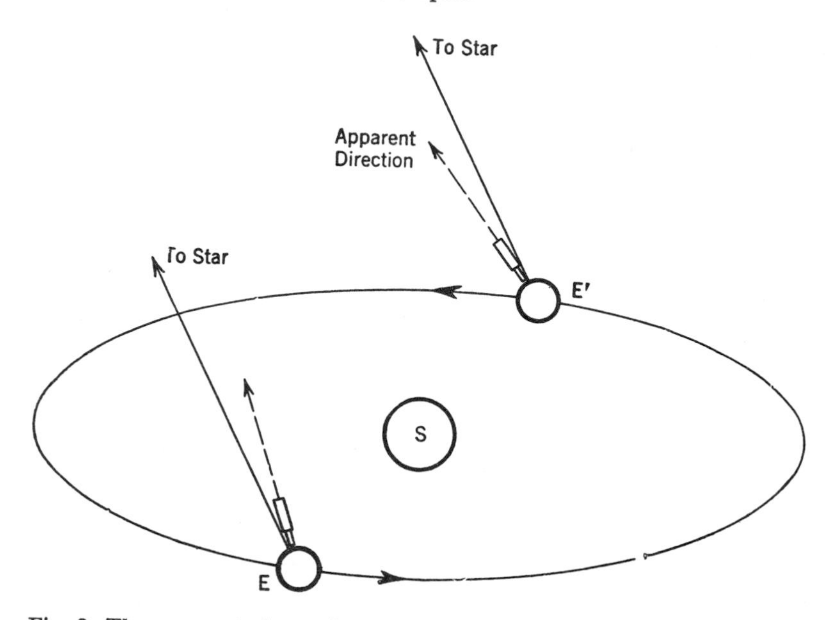

Fig. 2. The apparent change in position of a fixed star during a year (aberration). This change is exaggerated in the figure and amounts to not more than about 20″.5.

This explanation of aberration again contradicts the assumption that the transmitting medium is decisive for the speed of light. For if this assumption were true, the light rays, upon entering our atmosphere, would be "swept" along, and no aberration would take place.

The absolute frame of reference. All these arguments suggested the independence of the electromagnetic laws from the motion of either the source of radiation or the transmitting medium. The other alternative, it appeared, was to give up the principle of relativity and to assume that there existed a universal frame of reference with respect to which the speed of light was independent of the direction of propagation. As mentioned before, the equations of the electromagnetic field would have taken their standard form with respect to that frame. As the accelerations of charged particles are proportional to the field, this frame could be expected to be an inertial system, so that the accelerations would tend to zero with the field.

The experiment of Michelson and Morley. On the basis of this assumption, Michelson and Morley devised an experiment which was designed to determine the motion of the earth with respect to the privileged frame of reference in which the speed of light was to be uniform. The essential idea of their experiment was to compare the apparent speed of light in two different directions.

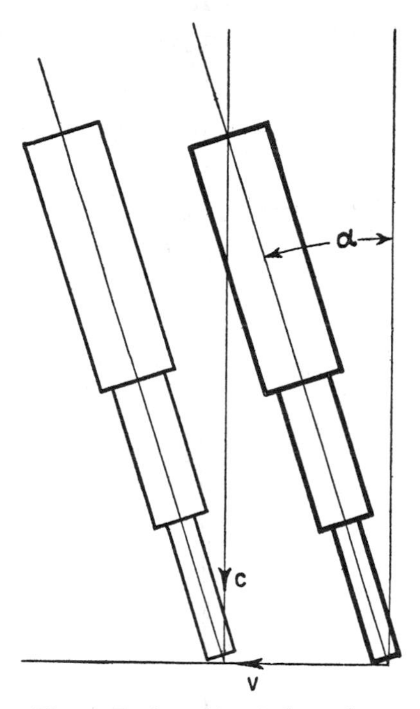

Fig. 3. Explanation of aberration.

Before studying their experimental set-up, let us discuss the expected results from the standpoint of this new hypothesis. The earth itself cannot be the privileged frame of reference with respect to which the equations of Maxwell hold, for it is continually subject to the gravitational action of the sun; and in a frame of reference connected with the center of gravitation of our solar system, the velocity of the earth is of the order of 3×10^6 cm sec^{-1}. It changes, therefore, about 6×10^6 cm sec^{-1} in the course of one half-year relatively to a frame of reference which approximates an inertial system better than the earth does. Therefore, even if the earth could at any time be identified with the state of motion of the privileged frame of reference, it would have a speed of 6×10^6 cm sec^{-1} half a year later relative to the privileged system.

In any case, it has a speed of at least 3×10^6 cm sec^{-1} relative to any inertial system through 6 months of the year. The speed of light is about 3×10^{10} cm sec^{-1}. If it is possible to compare the speed of light in two orthogonal directions with a relative accuracy better than 10^{-4}, and if the experiments are carried out over a period exceeding 6 months, the effects of the motion of the earth would become noticeable.

We proceed now to a description of Michelson and Morley's experiment (Fig. 4). Light from a terrestrial source L is separated into two parts by a thinly silvered glass plate P. At nearly equal distances from P, and at right angles to each other, two mirrors S_1 and S_2 are placed which reflect the light back to P. There, a part of each of the two rays reflected by S_1 and S_2, respectively, are reunited and are observed through a telescope F. Since the light emanating from L has

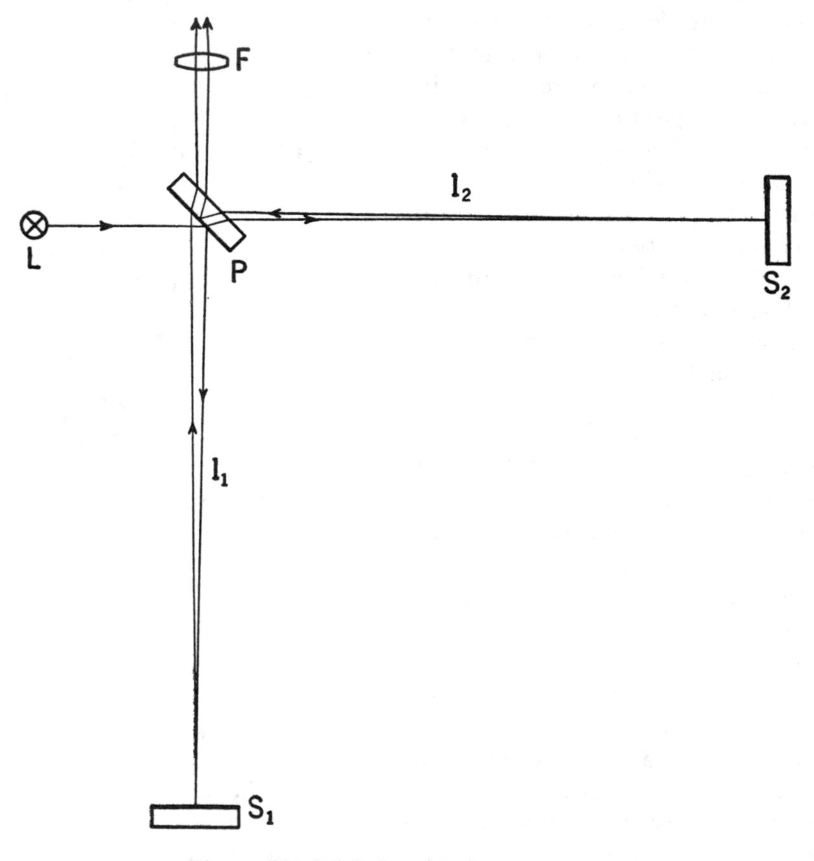

Fig. 4. The Michelson-Morley apparatus.

travelled almost equal distances, $L - P - S_1 - P - F$ and $L - P - S_2 - P - F$, respectively, interference fringes are observed, and their exact location depends on the difference between the distances l_1 and l_2.

So far, we have assumed that the speed of light is the same in all directions. If this assumption is dropped, the position of the interference fringes in F will also depend on the difference in the speeds along

l_1 and l_2. Let us assume that the earth, and with it the apparatus, is moving, relatively to the "absolute" frame of reference, along the direction of l_1 at the rate of speed v. With respect to the apparatus, the speed of light along the path $P - S_1$ equals $(c - v)$, and along the path $S_1 - P$ it is $(c + v)$. The time required to travel the path $P - S_1 - P$ will be

$$t_1 = \frac{l_1}{c - v} + \frac{l_1}{c + v} = \frac{2l_1/c}{1 - v^2/c^2}. \tag{3.7}$$

The relative speed of the light travelling along the path $P - S_2 - P$ will also be modified. While the light travels from P to S_2, the whole apparatus is moving sideways a distance δ,

$$\frac{\delta}{\sqrt{l_2^2 + \delta^2}} = \frac{v}{c}, \qquad \delta = \frac{v}{c}\frac{l_2}{\sqrt{1 - v^2/c^2}}, \tag{3.8}$$

and the actual distance travelled by the light is

$$l_2^* = \sqrt{l_2^2 + \delta^2} = \frac{l_2}{\sqrt{1 - v^2/c^2}}. \tag{3.9}$$

On the way back, the light has to travel an equal distance. The total time required by the light for the path $P - S_2 - P$ is, therefore,

$$t_2 = \frac{2l_2/c}{\sqrt{1 - v^2/c^2}}. \tag{3.10}$$

After the apparatus has been swung 90° about its axis, the times required to travel the paths $P - S_1 - P$ and $P - S_2 - P$ are, respectively,

$$\left.\begin{aligned} \bar{t}_1 &= \frac{2l_1/c}{\sqrt{1 - v^2/c^2}}, \\ \bar{t}_2 &= \frac{2l_2/c}{1 - v^2/c^2}. \end{aligned}\right\} \tag{3.11}$$

The time *differences* between the two alternative paths are, therefore, before and after the apparatus has been swung around,

$$\Delta t = t_1 - t_2 = \frac{2/c}{\sqrt{1 - v^2/c^2}}\left(\frac{l_1}{\sqrt{1 - v^2/c^2}} - l_2\right), \tag{3.12}$$

and

$$\Delta \bar{t} = \bar{t}_1 - \bar{t}_2 = \frac{2/c}{\sqrt{1 - v^2/c^2}}\left(l_1 - \frac{l_2}{\sqrt{1 - v^2/c^2}}\right). \tag{3.12a}$$

The change in Δt which is brought about by rotating the apparatus is, therefore,

$$\Delta \bar{t} - \Delta t = -\frac{2/c}{\sqrt{1 - v^2/c^2}} (l_1 + l_2) \left(\frac{1}{\sqrt{1 - v^2/c^2}} - 1 \right). \tag{3.13}$$

As $(v/c)^2$ is of the order of magnitude of 10^{-8}, we shall expand the right-hand side of eq. (3.13) into a power series in $(v/c)^2$ and consider only the first nonvanishing term. We obtain the approximate expression

$$\Delta \bar{t} - \Delta t = -\frac{1}{c} (l_1 + l_2) \frac{v^2}{c^2}. \tag{3.14}$$

We should expect the interference fringes in the telescope to shift because of this change in the time difference Δt. The amount of this shift, expressed in terms of the width of one fringe, would be equal to $(\Delta \bar{t} - \Delta t)$ divided by the time of one period of oscillation, $\frac{1}{\nu}$,

$$\frac{\Delta s}{s} = -\frac{\nu}{c} (l_1 + l_2) \frac{v^2}{c^2}. \tag{3.15}$$

v is the velocity of the earth relative to the "absolute" frame of reference, presumably at least of the order 3×10^6 cm sec^{-1}. $(v/c)^2$ is, therefore, of the order 10^{-8}. $\frac{\nu}{c}$ is the wave number, and for visible light, about 2×10^4 cm^{-1}. We have, therefore,

$$\frac{\Delta s}{s} \sim -\frac{l_1 + l_2}{5 \times 10^3 \text{ cm}}. \tag{3.15a}$$

By using multiple reflection, Michelson and Morley were able to work with effective lengths l_1 and l_2 of several meters. Any effect should have been clearly observable after all the usual sources of error, such as stresses, temperature effects, and so forth, had been eliminated. Nevertheless, no effect was observed.

An impasse was at hand: No consistent theory would agree with the results of Fizeau's experiment, the Michelson-Morley experiment, and the effect of aberration. A great number of additional experiments were performed along similar lines. Their discussion can be omitted here, because they did not change the situation materially. What was needed was not more experiments, but some new theory which would explain the apparent contradictions.

The ether hypothesis. Before launching into an explanation of that new theory, the theory of relativity, we mention a hypothesis which today has only historical significance. Physicists had been accustomed to think largely in terms of mechanics. When Faraday, Maxwell, and Hertz created the first field theory, it was only natural that attempts were made by many physicists to explain the new fields in terms of mechanical concepts. Maxwell and Hertz themselves contributed to these efforts. Within the realm of mechanics itself, there existed a branch which used concepts and methods resembling those of field physics, namely, the mechanics of continuous media. So the electromagnetic fields were explained as the stresses of a hypothetical material medium, the so-called ether.

There are many reasons why this interpretation of the electromagnetic field finally had to be abandoned. Among them are: the ether would have to be endowed with properties not shared by any known medium; it would have to penetrate all matter without exhibiting any frictional resistance; and it would have no mass and would not be affected by gravitation. Also, Maxwell's equations are different in many ways from the equations to which elastic waves are subject. There exists, for instance, no analogy in electrodynamics to the "longitudinal" elastic waves.

At the end of the nineteenth century, however, the ether was regarded as a most promising and even necessary hypothesis. Naturally, attempts were made to apply this concept to the problem discussed in this chapter, namely, to find that coördinate system in which the speed of light is equal to c in all directions. The idea of the ether suggested that it might be the coördinate system in which the ether is at rest. That theory, however, does little to solve the fundamental difficulty. All that it does is reword the problem; for, in order to find out what the state of motion of the ether really is, we would have no other means than to measure the speed of light. The outcome of the Michelson-Morley experiment would, therefore, suggest that the ether is dragged along with the earth, as far as the immediate neighborhood of the earth is concerned. The motion of small masses, such as in Fizeau's experiment, would carry the ether along, but not completely. But these hypotheses could not account for aberration. The existence of the aberration effect would be consistent with an ether hypothesis only if the earth could glide through the ether without carrying it along, even right on its surface, where our telescope picks up the light.

CHAPTER IV

The Lorentz Transformation

Several decades of experimental research showed that there was no way of determining the state of motion of the earth through the "ether." All the evidence seemed to point toward the existence of a "relativity principle" in optics and electrodynamics, even though the Galilean transformation equations ruled that out.

Nevertheless, Fitzgerald and especially H. A. Lorentz tried to preserve the traditional transformation equations and still account theoretically for the experimental results. Lorentz was able to show that the motion of a frame of reference through the ether with a velocity v would produce only "second-order effects"; that is, all observable deviations from the laws which were valid with respect to the frame connected with the ether itself would be proportional not to v/c, but to $(v/c)^2$.

One of these expected second-order effects was that, in a system moving relatively to the ether, a light ray would take longer to go out and back over a fixed distance parallel to the direction of the motion than over an equal distance perpendicular to the motion. The Michelson-Morley experiment was designed to measure that effect. In order to explain the negative outcome of the experiment, Fitzgerald and Lorentz assumed that scales and other "rigid" bodies moving through the ether contracted in the direction of the motion just sufficiently to offset this effect. This hypothesis preserved fully the privileged character of one frame of reference (the ether). The negative result of the Michelson-Morley experiment was not explained by the existence of an "optical relativity principle," but was attributed to an unfortunate combination of effects which made it impossible to determine experimentally the motion of the earth through the ether.

Einstein, on the contrary, accepted the experiments as conclusive evidence that the relativity principle was valid in the field of electrodynamics as well as in mechanics. Therefore, his efforts were directed toward an analysis and modification of the Galilean transformation equations so that they would become compatible with the relativity principle in optics. We shall now retrace this analysis in order to derive the new transformation laws.

In writing down transformation equations, we always made two assumptions, although we did not always stress them: That there exists a universal time t which is defined independently of the coördinate system or frame of reference, and that the distance between two simultaneous events is an invariant quantity, the value of which is independent of the coördinate system used.

The relative character of simultaneity. Let us take up the first assumption. As soon as we set out to define a universal time, we are confronted with the necessity of defining simultaneity. We can compare and adjust time-measuring devices in a unique way only if the statement "The two events A and B occurred simultaneously" can be given a meaning independent of a frame of reference. That this can be done is one of the most important assumptions of classical physics; and this assumption has become so much a part of our way of thinking, that almost everyone has great difficulty in analyzing its factual basis.

To examine this hypothesis, we must devise an experimental test which will decide whether two events occur simultaneously. Without such an experiment (which can be performed, at least in principle), the statement "The two events A and B occurred simultaneously" is devoid of physical significance.

When two events occur close together in space, we can set up a mechanism somewhat like the coincidence counters used in the investigation of cosmic rays. This mechanism will react only if the two events occur simultaneously.

If the two events occur a considerable distance apart, the coincidence apparatus is not adequate. In such a case, signals have to transmit the knowledge that each event has occurred, to some location where the coincidence apparatus has been set up. If we had a method of transmitting signals with infinite velocity, no great complication would arise. By "infinite velocity" we mean that the signal transmitted from a point P_1 to another point P_2 and then back to P_1 would return to P_1 at the same time as it started from there.

Unfortunately, no signal with this property is known. All actual signals take a finite time to travel out and back to the point of origin, and this time increases with the distance traversed. In choosing the type of signal, we should naturally favor a signal where the speed of transmission depends on as few factors as possible. Electromagnetic waves are most suitable, because their transmission does not require the presence of a material medium, and because their speed in empty space does not depend on their direction, their wave length, or their intensity. As the recording device, we can use a coincidence circuit with two photon counters.

To account for the finite time lost in transmission, we set up our apparatus at the midpoint of the straight line connecting the sites of the two events A and B. Each event, as it occurs, emits a light signal, and we shall call the events simultaneous if the two light signals arrive simultaneously at the midpoint. This experiment has been designed to determine the simultaneity of two events without the use of specific time-measuring devices. It is assumed that simultaneity as defined by this experiment is "transitive," meaning that if two events A and B occur simultaneously (by our definition), and if the two events A and C also occur simultaneously, then B and C are simultaneous. It must be understood that this assumption is a hypothesis concerning the behavior of electromagnetic signals.

Granted that this hypothesis is correct, we still have no assurance that our definition of simultaneity is independent of the frame of reference to which we refer our description of nature. Locating two events and constructing the point midway on the connecting straight line necessarily involves a particular frame and its state of motion.

Is our definition invariant with respect to the transition from one frame to another frame in a different state of motion? To answer this question, we shall consider two frames of reference: One connected with the earth (S), the other with a very long train (S^*) moving along a straight track at a constant rate of speed. We shall have two observers, one stationed on the ground alongside the railroad track, the other riding on the train. Each of the two observers is equipped with a recording device of the type described and a measuring rod. Their measuring rods need not be the same length; it is sufficient that each observer be able to determine the point midway between two points belonging to *his* reference body—ground or train.

Let us assume now that two thunderbolts strike, each hitting the train as well as the ground and leaving permanent marks. Also suppose that each observer finds afterwards that his recording apparatus was stationed exactly midway between the marks left on his reference body. In Fig. 5, the marks are denoted by A, B, A^*, and B^*, and the coincidence apparatus by C and C^*. Is it possible that the light signals issuing from A, A^* and from B, B^* arrive simultaneously at C and also simultaneously at C^*?

At the instant that the thunderbolt strikes at A and A^*, these two points coincide. The same is true of B and B^*. If eventually it turns out that the two bolts struck simultaneously as observed by the ground observer, then C^* must coincide with C at the same time that A coincides with A^* and B with B^* (that is, when the two thunderbolts strike).[1] It is understood that all these simultaneities are defined with respect to the frame S.

Because of the finite time needed by the light signals to reach C and C^*, C^* travels to the left (Fig. 5, stages b, c, d). The signal issuing from A, A^* reaches C, therefore, only after passing C^* (stages b, c); while the light signal from B, B^* reaches C before it gets to C^* (stages c, d). As a result, the train observer finds that the signal from A, A^* reaches his coincidence apparatus sooner than the signal from B, B^* (stages b, d).

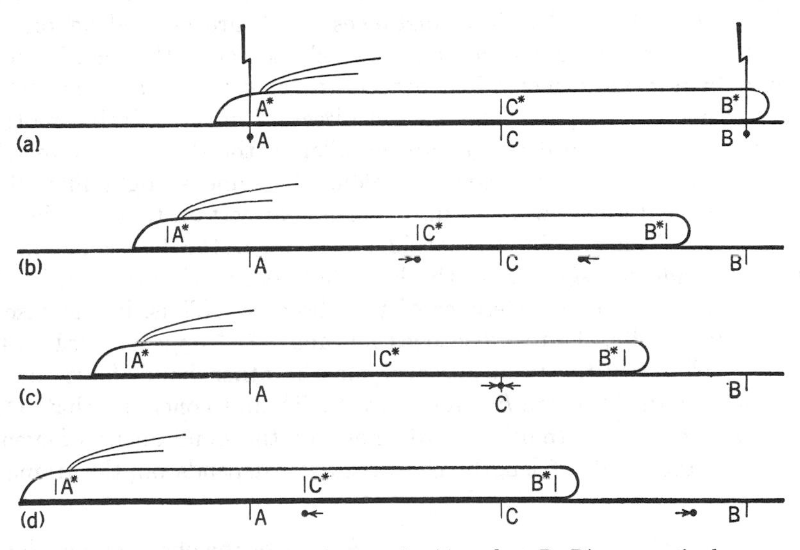

Fig. 5. The two events occurring at A, A^* and at B, B^*, respectively, appear simultaneous to an observer at rest relative to the ground (S), but not to an observer who is at rest relative to the train (S^*). At (a) the two events occur, (b) the light signal from A, A^* arrives at C^*, (c) the light signals from both events arrive at C, and (d) the light signal from B, B^* arrives at C^*.

This does not imply that the ground has a property not possessed by the train. It is possible for the thunderbolts to strike so that the light signals reach C^* simultaneously. But then the signal from A, A^* will arrive at C after the signal from B, B^*. In any case, it is impossible for both recording instruments, at C and at C^*, to indicate that the two thunderbolts struck simultaneously.

---

[1] Otherwise, the distances A^*C^* and B^*C^* would not appear equal from the point of view of the ground observer; we shall explain later why we do not assume anything of this kind.

We conclude, therefore, that two events which are simultaneous with respect to one frame of reference are in general not simultaneous with respect to another frame.[2]

The length of scales. Our conclusion affects the evaluation of length measurements. We have assumed that the ground observer and the train observer are able to carry out length measurements in their respective frames of reference. Two rods which are at rest relatively to the same frame of reference are considered equal in length if they can be placed alongside each other so that their respective end points E, E^* and F, F^* coincide. Two distances which are marked off on two different reference bodies moving relatively to each other can be compared by the same method, provided these distances are parallel to each other and perpendicular to the direction of the relative motion. However, if the two distances are parallel to the direction of relative motion, and if they are travelling along the same straight line, their respective end points will certainly coincide at certain times. The two distances EF and E^*F^* are considered equal if the two coincidences occur *simultaneously*. But whether they occur simultaneously depends on the frame of reference of the observer. Thus, in the case of the thunderbolts, the two distances AB and A^*B^* appear equal to the ground observer; the train observer, on the other hand, finds that A coincides with A^* before B coincides with B^*, and concludes that A^*B^* is longer than AB. In other words, not only the simultaneity of events, but also the result of length measurements, depends on the frame of reference.

The rate of clocks. The frame of reference of the observer also determines whether two clocks at a considerable distance from each other agree (that is, whether their hands assume equivalent positions simultaneously). Moreover, if the two clocks are in different states of motion, we cannot even compare their rates independently of the frame of reference. To illustrate this let us consider two clocks D and D^*, one stationed alongside the track and the other on the train. Let us assume that the two clocks happen to agree at the moment when D^* passes D. We can say that D^* and D go at the same rate if they continue to agree.

[2] Our definition of simultaneity is, of course, to a certain degree arbitrary. However, it is impossible to devise an experiment by means of which simultaneity could be defined independently of a frame of reference. From the outcome of the Michelson-Morley experiment, we conclude that the law of propagation of light takes the same form in all inertial systems. Had the outcome of the Michelson-Morley experiment been positive, in other words, if it were possible to determine the state of motion of the "ether," we should naturally have based our definition of simultaneity on the frame of reference connected with the ether, and thereby have given it absolute significance.

But after a while, D^* and D will be a considerable distance apart; and, as we know from our earlier considerations, their hands cannot assume equivalent positions simultaneously from the points of view both of the ground observer and of the train observer.

The Lorentz transformation. The above considerations help us to remove the apparent contradiction between the law of the propagation of electromagnetic waves and the principle of relativity. If it is impossible to define a universal time, and if the length of rigid rods cannot be defined independently of the frame of reference, it is quite conceivable that the speed of light is actually the same with respect to different frames of reference which are moving relatively to each other. We are now in a position to show that the classical transformations connecting two inertial systems (Galilean transformation equations) can be replaced by new equations which are not based on the assumptions of a universal time and the invariant length of scales, but which assume at the outset the invariant character of the speed of light.

In the derivation of these new transformation equations, we shall accept the principle of relativity as fundamental; that is, the transformation equations must contain nothing which would give one of the two coördinate systems a preferred position as compared with the other system. In addition, we shall assume that the transformation equations preserve the homogeneity of space; all points in space and time shall be equivalent from the point of view of the transformation. The equations must, therefore, be linear transformation equations. This is why we considered the two distances A^*C^* and B^*C^* equal in terms of S-coördinates as well as in S^*-coördinates (see page 31).

Let us consider two inertial coördinate systems, S and S^*. S^* moves relatively to S at the constant rate v along the X-axis; at the S-time $t = 0$, the points of origin of S and S^* coincide. The X^*-axis is parallel to the X-axis and, in fact, coincides with it. Points which are at rest relative to S^* will move with speed v relative to S in the X-direction. The first of our transformation equations will, thus, take the form

$$x^* = \alpha(x - vt), \tag{4.1}$$

where α is a constant to be determined later.

It is not quite obvious that a straight line which is perpendicular to the X-axis should also be perpendicular to the X^*-axis (the angles to be measured by observers in S and S^*, respectively). But if we did not assume that it was, the left-right symmetry with respect to the X-axis would be destroyed by the transformation. For similar reasons, we shall assume that the Y- and the Z-axes are orthogonal to each other, as observed from either system, and that the same is true of the Y^*- and Z^*-axes.

As mentioned before, we can compare the lengths of rods in different states of motion in an invariant manner if they are parallel to each other and orthogonal to the direction of relative motion. If their respective end points coincide, it follows from the principle of relativity that they are the same length. Otherwise, the relationship between S and S^* would not be reciprocal.

On the basis of this, we can formulate two further transformation equations,

$$\left.\begin{aligned} y^* &= y, \\ z^* &= z. \end{aligned}\right\} \tag{4.2}$$

To complete this set of equations, we have to formulate an equation connecting t^*, the time measured in S^*, with the time and space coördinates of S. t^* must depend on t, x, y, and z linearly, because of what we have called the "homogeneity" of space and time. For reasons of symmetry, we assume further that t^* does not depend on y and z. Otherwise, two S^*-clocks in the Y^*Z^*-plane would appear to disagree as observed from S. Choosing the point of time origin so that the inhomogeneous (constant) term in the transformation equation vanishes, we have

$$t^* = \beta t + \gamma x. \tag{4.3}$$

Finally, we must evaluate the constants α of eq. (4.1) and β and γ eq. (4.3). We shall find that they are determined by the two conditions that the speed of light be the same with respect to S and S^*, and that the new transformation equations go over into the classical equations when v is small compared with the speed of light, c.

Let us assume that at the time $t = 0$ an electromagnetic spherical wave leaves the point of origin of S, which coincides at that moment with the point of origin of S^*. The speed of propagation of the wave is the same in all directions and equal to c in terms of either set of coördinates. Its progress is therefore described by either of the two equations

$$x^2 + y^2 + z^2 = c^2t^2, \tag{4.4}$$

$$x^{*2} + y^{*2} + z^{*2} = c^2t^{*2}. \tag{4.5}$$

By applying eqs. (4.1), (4.2), and (4.3), we can replace the starred quantities in eq. (4.5) completely by unstarred quantities,

$$c^2(\beta t + \gamma x)^2 = \alpha^2(x - vt)^2 + y^2 + z^2. \tag{4.6}$$

By rearranging the terms, we obtain

$$(c^2\beta^2 - v^2\alpha^2)t^2 = (\alpha^2 - c^2\gamma^2)x^2 + y^2 + z^2 - 2(v\alpha^2 + c^2\beta\gamma)xt. \quad (4.7)$$

This equation goes over into eq. (4.4) only if the coefficients of t^2 and x^2 are the same in eqs. (4.7) as in eqs. (4.4), and if the coefficient of xt in eq. (4.7) vanishes. Therefore,

$$\left.\begin{aligned} c^2\beta^2 - v^2\alpha^2 &= c^2, \\ \alpha^2 - c^2\gamma^2 &= 1, \\ v\alpha^2 + c^2\beta\gamma &= 0. \end{aligned}\right\} \quad (4.8)$$

We solve these three equations for the three unknowns α, β, and γ by first eliminating α^2. We obtain the equations

$$\left.\begin{aligned} \beta(\beta + v\gamma) &= 1, \\ c^2\gamma(\beta + v\gamma) &= -v. \end{aligned}\right\} \quad (4.9)$$

Then we eliminate γ and obtain for β^2 the expression

$$\beta^2 = \frac{1}{1 - v^2/c^2}. \quad (4.10)$$

β is not equal to unity, as it is in the classical transformation theory. But by choosing the positive root of (4.10), we can make it nearly equal to unity for small values of v/c; its deviation from unity is of the second order. γ is given by the equation

$$\gamma = \frac{1 - \beta^2}{v\beta} = -\frac{\beta v}{c^2}; \quad (4.11)$$

and finally, α is obtained from the equation

$$\alpha^2 = -c^2\beta\gamma/v = \beta^2. \quad (4.12)$$

Again we choose the positive sign of the root.

By substituting all these values into eqs. (4.1), (4.3), we get the new transformation equations,

$$\left.\begin{aligned} x^* &= \frac{x - vt}{\sqrt{1 - v^2/c^2}}, \\ y^* &= y, \\ z^* &= z, \\ t^* &= \frac{t - \frac{v}{c^2}x}{\sqrt{1 - v^2/c^2}}. \end{aligned}\right\} \quad (4.13)$$

These equations are the so-called *Lorentz transformation equations.* For small values of v/c, they are approximated by the Galilean transformation equations,

$$\left.\begin{aligned} x^* &= x - vt, \\ y^* &= y, \\ z^* &= z, \\ t^* &= t. \end{aligned}\right\} \tag{4.14}$$

The deviations are all of the second order in v/c (or x/ct). We can therefore test the Lorentz transformation equations experimentally only if we are able to increase $(v/c)^2$ beyond the probable experimental error. Michelson and Morley, in their famous experiment, were able to increase the accuracy to such an extent that they could measure a second-order effect and prove experimentally the inadequacy of the Galilean transformation equations.

When we solve the equations (4.13) with respect to x, y, z, and t, we obtain

$$\left.\begin{aligned} x &= \frac{x^* + vt^*}{\sqrt{1 - v^2/c^2}}, \\ y &= y^*, \\ z &= z^*, \\ t &= \frac{t^* + \dfrac{v}{c^2}\, x^*}{\sqrt{1 - v^2/c^2}}. \end{aligned}\right\} \tag{4.15}$$

Comparing eqs. (4.15) with eqs. (4.13), we conclude that S has the relative velocity $(-v)$ with respect to S^*. This is not a trivial conclusion, for neither the unit length nor the unit time is directly comparable in S and S^*.

The velocity of a light signal emanating from any point at any time is equal to c with respect to any one system if it is equal to c in the other system, for the coördinate and time differences of two events transform exactly like x, y, z, and t themselves.

The Lorentz transformation equations do away with the classical notions regarding space and time. They extend the validity of the relativity principle to the law of propagation of light.

So far, we have fashioned our transformation theory to fit the outcome of the Michelson-Morley experiment. How does this new theory account for *aberration*? We have to compare the direction of the incoming light with respect to two frames of reference, that of the sun and that of the earth. The amount of aberration depends on the angle between the incoming light and the relative motion of these two frames of reference. We shall call that angle α. Both coördinate systems are to be arranged so that their relative motion is along their common X-axis, and that the path of the light ray lies entirely within the XY-plane. With respect to the sun, the path of the light ray is given by

$$x = ct\cdot\cos\alpha, \qquad y = ct\cdot\sin\alpha. \tag{4.16}$$

With respect to the moving earth, we find the equations of motion by applying the inverted equations of the Lorentz transformation, (4.15). Eq. (4.16) takes the form

$$\left.\begin{aligned} x^* + vt^* &= c(t^* + v/c^2\cdot x^*)\cos\alpha, \\ y^*\sqrt{1 - v^2/c^2} &= c(t^* + v/c^2\cdot x^*)\sin\alpha. \end{aligned}\right\} \tag{4.17}$$

By solving these equations with respect to x^* and y^*, we get

$$\left.\begin{aligned} x^* &= ct^*\,\frac{\cos\alpha - v/c}{1 - v/c\cdot\cos\alpha} = ct^*\cos\alpha^*, \\ y^* &= ct^*\sqrt{1 - v^2/c^2}\,\frac{\sin\alpha}{1 - v/c\cdot\cos\alpha} = ct^*\sin\alpha^*. \end{aligned}\right\} \tag{4.18}$$

The cotangent of the new direction is

$$\operatorname{ctg}\alpha^* = \frac{\operatorname{ctg}\alpha - v/c\cdot\operatorname{cosec}\alpha}{\sqrt{1 - v^2/c^2}}. \tag{4.19}$$

According to the classical explanation given on page 21, the angle would turn out to be

$$\operatorname{ctg}\alpha^* = \operatorname{ctg}\alpha - v/c\cdot\operatorname{cosec}\alpha. \tag{4.20}$$

If we wish to compare eq. (4.19) with eq. (4.20), we have to keep in mind that v/c is a small quantity (about 10^{-4}). Therefore, we expand both formulas into power series with respect to v/c. We get

$$\operatorname{ctg}\alpha^*_{\text{rel}} = \operatorname{ctg}\alpha - v/c\cdot\operatorname{cosec}\alpha + \tfrac{1}{2}(v/c)^2\operatorname{ctg}\alpha + \cdots, \tag{4.19a}$$

and

$$\operatorname{ctg}\alpha^*_{\text{class}} = \operatorname{ctg}\alpha - v/c\cdot\operatorname{cosec}\alpha. \tag{4.20a}$$

The observed effect is the first-order effect, while the relativistic second-order effect is far below the attainable accuracy of observation. The relativistic equation (4.19) is, therefore, in agreement with the observed facts.

We can explain *Fizeau's experiment* by connecting the coördinate system S with the earth and S^* with the flowing liquid. With respect to S^*, the liquid is at rest, and the equation of the light rays must be of the form

$$x^* = c/n(t^* - t_0^*). \tag{4.21}$$

Applying the Lorentz transformation equations (4.13), we obtain

$$x - vt = c/n \cdot [(t - v/c^2 \cdot x) - t_0^* \sqrt{1 - v^2/c^2}]. \tag{4.22}$$

We obtain the velocity of the light ray with respect to S by solving this equation with respect to x,

$$x = \left[\frac{c}{n} - \frac{1/n^2 - 1}{1 + v/nc} v\right] t + \text{const.} \tag{4.23}$$

Again the observable first-order effect is in agreement with the experiment.

The "kinematic" effects of the Lorentz transformation. We shall now study in more detail the effect of the Lorentz equations on length and time measurements in different frames of reference.

Let us consider a clock that is rigidly connected with the starred frame of reference, stationed at some point (x_0^*, y_0^*, z_0^*). Let us compare the time indicated by that clock with the time t measured in the unstarred system. According to eq. (4.15), the unstarred time coördinate of the clock is given by

$$t = \frac{v/c^2 \cdot x_0^* + t^*}{\sqrt{1 - v^2/c^2}}.$$

An S-time interval, $(t_2 - t_1)$, is therefore related to the readings t_2^* and t_1^* of the clock as follows:

$$t_2 - t_1 = (t_2^* - t_1^*)/\sqrt{1 - v^2/c^2}. \tag{4.24}$$

Thus, the rate of the clock appears slowed down, from the point of view of S, by the factor $\sqrt{1 - v^2/c^2}$. But not only that. Observed from the unstarred frame of reference, different S^*-clocks go at the same rate, but with a phase constant depending on their position. The farther away an S^*-clock is stationed from the point of origin along the positive X^*-axis, the slower it appears to be. Two events that occur simultaneously with respect to S are not in general simultaneous with respect to S^*, and vice versa.

We can reverse our setup and compare an S-clock with S^*-time.

The clock may be located at the point (x_1, y_1, z_1), and the starred time is connected with the time indicated by the S-clock through the equation

$$t^* = \frac{t - v/c^2 \cdot x_1}{\sqrt{1 - v^2/c^2}}.$$

Again the readings of the S-clock are related to an S^*-time interval as follows:

$$t_2^* - t_1^* = (t_2 - t_1)/\sqrt{1 - v^2/c^2}. \quad (4.25)$$

It appears that the S-clock is slowed down, measured in terms of S^*-time, and that it is ahead of an S-clock placed at the origin, if its own x-coördinate is positive.

How is it that an observer connected with either frame of reference finds the rate of the clocks in the other system slow? To measure the rate of a clock T which is not at rest relatively to his frame of reference, an observer compares it with all the clocks in his system which T passes in the course of time. That is to say, an S-observer compares *one* S^*-clock with a succession of S-clocks, while an S^*-observer compares *one* S-clock with several S^*-clocks. The S^*-clock passes, in the course of time, S-clocks which are farther and farther along the positive X-axis and therefore increasingly fast with respect to S^*; consequently, the rate of the S^*-clock appears slow in comparison. Conversely, an S-clock passes S^*-clocks farther and farther along the negative X^*-axis and therefore increasingly fast with respect to S. The rate of the S-clock appears slow compared with S^*-clocks.

In the case of length measurements, conditions are somewhat more involved, because the transformation equations contain y and z in a different way than x, the direction of relative motion. A rigid scale that is perpendicular to the direction of relative motion has the same length in either coördinate system. However, when the scale is parallel to the X-axis and the X^*-axis, we have to distinguish whether the scale is at rest relative to one coördinate system or to the other. Let us first consider a rod rigidly connected with S^*, the end points of which have the coördinates $(x_1^*, 0, 0)$ and $(x_2^*, 0, 0)$. Its length in its own system is

$$l^* = x_2^* - x_1^*. \quad (4.26)$$

An observer connected with S will consider as the length of the rod the coördinate difference $(x_2 - x_1)$ of its end points at the same time, t.

The coördinates x_2^* and x_1^* are related to x_2, x_1, and t by equations (4.13), yielding

$$\left.\begin{aligned} x_1^* &= \frac{x_1 - vt}{\sqrt{1 - v^2/c^2}}, \\ x_2^* &= \frac{x_2 - vt}{\sqrt{1 - v^2/c^2}}. \end{aligned}\right\} \tag{4.27}$$

Therefore, the coördinate differences are

$$x_2^* - x_1^* = \frac{x_2 - x_1}{\sqrt{1 - v^2/c^2}}. \tag{4.28}$$

If we denote the length $(x_2 - x_1)$ by l, we obtain

$$l = \sqrt{1 - v^2/c^2} \cdot l^*. \tag{4.29}$$

The rod appears contracted by the factor $\sqrt{1 - v^2/c^2}$. This effect is called the *Lorentz contraction.*

A calculation that reverses the roles of the two coördinate systems shows that a rod at rest in the unstarred system appears contracted in the starred system.

Thus, we have the rules: *Every clock appears to go at its fastest rate when it is at rest relatively to the observer. If it moves relatively to the observer with the velocity* v, *its rate appears slowed down by the factor* $\sqrt{1 - v^2/c^2}$. *Every rigid body appears to be longest when at rest relatively to the observer. When it is not at rest, it appears contracted in the direction of its relative motion by the factor* $\sqrt{1 - v^2/c^2}$, *while its dimensions perpendicular to the direction of motion are unaffected.*

The proper time interval. In contrast to the classical transformation theory, we no longer consider length and time intervals as invariants. But the invariant character of the speed of light gives rise to the existence of another invariant. Let us return to equations (4.1), (4.2), and (4.3), and conditions (4.8). We shall consider two events having the space and time coördinares (x_1, y_1, z_1, t_1) and (x_2, y_2, z_2, t_2), respectively. The difference between the squared time interval and the squared distance, divided by c^2, shall be called τ_{12}^2, or

$$\tau_{12}^2 = (t_2 - t_1)^2 - \frac{1}{c^2}[(x_2 - x_1)^2 + (y_2 - y_1)^2 + (z_2 - z_1)^2]. \tag{4.30}$$

Correspondingly, we define a similar quantity with respect to S^*,

$$\tau_{12}^{*2} = (t_2^* - t_1^*)^2 - 1/c^2[(x_2^* - x_1^*)^2 + (y_2^* - y_1^*)^2 + (z_2^* - z_1^*)^2]. \tag{4.31}$$

Now we express τ_{12}^{*2} in terms of S-quantities, according to eqs. (4.1), (4.2), and (4.3), just as we did in the discussion of eq. (4.5), and obtain

$$\left.\begin{aligned}\tau_{12}^{*2} &= (\beta^2 - \alpha^2 v^2/c^2)(t_2 - t_1)^2 - \frac{1}{c^2}[(\alpha^2 - c^2\gamma^2)(x_2 - x_1)^2 \\ &\quad + (y_2 - y_1)^2 + (z_2 - z_1)^2] + 2(\alpha^2 v/c^2 + \beta\gamma)(x_2 - x_1)(t_2 - t_1).\end{aligned}\right\} \quad (4.32)$$

Because the constants α, β, and γ satisfy conditions (4.8), we find that τ_{12}^2 is an invariant with respect to the transformation equations (4.13), or,

$$\tau_{12}^{*2} = \tau_{12}^2. \quad (4.33)$$

It is also invariant with respect to spatial orthogonal transformations (1.1).

Hereafter, we shall call all the linear transformations with respect to which τ_{12}^2 is invariant, *Lorentz transformations*, regardless of whether the relative motion of the two systems takes place along the common X-axis or not. Obviously, the invariance of τ_{12}^2 implies theinvariance of the speed of light, for the path of a light ray is characterized by the vanishing of τ_{12}^2 for all pairs of points along its path.

What is the physical significance of this quantity τ_{12}^2? If there exists a frame of reference with respect to which both events take place at the same space point, then τ_{12} (the positive square root of τ_{12}^2) is the time recorded by a clock at rest in that frame of reference. τ_{12} is therefore called the *proper time interval* (or *eigen time interval*).

Does there always exist a frame of reference with respect to which two events take place at the same space point? If we were dealing with the classical transformation equations, the answer would be yes, unless the two events took place "simultaneously." But the equations of the Lorentz transformation, (4.13), become singular when v, the relative velocity of the two frames, becomes equal to the speed of light. For values of v greater than c, equations (4.13) would lead to imaginary values of x^* and t^*. The Lorentz transformation equations are, thus, defined only for relative velocities of the two frames of reference smaller than c. Therefore, if two events occur in such rapid succession that the time difference is equal to or less than the time needed by a light ray to traverse the spatial distance between the two events, no frame of reference exists with respect to which the two events occur at the same spot.

Whenever the two events can be just connected by a light ray which leaves the site of one event at the time it occurs and arrives at the site of the other event as it takes place, the proper time interval τ_{12} between them vanishes. Whenever the sequence of two events is such that a light ray coming from either event arrives at the site of the other only after it has occurred, τ_{12}^2 is negative. Then we introduce instead of τ_{12}^2 the invariant $\sigma_{12}^2 = -c^2\tau_{12}^2$,

$$\sigma_{12}^2 = (x_2 - x_1)^2 + (y_2 - y_1)^2 + (z_2 - z_1)^2 - c^2(t_2 - t_1)^2. \qquad (4.34)$$

Either τ_{12} or σ_{12} is real for any two events. Whenever σ_{12} is real, we can carry out a Lorentz transformation so that $t_2^* - t_1^*$ vanishes. In other words, there exists a frame of reference with respect to which the events occur simultaneously. In that frame of reference, the spatial distance between the two events is simply σ_{12}.

Frequently, either τ_{12} or σ_{12} is referred to as the space-time interval between the two events. The interval is called *time-like* when τ_{12} is real, and *space-like* when σ_{12} is real. Whether the interval between the two events is time-like or space-like does not depend on the frame of reference or the coördinate system used, but is an invariant property of the two events.

We mentioned before that the Lorentz transformation is defined only for relative velocities smaller than the speed of light. If a frame of reference could move as fast as or faster than light, it would be, indeed, impossible for light to propagate at all in the forward direction, much less with the speed c.

The relativistic law of the addition of velocities. Is it possible to find two frames of reference which are moving relatively to each other with a velocity greater than c by carrying out a series of successive Lorentz transformations? To answer this question, we shall study the superposition of two (or more) Lorentz transformations. We shall introduce three frames of reference, S, S^*, S^{**}. S^* has the velocity v relative to S, and S^{**} has the velocity w relative to S^*. We want to find the transformation equations connecting S^{**} with S. Starting with the equations

$$\left.\begin{aligned} x^* &= \frac{x - vt}{\sqrt{1 - v^2/c^2}}, & y^* &= y, \\ t^* &= \frac{t - v/c^2 \cdot x}{\sqrt{1 - v^2/c^2}}, & z^* &= z, \\ x^{**} &= \frac{x^* - wt^*}{\sqrt{1 - w^2/c^2}}, & y^{**} &= y^*, \\ t^{**} &= \frac{t^* - w/c^2 \cdot x^*}{\sqrt{1 - w^2/c^2}}, & z^{**} &= z^*, \end{aligned}\right\} \qquad (4.35)$$

we have to substitute the first set of equations in the second set. The result of the straightforward calculation is

$$\left.\begin{aligned} x^{**} &= \frac{x - ut}{\sqrt{1 - u^2/c^2}}, \\ y^{**} &= y, \\ z^{**} &= z, \\ t^{**} &= \frac{t - u/c^2 \cdot x}{\sqrt{1 - u^2/c^2}}, \end{aligned}\right\} \tag{4.36}$$

with

$$u = \frac{v + w}{1 + vw/c^2}. \tag{4.37}$$

Thus, two Lorentz transformations, carried out one after the other, are equivalent to one Lorentz transformation. But the relative velocity of S^{**} with respect to S is not simply the sum of v and w. As long as both v/c and w/c are small compared with unity, u is very nearly equal to $v + w$; but as one of the two velocities approaches c, the deviation becomes important. Eq. (4.37) can be written in the form

$$u = c\left[1 - \frac{(1 - v/c)(1 - w/c)}{1 + vw/c^2}\right]. \tag{4.37a}$$

In this form, it is obvious that u cannot become equal to or greater than c, as long as both v and w are smaller than c. Therefore, it is impossible to combine several Lorentz transformations in one involving a relative velocity greater than c.

Eq. (4.37) can be interpreted in a slightly different way, for a body which has the velocity w with respect to S^* has the velocity u with respect to S. Then eq. (4.37) can be regarded as the transformation law for velocities (in the X-direction). In this case, it would be preferable to write it

$$u^* = \frac{u - v}{1 - uv/c^2}, \tag{4.38}$$

where w has been replaced by u^*. We conclude that a body has a velocity smaller than c in every inertial system if its velocity is less than c with respect to one inertial system.

The Lorentz transformation equations imply that no material body can have a velocity greater than c with respect to any inertial system. For each material body can be used as a frame of reference; and if it is removed from interaction with other bodies and does not rotate around its own center of gravity, it defines a new inertial system. Then, if the body could assume a velocity greater than c with respect to any inertial system, this system and the one connected with the body would have a relative velocity greater than c.

The proper time of a material body. We have spoken before of the space-time interval between two events. The application of this concept to the motion of a material body and to the space-time points along its path is particularly important for the development of relativistic mechanics. Since the velocity of a material body remains below c at all times, such an interval is always time-like. If the motion of the body is not straight-line and uniform, we can still define the parameter along its path by the differential equation

$$d\tau^2 = dt^2 - \frac{1}{c^2}(dx^2 + dy^2 + dz^2)$$

$$= \left\{1 - \frac{1}{c^2}\left[\left(\frac{dx}{dt}\right)^2 + \left(\frac{dy}{dt}\right)^2 + \left(\frac{dz}{dt}\right)^2\right]\right\} dt^2. \quad (4.39)$$

τ is the time shown by a clock rigidly connected with the moving body, really its "proper time" (its own time). When eq. (4.39) is divided by dt^2 and the root is taken, we obtain the relation between coördinate time and proper time,

$$\frac{d\tau}{dt} = \sqrt{1 - u^2/c^2}, \quad (4.40)$$

where u is the velocity of the body. This relation is valid for accelerated as well as unaccelerated bodies.

Both $d\tau$ and τ, which is defined by the integral

$$\tau = \int \sqrt{1 - u^2/c^2}\, dt, \quad (4.40a)$$

are invariant with respect to Lorentz transformations, though dt and u are not.

PROBLEMS

1. On page 39 we have discussed one method of measuring the length of a moving rod. We could also define that length as the product of the velocity of the moving rod by the time interval between the instant when one end point of the moving rod passes a fixed marker and the instant when the other end point passes the same marker.

Show that this definition leads also to the Lorentz contraction formula, equation (4.29).

2. Two rods which are parallel to each other move relatively to each other in their length directions. Explain the apparent paradox that either rod may appear longer than the other, depending on the state of motion of the observer.

3. Suppose that the frequency of a light ray is ν with respect to a frame of reference S. Its frequency ν^* in another frame of reference, S^*, depends on the angle α between the direction of the light ray and the direction of relative motion of S and S^*. Derive both the classical and the relativistic equations stating how ν^* depends on ν and the angle α.

For this purpose, the light may be treated as a plane scalar wave moving with the velocity c.

Answer:

$$\nu^*_{\text{cl.}} = \nu(1 - \cos\alpha\cdot v/c),$$

$$\nu^*_{\text{rel.}} = \nu\,\frac{1 - \cos\alpha\cdot v/c}{\sqrt{1 - v^2/c^2}} = \nu(1 - \cos\alpha\cdot v/c + \tfrac{1}{2}(v/c)^2 - \cdots).$$

The first-order effect common to both formulas is the "classical" Doppler effect, the second-order term is called the "relativistic" Doppler effect. It is independent of the angle α.

4. H. A. Lorentz created a theory which was the forerunner to the relativity theory as we know it today. Instead of trying to extend the relativity principle to electrodynamics, he assumed that there exists one privileged frame of reference, with respect to which the ether was to be at rest. In order to account for the outcome of the Michelson-Morley experiment, he assumed that the ether affects scales and clocks which are moving through it. According to this hypothesis, clocks are slowed down and scales are contracted in the direction of their motion. It is possible to derive the quantitative expressions for the factors of time- and length-contraction with the help of these notions.

(*a*) Assuming that the Galilean transformation equations are applicable, derive the rigorous expression for the time that a light ray needs to travel a measured distance l in both directions along a straight path in a Michelson-Morley apparatus, provided that the velocity of the apparatus relative to the privileged system is v and that the angle between the path and the direction of v is α.

Answer:

$$t = \frac{2l}{c}\cdot\frac{\sqrt{1 - v^2/c^2\cdot\sin^2\alpha}}{1 - v^2/c^2}. \qquad (4.\text{p}1)$$

(*b*) Now we introduce Lorentz' hypothesis and assume that equation (4.p1) holds for the true, contracted length l and the true, distorted angle α. The time indicated by the observer's clock is not the real time t, but the clock time, t^*. Furthermore, we measure the length with scales that are contracted themselves; that is, what we measure is not the true, contracted length l, but the apparent, uncontracted length l^*. The relation between the clock-time t^* and the apparent length l^* is

$$t^* = \frac{2l^*}{c}, \qquad (4.\text{p}2)$$

according to the outcome of the Michelson-Morley experiment. We call the factor of time-contraction θ and the factor of length-contraction in the direction of v, Λ. Derive the relations between t and t^*, l and l^*, and determine Λ and θ so that eqs. (4.p1) and (4.p2) become equivalent.

Answer:

$$\left.\begin{aligned} t^* &= \theta t, \\ l^* &= l\sqrt{\sin^2 \alpha + \Lambda^{-2} \cos^2 \alpha}, \\ \Lambda = \theta &= \sqrt{1 - v^2/c^2}. \end{aligned}\right\} \qquad (4.\text{p}3)$$

(*c*) In order to obtain the complete Lorentz transformation equations (4.13), introduce two coördinate systems, one at rest and one moving through the ether (S and S^*). Determine the apparent distances of points on the starred coördinate axes from the starred point of origin. Finally, find out how moving clocks must be adjusted so that a signal spreading in all directions from the starred point of origin and starting at the time $t = t^* = 0$ has the apparent speed c in all directions.

CHAPTER V

Vector and Tensor Calculus in an n Dimensional Continuum

The classical transformation theory draws a sharp dividing line between space and time coördinates. The time coördinate is always transformed into itself, because time intervals are considered in classical physics to be invariant.

The relativistic transformation theory destroys this detached position of the time coördinate in that the time coördinate of one coördinate system depends on both the time and space coördinates of another system whenever the two systems considered are not at rest relative to each other.

The laws of classical physics are always formulated so that the time coördinate is set apart from the spatial coördinates, and this is quite appropriate because of the character of the transformations with respect to which these laws are covariant. It is possible to formulate relativistic physics so that the time coördinate retains its customary special position, but we shall find that in this form the relativistic laws are cumbersome and often difficult to apply.

A proper formalism must be adapted to the theory which it is to represent. The Lorentz transformation equations suggest the uniform treatment of the *four* coördinates x, y, z, and t. How this might be done was shown by H. Minkowski. We shall find that the application of his formalism will simplify many problems, and that with its help many relativistic laws and equations turn out to be more lucid than their non-relativistic analogues.

Classical physics is characterized by the invariance of length and time. We can formally characterize relativistic physics by the invariance of the expression

$$\tau_{12}{}^2 = (t_2 - t_1)^2 - \frac{1}{c^2}\,[(x_2 - x_1)^2 + (y_2 - y_1)^2 + (z_2 - z_1)^2]. \qquad (5.1)$$

The invariance of this quadratic form of the coördinate differences restricts the group of all conceivable linear transformations of the four coördinates x, y, z, and t to that of the Lorentz transformations, just as the invariance of the expression

$$s_{12}^2 = (x_2 - x_1)^2 + (y_2 - y_1)^2 + (z_2 - z_1)^2 \tag{5.2}$$

defines the group of three dimensional orthogonal transformations. The four dimensional continuum (x, y, z, t), with its invariant form τ_{12}^2, can be treated as a four dimensional "space," in which τ_{12} is the "distance" between the two "points" (x_1, y_1, z_1, t_1) and (x_2, y_2, z_2, t_2). This procedure permits the development of a sort of generalized vector calculus in the "Minkowski world," and the formulation of all invariant relations in a clear and concise way.

We shall begin the study of this mathematical method with a recapitulation of elementary vector calculus, focusing our attention on its formal aspects. Then we shall generalize the formalism so that it becomes applicable to the space-time continuum.

Orthogonal transformations. Let us start with a rectangular Cartesian coördinate system and call its three coördinates x_1, x_2, and x_3 (instead of x, y, and z). Call the coördinate differences between two points P and P', Δx_1, Δx_2, and Δx_3. The distance between the two points is given by

$$s^2 = \sum_{i=1}^{3} \Delta x_i^2. \tag{5.2a}$$

If we carry out a linear coördinate transformation,

$$x'_i = \sum_{k=1}^{3} c_{ik} x_k + \mathring{x}'_i, \qquad i = 1, 2, 3, \tag{5.3}$$

the new coördinate differences are

$$\Delta x'_i = \sum_{k=1}^{3} c_{ik} \Delta x_k, \qquad i = 1, 2, 3. \tag{5.4}$$

These equations can be solved with respect to the Δx_k;

$$\Delta x_k = \sum_{i=1}^{3} c'_{ki} \Delta x'_i, \qquad k = 1, 2, 3. \tag{5.5}$$

Eq. (5.2a) expresses itself in terms of the new coördinates thus:

$$s^2 = \sum_{i,k,l=1}^{3} c'_{ik} c'_{il} \Delta x'_k \Delta x'_l. \tag{5.6}$$

The new coördinate system is a rectangular Cartesian system only if eq. (5.6) is formally identical with eq. (5.2a), that is, if

$$\sum_{i=1}^{3} c'_{ik} c'_{il} = \begin{Bmatrix} 0 & \text{if} & k \neq l, \\ 1 & \text{if} & k = l. \end{Bmatrix} \tag{5.7}$$

These equations take a more concise form if we use the so-called Kronecker symbol δ_{kl}, which is defined by the equations

$$\left.\begin{aligned} \delta_{kl} &= 0, \qquad k \neq l, \\ \delta_{kl} &= 1, \qquad k = l. \end{aligned}\right\} \tag{5.8}$$

Eq. (5.7) takes the form

$$\sum_{i=1}^{3} c'_{ik} c'_{il} = \delta_{kl}, \qquad k, l = 1, 2, 3. \tag{5.7a}$$

Eq. (5.7a) is the condition which must be satisfied if the transformation equations (5.3) are to represent the transition from one Cartesian coördinate system to another.

We can easily formulate the condition to be satisfied by the c_{ik} themselves. By substituting eqs. (5.4) in eqs. (5.5) we obtain

$$\Delta x_k = \sum_{i,l=1}^{3} c'_{ki} c_{il} \Delta x_l, \qquad k = 1, 2, 3, \tag{5.9}$$

and because this equation holds for arbitrary Δx_k, we find

$$\sum_{i=1}^{3} c'_{ki} c_{il} = \delta_{kl}, \qquad k, l = 1, 2, 3. \tag{5.10}$$

Now we can multiply eqs. (5.7a) by c_{lm} and sum over the three possible values of l. We obtain, because of (5.10),

$$\sum_{i,l=1}^{3} c'_{ik} c'_{il} c_{lm} = \underline{c'_{mk}} = \sum_{l=1}^{3} \delta_{kl} c_{lm} = \underline{c_{km}}. \tag{5.11}$$

By substituting c_{ki} for c'_{ik}, and so forth, in eqs. (5.7a), we obtain

$$\sum_{i=1}^{3} c_{ki} c_{li} = \delta_{kl}, \qquad k, l = 1, 2, 3, \tag{5.7b}$$

and eqs. (5.10) take the form

$$\sum_{i=1}^{3} c_{ik} c_{il} = \delta_{kl}, \qquad k, l = 1, 2, 3. \tag{5.10a}$$

Either eqs. (5.7b) or (5.10a), together with eqs. (5.3), define the group of orthogonal transformations.

Transformation determinant. We shall now investigate the transformations (5.3) and (5.7b) a little further.

The determinant of the coefficients c_{ik},

$$\begin{vmatrix} c_{11}, & c_{12}, & c_{13} \\ c_{21}, & c_{22}, & c_{23} \\ c_{31}, & c_{32}, & c_{33} \end{vmatrix},$$

is equal to ± 1. To prove this statement, we make use of the multiplication law of determinants, which states that the product of two determinants $|a_{ik}|$ and $|b_{ik}|$ is equal to the determinant $|\sum_l a_{il}b_{lk}|$. Now we form the determinant of both sides of eqs. (5.7b),

$$\left|\sum_{i=1}^{3} c_{ki}\,c_{li}\right| = |\delta_{kl}|. \tag{5.12}$$

According to the above-mentioned multiplication law, the left-hand side can be written

$$\left|\sum_{i=1}^{3} c_{ki}\,c_{li}\right| = \left|\sum_{i=1}^{3} c_{ki}\,c'_{il}\right| = |c_{ki}|\cdot|c'_{il}| = |c_{ki}|\cdot|c_{li}| = |c_{ki}|^2. \tag{5.13}$$

The value of the right-hand side of eq. (5.12) is equal to unity, since

$$|\delta_{mn}| = \begin{vmatrix} 1 & 0 & 0 \\ 0 & 1 & 0 \\ 0 & 0 & 1 \end{vmatrix} = 1. \tag{5.14}$$

Therefore, we really have

$$|c_{ki}| = \pm 1. \tag{5.15}$$

The value $+1$ of the determinant belongs to the "proper" rotations, while the value -1 belongs to orthogonal transformations involving a reflection.

Improved notation. In the great majority of equations occurring in three dimensional vector (and tensor) calculus, every literal index which occurs once in a product assumes any of the three values 1, 2, 3, and every literal index which occurs twice in a product is a summation index. From now on, therefore, we shall omit all summation signs and all remarks of the type $(i, k = 1, 2, 3)$, and it shall be understood that:

(1) *Each literal index which occurs once in a product assumes all its possible values*;

(2) *Each literal index which occurs twice in a product is a summation index, where the summation is to be carried out over all possible values.*

Thus, we write eqs. (5.3) and (5.6) like this:

$$x'_i = c_{ik}x_k + \overset{\circ}{x}'_i\,,$$

$$s^2 = c'_{ik}c'_{il}\Delta x'_k \Delta x'_l\,.$$

Summation indices are often called *dummy indices* or simply *dummies*. The significance of an expression is not changed if a pair of dummies is replaced by some other letter, for example,

$$c_{ik}x_k = c_{il}x_l\,.$$

Vectors. The transformation law of the Δx_k, (5.4), is the general transformation law of vectors with respect to orthogonal transformations, or, rather: A vector is defined as a set of three quantities which transform like coördinate differences:

$$a'_k = c_{ki}a_i\,. \tag{5.16}$$

When the vector components are given with respect to any one Cartesian coördinate system, they can be computed with respect to every other Cartesian coördinate system.

The norm of a vector is defined as the sum of the squared vector components.

We shall prove that the norm is an invariant with respect to orthogonal transformations, or, that

$$a'_k a'_k = a_i a_i\,. \tag{5.17}$$

Substituting for a'_k its expression (5.16), and making use of eq. (5.10a), we obtain

$$a'_k a'_k = c_{ki}a_i c_{kl}a_l = \delta_{il}a_i a_l = a_i a_i\,,$$

which proves that eq. (5.17) holds for orthogonal transformations.

The scalar product of two vectors is defined as the sum of the products of corresponding vector components,

$$(\mathbf{a}\cdot\mathbf{b}) \equiv a_i b_i\,. \tag{5.18}$$

That this expression is an invariant with respect to orthogonal transformations is shown by a computation analogous to the proof of eq. (5.17). The norm of a vector is the scalar product of the vector by itself.

The word *scalar* is frequently used in vector and tensor calculus instead of *invariant*. "Scalar product" means "invariant product." Sums and differences of vectors are, again, vectors,

$$\left.\begin{aligned} a_i + b_i &= s_i\,, \\ a_i - b_i &= d_i\,. \end{aligned}\right\} \tag{5.19}$$

That the new quantities s_i and d_i really transform according to eq. (5.16) follows from the linear, homogeneous character of that transformation law.

The product of a vector and a scalar (invariant) is a vector,

$$s \cdot a_i = b_i\,. \tag{5.20}$$

The proof is left to the reader.

The discussion of the remaining algebraic vector operation, the vector product, must be deferred until later in this chapter, because its transformation properties are not quite like those of a vector.

Vector analysis. We are now ready to go on to the simplest differential operations, the gradient and the divergence. In the three dimensional space of the three coördinates x_i, let us take a scalar field V, that is, a function of the three coördinates x_i which is invariant with respect to coördinate transformations. The form of the function V of the coördinates will, of course, depend on the coördinate system used, but in such a way that its value at a fixed point P is not changed by the transformation.

What is the transformation law of the derivatives of V with respect to the three coördinates,

$$V_{,i} = \frac{\partial V}{\partial x_i}\,? \tag{5.21}$$

We must express the derivatives with respect to x'_k in terms of the derivatives with respect to x_i,

$$\frac{\partial V}{\partial x'_k} = \frac{\partial x_i}{\partial x'_k}\,\frac{\partial V}{\partial x_i}\,. \tag{5.22}$$

According to eq. (5.3), the x'_k are linear functions of the x_i, and vice versa. Therefore, the $\partial x_i / \partial x'_k$ are constants, and they are the constants c'_{ik} defined by eqs. (5.5). We have, therefore,

$$V_{,k'} = c'_{ik} \cdot V_{,i}$$

and, according to eq. (5.11),

$$V_{,k'} = c_{ki} V_{,i}\,. \tag{5.23}$$

The three quantities $V_{,i}$ transform according to eq. (5.16); therefore, they are the components of a vector, which is called the gradient of the scalar field V.

Three functions of the coördinates, $V_i(x_1, x_2, x_3)$, are the components of a vector field if at each space-point they transform as the components of a vector. The functions V'_i of the coördinates x'_r are, thus, given by the equations

$$V'_i(x'_r) = c_{ik}V_k(x_s), \tag{5.11a}$$

where the x_s are connected with the x'_r by the transformation equations. The gradient creates a vector field out of a scalar field.

The divergence does the opposite. Given a vector field V_i, we form the sum of the three derivatives of each component with respect to the coördinate with the same index,

$$\text{div } \mathbf{V} \equiv V_{i,i}. \tag{5.24}$$

We have to show that this expression is an invariant (or scalar),

$$V'_{k,k'} = V_{i,i}. \tag{5.25}$$

The procedure is exactly the same as before. We replace the primed quantities and derivatives by the unprimed quantities,

$$V'_{k,k'} = c'_{mk}(c_{kl}V_l)_{,m} = c_{kl}c'_{mk}V_{l,m}. \tag{5.26}$$

Because of eq. (5.10), this last expression is equal to the right-hand side of eq. (5.25).

The divergence of a gradient of a scalar field is the Laplacian of that scalar field and, of course, is itself a scalar field,

$$\text{div grad } V \equiv V_{,ss} \equiv \nabla^2 V. \tag{5.27}$$

Tensors. In many parts of physics we encounter quantities whose transformation laws are somewhat more involved than those of vectors. As an example, let us consider the so-called "vector gradient." When a vector field V_i is given, we can obtain a set of quantities which determine the change of each component of V_i as we proceed from a point with the coördinates x_i in an arbitrary direction to the infinitesimally near point with the coördinates $x_i + \delta x_i$. The increments of the three quantities V_i are

$$\delta V_i = V_{i,k}\delta x_k, \tag{5.28}$$

and the nine quantities $V_{i,k}$ are called the vector gradient of V_i. We can easily derive its transformation law in the usual manner:

$$V'_{m,n'} = c'_{kn}(c_{mi}V_i)_{,k} = c_{mi}c_{nk}V_{i,k}. \tag{5.29}$$

The vector gradient is one example of the new class of quantities which we are now going to treat, the tensors. *In general, a tensor has N indices, all of which take all values* 1 *to* 3. *The tensor has, therefore,* 3^N *components. These* 3^N *components transform according to the transformation law*

$$T'_{mns\ldots} = c_{mi}c_{nk}c_{sl} \cdots T_{ikl\ldots}. \tag{5.30}$$

The number of indices, N, is called its rank. The vector gradient is a tensor of rank 2, vectors are tensors of rank 1, and scalars may be called tensors of rank 0.

One very important tensor is the Kronecker symbol. Its values in one coördinate system, when substituted into eq. (5.30), yield the same values in another coördinate system,

$$\delta'_{kl} = c_{ki}c_{lj}\delta_{ij} = c_{ki}c_{li} = \delta_{kl}, \tag{5.31}$$

according to eq. (5.7b).

The sum or difference of two tensors of equal rank is a tensor of the same rank. We formulate this law for tensors of rank 3:

$$T_{ikl} + U_{ikl} = V_{ikl}, \tag{5.32}$$

$$T_{ikl} - U_{ikl} = W_{ikl}. \tag{5.33}$$

The proof is the same as for the corresponding law for vectors, eq. (5.19).

The product of two tensors of ranks M and N is a new tensor of rank $(M + N)$,

$$T_{ik\ldots}U_{lm\ldots} = V_{ik\ldots lm\ldots}. \tag{5.34}$$

The rank of a tensor may be lowered by 2 (or by any even number) by an operation called "contraction." Any two indices are converted into a pair of dummy indices. For instance, we can contract the tensor $T_{ikl\ldots}$ to obtain the tensors $T_{ssl\ldots}$ or $T_{irr\ldots}$. The proof that these new contracted tensors are again tensors is very simple. For the first example given here, it runs as follows.

$$T'_{ssl\ldots} = c_{si}c_{sk}c_{lm} \cdots T_{ikm\ldots}.$$

Because of eq. (5.10a), the right-hand side is equal to

$$T'_{ssl\ldots} = \delta_{ik}c_{lm} \cdots T_{ikm\ldots} = c_{lm} \cdots T_{iim\ldots}. \tag{5.35}$$

When we contract the vector gradient (tensor of rank 2), we obtain the divergence (tensor of rank 0). The operations product, (5.34), and the contraction can be combined so that they yield tensors such as $T_{ik}U_{lk}$, $T_{ik}U_{km}$, $T_{ik}U_{ik}$, $T_{ik}U_{ki}$.

Tensors may have *symmetry properties* with respect to their indices

If a tensor is not changed when two or more indices are exchanged, then it is *symmetric* in these indices. Instances are

$$t_{ikl} = t_{kil} \,,$$

$$t_{iklm} = t_{ilkm} = t_{kilm} = t_{lkim} = t_{klim} = t_{likm} \,.$$

The first tensor is symmetric in its first two indices; the second tensor is symmetric in its first three indices.

When a tensor remains the same or changes the sign of every component upon the permutation of certain indices, the sign depending on whether it is an even or an odd permutation, we say that the tensor is *antisymmetric* (also *skewsymmetric* or *alternating*) with respect to these indices. Instances are

$$t_{ikl} = -t_{kil} \,,$$

$$t_{iklm} = t_{klim} = t_{likm} = -t_{ilkm} = -t_{kilm} = -t_{lkim} \,.$$

All such symmetry properties of a tensor are invariant. The proof is extremely simple and shall be left to the student.

The Kronecker tensor is symmetric in its two indices.

Tensor analysis. When a tensor is differentiated with respect to the coördinates, a new tensor is obtained, the rank of which is greater by 1. The proof again consists of simple computation:

$$T'_{mn\cdots,s'} = c'_{ls}(c_{mi}c_{nk} \cdots T_{ik\cdots})_{,l} = c_{mi}c_{nk} \cdots c_{sl}T_{ik\cdots,l} \,. \quad (5.36)$$

When the resulting tensor is contracted with respect to the index of differentiation and another index, for example, $T_{ik\cdots,k}$, it is often called a divergence.

Tensor densities. The "vector product" of two vectors **a** and **b** is usually defined as a vector which is perpendicular to **a** and to **b** and which has the magnitude $|\mathbf{a}|\cdot|\mathbf{b}|\cdot\sin(\mathbf{a}, \mathbf{b})$. As there are always two vectors satisfying these conditions, viz., $\mathbf{P}_1$ and $\mathbf{P}_2$ in Fig. 6, a choice is made between these two vectors by the further condition that **a**, **b**, and **P** shall form a "screw" of the same type as the coördinate axes in the sequence x, y, z. In Fig. 6, the vector $\mathbf{P}_1$ satisfies this condition, but only because the chosen coördinate system is a "right-handed" coördinate system. If we carry out a "reflection" (for example, give the positive X-axis the direction to the rear of the figure instead of to the front), $\mathbf{P}_2$ becomes automatically the vector product of **a** and **b**.

The vector product is, thus, not an ordinary vector, but changes its sign when we transform a right-handed coördinate system into a left-handed system, or vice versa. Such quantities are called "*axial vectors*," while ordinary vectors are called "*polar vectors*."

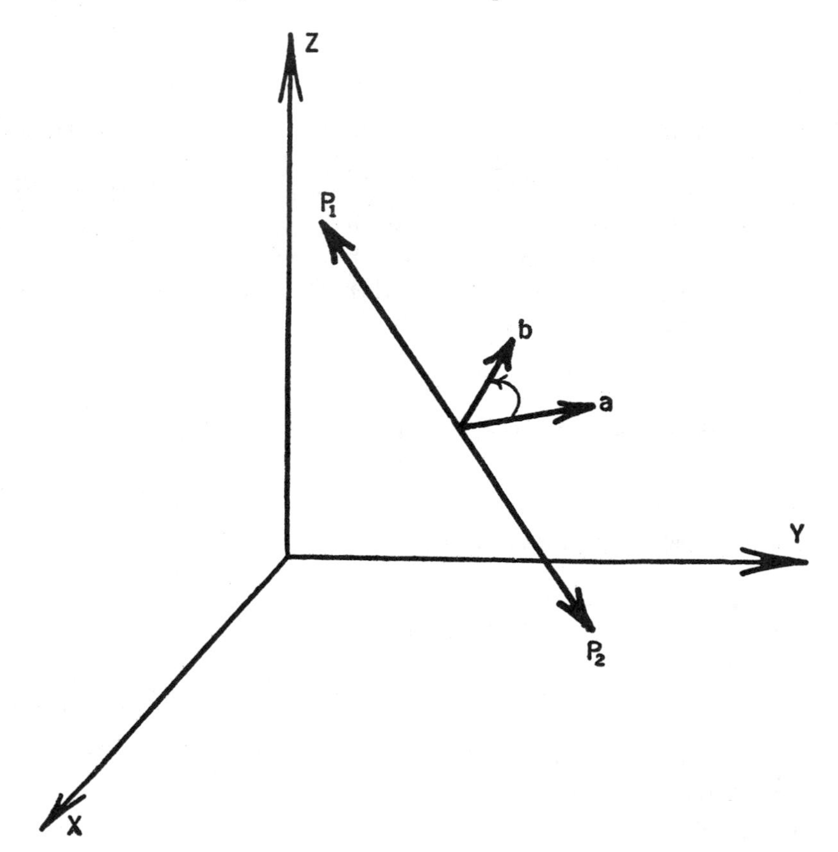

Fig. 6. The vector product. In a right-handed coördinate system, $\mathbf{P}_1$ represents the vector product of **a** by **b**.

With respect to a Cartesian coördinate system, the components of **P** are given by the expressions

$$\left.\begin{aligned} P_1 &= a_2 b_3 - a_3 b_2, \\ P_2 &= a_3 b_1 - a_1 b_3, \\ P_3 &= a_1 b_2 - a_2 b_1. \end{aligned}\right\} \qquad (5.37)$$

Similarly, the *curl* of a vector field V_i is defined as an "axial vector" with the components

$$\left.\begin{aligned} C_1 &= V_{3,2} - V_{2,3}, \\ C_2 &= V_{1,3} - V_{3,1}, \\ C_3 &= V_{2,1} - V_{1,2}. \end{aligned}\right\} \tag{5.38}$$

From the point of view of tensor calculus, we can avoid the concept of "axial vector" by introducing vector product and curl as skewsymmetric tensors of rank 2,

$$P_{ik} = a_i b_k - a_k b_i , \tag{5.37a}$$

and

$$C_{ik} = V_{k,i} - V_{i,k} . \tag{5.38a}$$

It can be shown that all equations in which "axial vectors" appear can be written in the covariant manner with the help of such skewsymmetric tensors. Nevertheless, this treatment does not show very clearly the connection between the transformation law of a skewsymmetric tensor of rank 2 and that of an "axial vector." We can conform closely to the methods of elementary vector calculus by introducing in addition to tensors a new type of quantity, the "*tensor densities.*"

The tensor densities transform like tensors, except that they are also multiplied by the transformation determinant (5.15). As long as this determinant equals +1, that is, when the transformation is a "proper orthogonal transformation" without reflection, there is no difference between a tensor and a tensor density. But a density undergoes a change of sign (compared with a tensor) when a reflection of the coördinate system is carried out. The tensor densities have, thus, the same relationship to tensors as the "axial vectors" have to the "polar vectors." Their transformation law can be written thus:

$$\mathfrak{T}'_{mn\cdots} = c_{mi} c_{nk} \cdots \mid c_{ab} \mid \mathfrak{T}_{ik\cdots} . \tag{5.39}$$

The laws of tensor density algebra and calculus are: The sum or difference of two tensor densities of equal rank is again a tensor density of the same rank. The product of a tensor and a tensor density is a tensor density. The product of two tensor densities is a tensor. The contraction of a tensor density yields a new tensor density of lower rank. The derivatives of the components of a tensor density are the components of a new tensor density, the rank of which is greater by 1 than the rank of the original density.

The tensor density of Levi-Civita. We found that the Kronecker symbol is a tensor, the components of which take the same constant values in every coördinate system. Likewise, there exists a constant tensor density of rank 3, the Levi-Civita tensor density, defined as follows. *δ_{ikl} is skewsymmetric in its three indices; therefore, all those components which have at least two indices equal vanish. The values of the nonvanishing components are ± 1, the sign depending on whether (i, k, l) is an even or an odd permutation of* $(1, 2, 3)$.

We have yet to show that δ_{ikl} are really the components of a tensor density. To do that, let us consider a tensor density D_{ikl} which has the components δ_{ikl} in one coördinate system. If it turns out that its components in some other coördinate system are again δ_{ikl}, our assertion is proved.

The components of D_{ikl} in another coördinate system are

$$D'_{mns} = | c_{ab} | \, c_{mi} c_{nk} c_{sl} \delta_{ikl} . \tag{5.40}$$

As the skewsymmetry of D_{ikl} is preserved by the coördinate transformation, we know that all components D'_{mns} with at least two equal indices vanish. We have to compute only components with all three indices different. The component D'_{123} is given by the expression

$$D'_{123} = | c_{ab} | \, c_{1i} c_{2k} c_{3l} \delta_{ikl} . \tag{5.41}$$

The right-hand side is simply the square of $| c_{ab} |$, and equal to unity. For δ_{ikl} is defined so that $c_{1i} c_{2k} c_{3l} \delta_{ikl}$ is just the determinant $| c_{ab} |$.

Now that we know that D'_{123} is equal to unity, the remaining components are obtained simply by using the symmetry properties. They are

$$D'_{123} = D'_{231} = D'_{312} = -D'_{132} = -D'_{213} = -D'_{321} . \tag{5.42}$$

In other words, the D'_{mns} are again equal to δ_{mns}, and the proof is completed.

Vector product and curl. With the help of the Levi-Civita tensor density, we can associate skewsymmetric tensors of rank 2 with vector densities:

$$\mathfrak{w}_i = \tfrac{1}{2} \delta_{ikl} w_{kl} . \tag{5.43}$$

The converse relation is

$$w_{kl} = \delta_{kli} \mathfrak{w}_i . \tag{5.44}$$

Applying eq. (5.43) to the vector product and to the curl, defined by eqs. (5.37) and (5.38), respectively, we obtain

$$\mathfrak{P}_i = \delta_{ikl} a_k b_l , \tag{5.37b}$$

$$\mathfrak{C}_i = \delta_{ikl} a_{l,k} . \tag{5.38b}$$

Because these two vector densities $\mathfrak{P}_i$ and $\mathfrak{C}_i$ transform like vectors except for the change of sign in the case of coördinate reflections, they are treated as vectors in vector calculus, but they are referred to as "axial" vectors, implying that they have something to do with "rotation."

They really do have something to do with rotation. The angular momentum, for instance, is the vector product of the radius vector and the ordinary momentum,

$$\mathfrak{J}_i = \delta_{ikl} x_k p_l \,. \tag{5.45}$$

In the case of a reflection, it transforms as an ordinary vector would, except for its sign. Assume that of the x_k only x_1 does not vanish, and that $\mathbf{p}$ has only the component p_2. Then the angular momentum has only the component $\mathfrak{J}_3$. We can carry out a reflection in three different ways: We can replace x_1 by $(-x_1)$ or we can do the same thing with x_2 or with x_3, the other two coördinates remaining unchanged in every case. $\mathfrak{J}_3$ changes its sign in the first two cases, and it remains unchanged when x_3 is replaced by $(-x_3)$. A genuine vector would change its sign only when x_3 is replaced by $(-x_3)$.

Generalization. Now that we have reviewed briefly vector and tensor calculus in three dimensions with respect to orthogonal transformations, we are in a position to generalize the concepts obtained so that they will be applicable to the problems we shall discuss later. The generalization is to be carried out in two steps. First, we have to extend the formalism so that it applies to any positive integral number of dimensions; second, we shall have to consider coördinate transformations other than orthogonal transformations.

***n* dimensional continuum.** The first generalization is almost trivial. Instead of three coördinates x_1, x_2, x_3, we have n coördinates, $x_1 \cdots x_n$, describing an n dimensional manifold. We assume, again, that there exists an invariant distance between two points,

$$s^2 = \Delta x_i \Delta x_i \,, \tag{5.2b}$$

where the summation is to be carried out over all n values of the index i. Eq. (5.2b) is invariant with respect to the group of n dimensional, orthogonal transformations,

$$x_i' = c_{ik} x_k + \mathring{x}_i' \,, \tag{5.3a}$$

where the c_{ik} have to satisfy the conditions

$$c_{ik} c_{il} = \delta_{kl} \tag{5.10b}$$

All indices take all values $1 \cdots n$, and summations are to be carried out from 1 to n. The determinant $| c_{ab} |$ is again equal to ± 1.

Vectors are defined by the transformation law

$$a_k' = c_{ki} a_i \,, \tag{5.16a}$$

and their algebra and analysis are identical with the algebra and analysis of three dimensional vectors.

Tensors and tensor densities are defined as in three dimensional space, except that all indices run from 1 to n. δ_{ik} is again a symmetric tensor.

The Levi-Civita tensor density is defined as follows. $\delta_{ik \cdots s}$ is a tensor density with n indices (of rank n), skewsymmetric in all of them. The nonvanishing components are ± 1, the sign depending on whether $(i, k, \cdots, s)$ is an even or an odd permutation of $(1, 2, \cdots, n)$. The "vector product" is no longer a vector density. With the help of the Levi-Civita tensor density, we can form from a skewsymmetric tensor of rank m $(m \leqslant n)$ a skewsymmetric tensor density of rank $(n - m)$. Only when n is 3 is the "conjugate" tensor density to a tensor of rank 2 a vector density.

General transformations. The "length" defined in the Minkowski space, (5.1), does not have the form (5.2b). We shall, therefore, no longer restrict ourselves to transformations which leave eqs. (5.2b) invariant, but shall take up general coördinate transformations. Since the Lorentz transformations are much less general than the coördinate transformations which we are about to consider, it may appear that we are deviating from our main purpose. But we shall need the general coördinate transformations in the general theory of relativity; and, since they are as simple in most respects as the more restricted group of Lorentz transformations, we shall thus avoid needless repetition.

Let us consider a space in which we can introduce Cartesian coördinate systems so that the length is defined by eq. (5.2b). Then let us pass from a Cartesian coördinate system to another coördinate system which is not Cartesian. The new coördinates may be called $\xi^1, \xi^2, \cdots, \xi^n$ (the superscripts are not to be mistaken for power exponents). We have, then,

$$\xi^i = f^i(x_1, \cdots x_n), \qquad i = 1 \cdots n, \tag{5.46}$$

where the n functions f^i are arbitrary, except that we shall assume that their derivatives exist up to the order needed in any discussion; that the Jacobian of the transformation,

$$\det \left| \frac{\partial \xi^s}{\partial x_r} \right|$$

vanishes nowhere; and that the ξ^i are real for all real values of the $x_1 \cdots x_n$.

s^2 is not, in general, a quadratic form of the $\Delta\xi^i$, as it is of the Δx_i. But the square of the distance between two infinitesimally near points remains a quadratic form of the coördinate differentials. In terms of Cartesian coördinates, this infinitesimal distance is given by

$$ds^2 = dx_k dx_k, \tag{5.47}$$

and dx_k can be expressed in terms of the $d\xi^i$,

$$dx_k = \frac{\partial x_k}{\partial \xi^i} d\xi^i. \tag{5.48}$$

Substitution into eq. (5.47) yields

$$ds^2 = \frac{\partial x_k}{\partial \xi^i} \frac{\partial x_k}{\partial \xi^l} d\xi^i d\xi^l. \tag{5.49}$$

ds^2 is a quadratic form of the $d\xi^i$, regardless of the coördinate system used. This suggests that the coördinate differentials $d\xi^i$ and the distance differential ds will, in the field of general coördinate transformations, take the place of the coördinate differences Δx_i and the distance s, which are adapted to Cartesian coördinates and orthogonal transformations.

Vectors. Let us see how the coördinate differentials transform in the case of a general coördinate transformation. Let ξ^i, ξ'^l be two sets of non-Cartesian coördinates. Then the coördinate differentials are connected by the equations

$$d\xi'^l = \frac{\partial \xi'^l}{\partial \xi^i} d\xi^i. \tag{5.50}$$

The transformed coördinate differentials $d\xi'^l$ are linear, homogeneous functions of the $d\xi^i$; but the transformation coefficients $(\partial\xi'^l/\partial\xi^i)$ are not constant, but functions of the ξ^i. Neither is their determinant, $|\partial\xi'^a/\partial\xi^b|$, a constant. We shall use them for the purpose of defining a type of geometrical quantity, the "contravariant vector": *A contravariant vector has n components, which transform like coordinate differentials,*

$$a'^l = \frac{\partial \xi'^l}{\partial \xi^i} a^i. \tag{5.51}$$

The sum or the difference of contravariant vectors, and the product of a contravariant vector and a scalar are also contravariant vectors.

It is impossible to form scalar products of contravariant vectors alone.

To find out what corresponds to the scalar product in our formalism, we shall consider a scalar field $V(\xi^1, \cdots, \xi^n)$. The change of V along an infinitesimal displacement $\delta\xi^i$ is given by

$$\delta V = V_{,i}\,\delta\xi^i, \qquad V_{,i} = \frac{\partial V}{\partial \xi^i}. \tag{5.52}$$

The left-hand side is obviously invariant. The right-hand side has the form of an inner product; one factor is the contravariant vector $\delta\xi^i$, the other is the gradiant of V, $V_{,i}$.

The components of the gradient of V transform according to the law

$$V_{,l'} = \frac{\partial \xi^i}{\partial \xi'^l} V_{,i}. \tag{5.53}$$

The $V_{,l'}$ are linear, homogeneous functions of the $V_{,i}$. The transformation law (5.53) is not that of a contravariant vector. We call the gradient of a scalar field a *covariant vector* and define in general a covariant vector as *a set of n quantities which transform according to the law*

$$a_l' = \frac{\partial \xi^i}{\partial \xi'^l} a_i. \tag{5.54}$$

The sum or difference of covariant vectors, and the product of a covariant vector and a scalar are, again, covariant vectors.

In order to distinguish between contravariant and covariant vectors, we shall always write contravariant vectors with superscripts, and covariant vectors with subscripts.

The transformation coefficients of contravariant and covariant vectors are different, but they are related. The $(\partial\xi'^l/\partial\xi^i)$ of eq. (5.51) and the $(\partial\xi^i/\partial\xi'^l)$ of eq. (5.54) are connected by the n^2 equations

$$\frac{\partial \xi'^l}{\partial \xi^i}\frac{\partial \xi^k}{\partial \xi'^l} = \delta_i^k, \tag{5.55}$$

where δ_i^k is, again, the Kronecker symbol. Because of eq. (5.55), the inner product of a covariant and a contravariant vector is an invariant,

$$a_l' b'^l = a_i b^i. \tag{5.56}$$

Let us return for a moment to the orthogonal transformations. Their transformation coefficients, c_{ik}, satisfy the equations (5.7b) and (5.10a). In the case of orthogonal transformations, the coördinate transformation derivatives are

$$\frac{\partial x_l'}{\partial x_i} = c_{li};$$

and, because eq. (5.55) holds for all transformations, it follows from eq. (5.7b) that $(\partial x_i/\partial x'_l)$, too, is

$$\frac{\partial x_i}{\partial x'_l} = c_{li}. \tag{5.57}$$

That is why the distinction between contravariant and covariant vectors does not exist in the realm of orthogonal transformations.

Tensors. *Tensors are defined as forms with n^N components (N being the rank of the tensor), which transform with respect to each index like a vector. They may be covariant in all indices, contravariant in all indices, or mixed, that is, contravariant in some indices, covariant in others.* The contravariant indices are written as superscripts, the covariant indices as subscripts. The example of a mixed tensor of rank three may illustrate the definition:

$$t'^{\cdot\cdot s}_{mn\cdot} = \frac{\partial \xi^i}{\partial \xi'^m} \frac{\partial \xi^k}{\partial \xi'^n} \frac{\partial \xi'^s}{\partial \xi^l} t^{\cdot\cdot l}_{ik\cdot}. \tag{5.58}$$

Symmetry properties of tensors are invariant with respect to coördinate transformations only if they exist with respect to indices of the same type (covariant or contravariant).

The Kronecker symbol is a mixed tensor,

$$\delta'^i_k = \frac{\partial \xi'^i}{\partial \xi^m} \frac{\partial \xi^n}{\partial \xi'^k} \delta^m_n = \frac{\partial \xi'^i}{\partial \xi^m} \frac{\partial \xi^m}{\partial \xi'^k} = \delta^i_k. \tag{5.59}$$

The product of two tensors of ranks M and N is a tensor of rank $(M + N)$, and every index of either factor keeps its character as a contravariant or covariant index. Furthermore, just as the scalar product of a covariant and a contravariant vector is a scalar, so any two indices of different position can be used as a pair of dummies, and the result is a lowering of the tensor rank by 2. Examples of such products are:

$$a^{\cdot\cdot l}_{ik\cdot} b^{mn}_{\cdot\cdot s}, \qquad a^{\cdot\cdot l}_{ik\cdot} b^{kn}_{\cdot\cdot s}, \qquad a^{\cdot\cdot l}_{ik\cdot} b^{kn}_{\cdot\cdot l}.$$

When the corresponding components of two tensors of equal rank are added or subtracted, the sums or differences are components of a new tensor, provided that the two original tensors have equal numbers of indices of the same type.

These are the simple rules of tensor algebra. They indicate how new quantities may be formed which transform according to laws of the pattern (5.58).

Metric tensor, Riemannian spaces. The expression which occurs in eq. (5.49), $g_{il} = \frac{\partial x_k}{\partial \xi^i} \frac{\partial x_k}{\partial \xi^l}$, transforms as a tensor when we change from a coördinate system ξ^i to ξ'^m,

$$\left.\begin{aligned} \frac{\partial x_k}{\partial \xi'^m} \frac{\partial x_k}{\partial \xi'^n} &= \frac{\partial x_k}{\partial \xi^i} \frac{\partial \xi^i}{\partial \xi'^m} \cdot \frac{\partial x_k}{\partial \xi^l} \frac{\partial \xi^l}{\partial \xi'^n}; \\ g'_{mn} &= \frac{\partial \xi^i}{\partial \xi'^m} \frac{\partial \xi^l}{\partial \xi'^n} g_{il}\,. \end{aligned}\right\} \tag{5.60}$$

In other words, g_{il} is a covariant symmetric tensor of rank 2. It is called the *metric tensor*.

There exist "spaces" where it is not possible to introduce a Cartesian coördinate system. Two dimensional "spaces" of that kind include the surface of a sphere. If we introduce as a coördinate system the latitude and longitude, φ and ϑ, it is possible to express the distance between two infinitesimally near points on the spherical surface in terms of their coördinate differentials,

$$ds^2 = R^2(d\varphi^2 + \cos^2 \varphi \, d\vartheta^2).$$

In order to include such continuous manifolds within the scope of our investigations, we shall consider spaces with a metric tensor, without raising the question, for the time being, of whether a Cartesian coördinate system can be introduced or not. Whenever a "squared infinitesimal distance," that is, an invariant homogeneous quadratic function of the coördinate differentials, is defined, we call the manifold a "metric space" or a "Riemannian space." If it is possible to introduce in a Riemannian space a coördinate system with respect to which the components of the metric tensor assume the values δ_{ik} at every point, such a coördinate system is a Cartesian coördinate system, and the space is called a Euclidean space. Euclidean spaces are, therefore, a special case of Riemannian spaces.

Whenever the infinitesimal distance is given by equations of the form

$$ds^2 = g_{il} \, d\xi^i \, d\xi^l, \tag{5.61}$$

where ds^2 is an invariant, g_{il} is a covariant tensor. Our previous proof was based on the assumption that the g_{il} were given by the expression $\frac{\partial x_k}{\partial \xi^i} \frac{\partial x_k}{\partial \xi^l}$; in other words, we implied the possibility of introducing Cartesian coördinates. To show that the transformation properties of g_{il} are independent of this assumption, we shall consider the equation

$$g_{il} \, d\xi^i \, d\xi^l = g'_{mn} \, d\xi'^m \, d\xi'^n, \tag{5.62}$$

which expresses the invariance of ds^2. When we replace the $d\xi^i$ on the left-hand side by $(\partial\xi^i/\partial\xi'^m)\cdot d\xi'^m$, we obtain

$$g_{il}\frac{\partial\xi^i}{\partial\xi'^m}\frac{\partial\xi^l}{\partial\xi'^n}d\xi'^m\,d\xi'^n = g'_{mn}\,d\xi'^m\,d\xi'^n. \tag{5.63}$$

Because the $d\xi'^m$ are arbitrary, it follows that the coefficients on both sides are equal, that is, eq. (5.60) is satisfied.

If the determinant of the components of g_{il} does not vanish, it is possible to define new quantities g^{il} by the equations

$$g_{ik}g^{kl} = \delta^l_i\,. \tag{5.64}$$

In order to determine the transformation law of these quantities, we transform first the g_{ik}. We replace them by the expression

$$g_{ik} = \frac{\partial\xi'^m}{\partial\xi^i}\frac{\partial\xi'^n}{\partial\xi^k}g'_{mn}\,, \tag{5.65}$$

so that we get

$$\frac{\partial\xi'^m}{\partial\xi^i}g'_{mn}\frac{\partial\xi'^n}{\partial\xi^k}g^{kl} = \delta^l_i\,;$$

then multiply the latter equation by $\dfrac{\partial\xi^i}{\partial\xi'^r}\cdot\dfrac{\partial\xi'^s}{\partial\xi^l}$. We know from eq. (5.59) that the right-hand side becomes δ^s_r; and the left-hand side becomes

$$\frac{\partial\xi'^m}{\partial\xi^i}\frac{\partial\xi^i}{\partial\xi'^r}g'_{mn}\frac{\partial\xi'^n}{\partial\xi^k}g^{kl}\frac{\partial\xi'^s}{\partial\xi^l} = \delta^m_r\,g'_{mn}\frac{\partial\xi'^n}{\partial\xi^k}\frac{\partial\xi'^s}{\partial\xi^l}g^{kl}$$

$$= g'_{rn}\frac{\partial\xi'^n}{\partial\xi^k}\frac{\partial\xi'^s}{\partial\xi^l}g^{kl},$$

so that we get

$$g'_{rn}\frac{\partial\xi'^n}{\partial\xi^k}\frac{\partial\xi'^s}{\partial\xi^l}g^{kl} = \delta^s_r\,. \tag{5.66}$$

By comparing eq. (5.66) with eq. (5.64), we find that

$$\frac{\partial\xi'^n}{\partial\xi^k}\frac{\partial\xi'^s}{\partial\xi^l}g^{kl} = g'^{ns}; \tag{5.67}$$

that is, the g^{kl} are the components of a contravariant tensor. This tensor is symmetric. We can show this by multiplying eq. (5.64) by $g_{ls}g^{ir}$. The left-hand side becomes

$$g_{ik}g^{kl}g_{ls}g^{ir} = g_{ki}g^{ir}g^{kl}g_{sl} = \delta^r_kg^{kl}g_{sl} = g_{sl}g^{rl}.$$

while the right-hand side is equal to

$$\delta^l_i g_{ls} g^{ir} = g_{is} g^{ir} = g_{si} g^{ir} = \delta^r_s .$$

We obtain

$$g_{sl} g^{rl} = \delta^r_s .$$

Comparing this equation with eq. (5.64), we find that

$$g^{kl} = g^{lk}. \tag{5.68}$$

g^{kl} is called the contravariant metric tensor. The values of its components are equal to the cofactor of the g_{kl}, divided by the full determinant $g = | g_{ab} |$,

$$g^{kl} = g^{-1} \cdot \text{cofactor } (g_{kl}). \tag{5.69}$$

In the case of Cartesian coördinate systems, g^{kl} equals δ_{kl}.

Raising and lowering of indices. A covariant vector can be obtained from a contravariant vector by multiplying it by the covariant metric tensor and summing over a pair of indices,

$$a_i = g_{ik} a^k. \tag{5.70}$$

This process can be reversed by multiplying a_i by the contravariant metric tensor

$$a^k = g^{ik} a_i . \tag{5.71}$$

From the definition of the contravariant metric tensor, (5.64), it follows that eq. (5.71) is equivalent to eq. (5.70); in other words, eq. (5.71) leads back to the same contravariant vector which appears in eq. (5.70). The two vectors a_i and a^k can, therefore, be properly considered as two equivalent representations of the same geometrical object. The operations (5.70) and (5.71) are called lowering and raising of indices. It is possible, of course, to raise and lower any tensor index in the same way.

The norm of a vector is defined as the scalar

$$a^2 = g_{ik} a^i a^k = g^{ik} a_i a_k = a^i a_i , \tag{5.72}$$

while the scalar product of two vectors can be written in the alternative forms

$$a_i b^i = a^i b_i = g_{ik} a^i b^k = g^{ik} a_i b_k . \tag{5.73}$$

Tensor densities, Levi-Civita tensor density. *We call a tensor density a quantity which transforms according to the law*

$$\mathfrak{T}'^{m\cdots}_{\cdots n\cdots} = \frac{\partial \xi'^m}{\partial \xi^i} \cdots \frac{\partial \xi^k}{\partial \xi'^n} \cdots \left| \frac{\partial \xi^a}{\partial \xi'^b} \right|^W \mathfrak{T}^{i\cdots}_{\cdots k\cdots} . \tag{5.74}$$

W is a constant the value of which is characteristic for any given tensor density; this constant is called the weight of the tensor density. Tensors are tensor densities of weight zero. Depending on the number of indices, we speak also of scalar densities and vector densities.

The sum of two densities with the same numbers of like indices and the same weight is a density with the same characteristics. In the multiplication, the weights are added.

The Levi-Civita symbols $\delta_{a_1 \cdots a_n}$ and $\delta^{a_1 \cdots a_n}$ are densities of weight (-1) and $(+1)$, respectively. (n is the number of dimensions.) The proof is simple and analogous to the one given for orthogonal transformations.

The determinant of the covariant metric tensor,

$$g = | g_{ik} |, \tag{5.75}$$

is a scalar density of weight 2.

We shall have very little occasion to work with tensor densities.

Tensor analysis.[1] The consideration of tensor densities completes the discussion of tensor algebra as far as it is needed in this book. As for tensor analysis, we have already found that the ordinary derivatives of a scalar field are the components of a covariant vector field. In general, however, the derivatives of a tensor field do not form a new tensor field.

Let us take the derivative of a vector. The derivative compares the value of a vector at one point with its value at another infinitesimally near point, in a given direction. In the case of a coördinate transformation, the vectors at the two points do not transform with the same transformation coefficients, for the coefficients of the transformation are themselves functions of the coördinates. Therefore, the derivatives of the transformation coefficients enter into the transformation law of the derivatives of the vector.

However, there is a way which enables us to obtain new tensors by differentiation. The method is suggested by our experience in the realm of Cartesian coördinates. There we can describe the derivative of a vector or tensor thus: The vector is first carried to the "neighboring" point without changing the values of its components: it is displaced *parallel to itself.* (As long as we use Cartesian coördinates, that statement has an invariant meaning, because the transformation coefficients are the same at both points.) Then this parallel displaced vector is compared with the value of the vector (as a function of the coördinates)

---

[1] Tensor analysis with respect to general coördinates is discussed here because it is necessary to an understanding of the general theory of relativity. It is not needed anywhere in the special theory of relativity, and may be omitted by those who are not interested in the general theory of relativity.

at the same point. The difference is given by $A_{i,s}\delta x_s$. If it were possible to define "the same vector" or "the parallel vector" at a neighboring point, the difference between the parallel displaced vector and the actual vector at a neighboring point would be subject only to the transformation law at that point.

A definition of parallel displacement is actually possible in a comparatively simple way. Of course, the value of the displaced vector depends on the original vector itself and on the direction of displacement. Let us first consider a Euclidean space, where we can introduce a Cartesian coördinate system. With respect to such a coördinate system, the law of parallel displacement takes the form

$$\delta a_i \equiv a_{i,k}\delta x_k = 0, \tag{5.76}$$

where δx_k represents the infinitesimal displacement. Let us now introduce an arbitrary coördinate transformation (5.46). The vector components with respect to that new coördinate system may be denoted by a prime. Then we have

$$a_i = \frac{\partial \xi^r}{\partial x^i} a'_r,$$

$$\frac{\partial a_i}{\partial x_k} = \frac{\partial \xi^s}{\partial x_k}\frac{\partial}{\partial \xi^s}\left(\frac{\partial \xi^r}{\partial x_i} a'_r\right) = \frac{\partial \xi^s}{\partial x_k}\frac{\partial \xi^r}{\partial x_i}\frac{\partial a'_r}{\partial \xi^s} + \frac{\partial \xi^s}{\partial x_k}\frac{\partial x_l}{\partial \xi^s}\frac{\partial^2 \xi^r}{\partial x_i \partial x_l} a'_r .$$

δx_k transforms according to eq. (5.48), and we obtain

$$0 = a_{i,k}\,\delta x_k = \left\{\frac{\partial \xi^r}{\partial x_i}\frac{\partial a'_r}{\partial \xi^s} + \frac{\partial^2 \xi^r}{\partial x_i \partial x_l}\frac{\partial x_l}{\partial \xi^s} a'_r\right\}\delta \xi^s. \tag{5.76a}$$

$\dfrac{\partial a'_r}{\partial \xi^s}\delta\xi^s$ is the actual increment of a'_r as a result of the displacement, and shall be denoted by $\delta a'_r$. Multiplying the right-hand side of eq. (5.76a) by $\dfrac{\partial x_i}{\partial \xi^t}$, we get finally

$$\delta a'_t = -\frac{\partial x_l}{\partial \xi^s}\frac{\partial x_i}{\partial \xi^t}\frac{\partial^2 \xi^r}{\partial x_i \partial x_l} a'_r\,\delta\xi^s. \tag{5.77}$$

When no Cartesian coördinate system can be introduced, we shall retain the linear form of the last equation and assume that, because of a parallel displacement, the infinitesimal changes of the vector components are bilinear functions of the vector components and the components of the infinitesimal displacement,

$$\delta a^i = -\overset{\mathrm{I}}{\Gamma}{}^i_{kl}\, a^k\,\delta\xi^l, \tag{5.78}$$

$$\delta a_k = +\overset{\mathrm{II}}{\Gamma}{}^i_{kl}\, a_i\,\delta\xi^l. \tag{5.79}$$

The coefficients $\overset{\mathrm{I}}{\Gamma}{}^{i}_{kl}$ and $\overset{\mathrm{II}}{\Gamma}{}^{i}_{kl}$ of these new tentative laws are, so far, entirely unknown quantities. But we can determine their transformation laws. δa^i is the difference between two vectors at two points, characterized by the coördinate values ξ^l and $\xi^l + \delta\xi^l$. In the case of a coördinate transformation, the new $\delta a'^k$ are given by the following expressions:

$$\left.\begin{aligned}\delta a'^k &= \left(\frac{\partial \xi'^k}{\partial \xi^s} a^s\right)_{\xi^l+\delta\xi^l} - \left(\frac{\partial \xi'^k}{\partial \xi^s} a^s\right)_{\xi^l} = \frac{\partial}{\partial \xi^l}\left(\frac{\partial \xi'^k}{\partial \xi^s} a^s\right)\delta\xi^l \\ &= \frac{\partial^2 \xi'^k}{\partial \xi^s \partial \xi^l} a^s \delta\xi^l + \frac{\partial \xi'^k}{\partial \xi^s} a^s{}_{,l}\,\delta\xi^l \\ &= \frac{\partial^2 \xi'^k}{\partial \xi^s \partial \xi^l} a^s \delta\xi^l + \frac{\partial \xi'^k}{\partial \xi^s} \delta a^s.\end{aligned}\right\} \tag{5.80}$$

As stated before, δa^s does not transform like a vector; this was the original source of our difficulty.

We substitute the expression (5.80) into the left-hand side of the equation

$$\delta a'^k = -\overset{\mathrm{I}}{\Gamma}{}'^k_{mn}\, a'^m\, \delta\xi'^n$$

and replace a'^m and $\delta\xi'^n$ by the expressions from eqs. (5.51) and (5.50). We obtain

$$\frac{\partial^2 \xi'^k}{\partial \xi^s \partial \xi^l} a^s \delta\xi^l + \frac{\partial \xi'^k}{\partial \xi^s} \delta a^s = -\overset{\mathrm{I}}{\Gamma}{}'^k_{mn} \frac{\partial \xi'^m}{\partial \xi^s} a^s \frac{\partial \xi'^n}{\partial \xi^l} \delta\xi^l.$$

Substituting δa^s from eq. (5.78), we get

$$\left(\frac{\partial^2 \xi'^k}{\partial \xi^s \partial \xi^l} - \frac{\partial \xi'^k}{\partial \xi^r} \overset{\mathrm{I}}{\Gamma}{}^r_{sl}\right) a^s \delta\xi^l = -\overset{\mathrm{I}}{\Gamma}{}'^k_{mn} \frac{\partial \xi'^m}{\partial \xi^s} \frac{\partial \xi'^n}{\partial \xi^l} a^s \delta\xi^l.$$

As both a^s and $\delta\xi^l$ are arbitrary quantities, their coefficients on both sides must be equal,

$$\frac{\partial \xi'^m}{\partial \xi^s} \frac{\partial \xi'^n}{\partial \xi^l} \overset{\mathrm{I}}{\Gamma}{}'^k_{mn} = \frac{\partial \xi'^k}{\partial \xi^r} \overset{\mathrm{I}}{\Gamma}{}^r_{sl} - \frac{\partial^2 \xi'^k}{\partial \xi^s \partial \xi^l}.$$

The transformation law of the $\overset{\mathrm{I}}{\Gamma}{}^r_{sl}$ is obtained by multiplication by $\dfrac{\partial \xi^s}{\partial \xi'^a} \dfrac{\partial \xi^l}{\partial \xi'^b}$,

$$\overset{\mathrm{I}}{\Gamma}{}'^k_{ab} = \frac{\partial \xi^s}{\partial \xi'^a} \frac{\partial \xi^l}{\partial \xi'^b}\left(\frac{\partial \xi'^k}{\partial \xi^r} \overset{\mathrm{I}}{\Gamma}{}^r_{sl} - \frac{\partial^2 \xi'^k}{\partial \xi^s \partial \xi^l}\right). \tag{5.81}$$

The last term on the right-hand side can be written in a slightly different way, by shifting the second derivative. It is

$$-\frac{\partial \xi^s}{\partial \xi'^a}\frac{\partial \xi^l}{\partial \xi'^b}\frac{\partial^2 \xi'^k}{\partial \xi^s \partial \xi^l} = -\frac{\partial \xi^s}{\partial \xi'^a}\frac{\partial}{\partial \xi^s}\left(\frac{\partial \xi^l}{\partial \xi'^b}\frac{\partial \xi'^k}{\partial \xi^l}\right) + \frac{\partial \xi^s}{\partial \xi'^a}\frac{\partial \xi'^k}{\partial \xi^l}\frac{\partial}{\partial \xi^s}\left(\frac{\partial \xi^l}{\partial \xi'^b}\right)$$

$$= -\frac{\partial \xi^s}{\partial \xi'^a}\frac{\partial}{\partial \xi^s}(\delta^k_b) + \frac{\partial \xi'^k}{\partial \xi^l}\frac{\partial^2 \xi^l}{\partial \xi'^a \partial \xi'^b}.$$

The first term on the right-hand side vanishes, because the parenthesis in that term is constant. Eq. (5.81) thus becomes

$$\overset{\mathrm{I}}{\Gamma}{}'^k_{ab} = \frac{\partial \xi'^k}{\partial \xi^l}\left(\frac{\partial \xi^r}{\partial \xi'^a}\frac{\partial \xi^s}{\partial \xi'^b}\overset{\mathrm{I}}{\Gamma}{}^l_{rs} + \frac{\partial^2 \xi^l}{\partial \xi'^a \partial \xi'^b}\right). \tag{5.82}$$

By carrying out the same computation with eq. (5.79), we obtain the transformation law for $\overset{\mathrm{II}}{\Gamma}{}^k_{ab}$. It is identical with that of $\overset{\mathrm{I}}{\Gamma}{}^k_{ab}$.

We can now subject $\overset{\mathrm{I}}{\Gamma}{}^k_{ab}$ and $\overset{\mathrm{II}}{\Gamma}{}^k_{ab}$ to conditions which are compatible with their transformation law. The transformation law consists of two terms. One term depends on the Γ^k_{ab} in the old coördinate system and has exactly the same form as the transformation law of a tensor. The second term does not depend on the Γ^k_{ab} and adds an expression which is symmetric in the two subscripts. So, even though the Γ^k_{ab} may vanish in one coördinate system, they do not vanish in other systems. But if the Γ^k_{ab} were symmetric in their subscripts in one coördinate system, they would be symmetric in every other coördinate system as well. This would be particularly true if the Γ^k_{ab} were to vanish in one system. Furthermore, if the $\overset{\mathrm{I}}{\Gamma}{}^k_{ab}$ were equal to the $\overset{\mathrm{II}}{\Gamma}{}^k_{cb}$ in one coördinate system, this equality would be preserved by arbitrary coördinate transformations. We shall find that geometrical considerations lead us to treat only systems of Γ^k_{ab} which satisfy both these conditions.

Let us displace two vectors a_i and b^i parallel to themselves along an infinitesimal path, $\delta\xi^i$. The change of their scalar product, a_ib^i, is given by

$$\delta(a_ib^i) = a_i\delta b^i + b^i\delta a_i = a_ib^k(\overset{\mathrm{II}}{\Gamma}{}^i_{kl} - \overset{\mathrm{I}}{\Gamma}{}^i_{kl})\delta\xi^l. \tag{5.83}$$

When two vectors are displaced parallel to themselves, their scalar product always remains constant if and only if the $\overset{\mathrm{I}}{\Gamma}{}^i_{kl}$ *are equal to the corresponding* $\overset{\mathrm{II}}{\Gamma}{}^i_{kl}$.

Actually, the assumption that the two types of Γ^i_{kl} are equal is strongly suggested not only because in Cartesian coördinates the inner product of two constant vectors is itself a constant, but because of another consideration which does not refer to Cartesian coördinates.

By extending the law or definition of "parallel" displacement (5.78), (5.79) from vectors to tensors, we can displace any tensor "parallel" to itself according to the rule

$$\delta t_{ik\cdot}^{\ \ l} = (\overset{\text{II}}{\Gamma}{}^{r}_{is} t_{rk\cdot}^{\ \ l} + \overset{\text{II}}{\Gamma}{}^{r}_{ks} t_{ir\cdot}^{\ \ l} - \overset{\text{I}}{\Gamma}{}^{l}_{rs} t_{ik\cdot}^{\ \ r})\delta\xi^{s}. \tag{5.84}$$

This rule is derived from the postulate that the "parallel displacement" of a product is given by the same law that applies to the differentiation of products,

$$\delta(abc) = ab\ \delta c + ac\ \delta b + bc\ \delta a. \tag{5.85}$$

Applying the law (5.84) to the parallel displacement of the Kronecker tensor, we obtain

$$\delta(\delta_i^k) = (\overset{\text{II}}{\Gamma}{}^{r}_{is}\delta_r^k - \overset{\text{I}}{\Gamma}{}^{k}_{rs}\delta_i^r)\delta\xi^s = (\overset{\text{II}}{\Gamma}{}^{k}_{is} - \overset{\text{I}}{\Gamma}{}^{k}_{is})\delta\xi^s. \tag{5.86}$$

Now apply eq. (5.86) and the product rule (5.85) to the "parallel displacement" of the product $a^i\delta_i^k$. The result is

$$\delta(a^i\delta_i^k) = \delta_i^k\delta a^i + a^i\delta(\delta_i^k) = \delta a^k + a^i\delta(\delta_i^k).$$

On the other hand, the product $a^i\delta_i^k$ is equal to a^k. We have, therefore,

$$\delta a^k = \delta a^k + a^i\delta(\delta_i^k).$$

Therefore, $\delta(\delta_i^k)$ must vanish. Accordingly, we have

$$\overset{\text{II}}{\Gamma}{}^{k}_{is} = \overset{\text{I}}{\Gamma}{}^{k}_{is}. \tag{5.87}$$

Henceforth, we shall omit the distinguishing marks I and II.

As mentioned before, the Γ^k_{is} are symmetric in their subscripts if it is possible to introduce a coördinate system in which they vanish at least locally. *From now on, let it be understood that we shall consider only symmetric* Γ^k_{is}. The Γ^k_{is} are still, to a high degree, arbitrary. They are, however, uniquely determined if we connect them with the metric tensor g_{ik} by the following condition: The result of the parallel displacement of a vector **a** shall not depend on whether we apply the law of parallel displacement (5.78) to its contravariant representation or the law (5.79) to its covariant representation. The two representations a^i and a_k have, at the point $(\xi^s + \delta\xi^s)$, the components $(a^i + \delta a^i)$ and $(a_k + \delta a_k)$, respectively, where δa^i and δa_k are given by equations (5.78) and (5.79). That these two vectors are again to be representations of the same vector $(\mathbf{a} + \delta\mathbf{a})$, at the point $(\xi^s + \delta\xi^s)$, is expressed by the equation

$$a_k + \delta a_k = (g_{ik} + \delta g_{ik})(a^i + \delta a^i), \tag{5.88}$$

where δg_{ik} is

$$\delta g_{ik} = g_{ik,l}\,\delta\xi^l.$$

Eq. (5.88) must be satisfied up to linear terms in the differentials and for arbitrary a^i and $\delta\xi^s$. If we multiply out the right-hand side of eq. (5.88), we obtain

$$\delta a_k = a^i g_{ik,l}\delta\xi^l + g_{ik}\delta a^i.$$

Substituting δa_k and δa^i from eqs. (5.78) and (5.79), we obtain

$$\Gamma^i_{kl}a_i\delta\xi^l = a^i g_{ik,l}\delta\xi^l - g_{ik}\Gamma^i_{sl}a^s\delta\xi^l$$

or

$$a^s\delta\xi^l(g_{si}\Gamma^i_{kl} + g_{ik}\Gamma^i_{sl} - g_{sk,l}) = 0. \tag{5.88a}$$

a^s and $\delta\xi^l$ are arbitrary; therefore, the contents of the parenthesis vanish.

Now we make use of the symmetry condition and write the vanishing bracket down three times, with different index combinations:

$$\Gamma^r_{is}g_{kr} + \Gamma^r_{ks}g_{ir} - g_{ik,s} = 0,$$

$$\Gamma^r_{ik}g_{rs} + \Gamma^r_{sk}g_{ir} - g_{is,k} = 0,$$

$$\Gamma^r_{ki}g_{rs} + \Gamma^r_{si}g_{rk} - g_{ks,i} = 0.$$

We subtract the first of these equations from the sum of the other two equations. Several terms cancel, and we obtain the equation

$$g_{rs}\Gamma^r_{ik} = \tfrac{1}{2}(g_{is,k} + g_{ks,i} - g_{ik,s}). \tag{5.89}$$

We multiply this equation by g^{sl} to obtain the final expression for Γ^l_{ik},

$$\Gamma^l_{ik} = \tfrac{1}{2}g^{ls}(g_{is,k} + g_{ks,i} - g_{ik,s}). \tag{5.90}$$

This expression is usually referred to as the Christoffel three-index symbol of the second kind, and it is denoted by the symbol $\left\{ {l \atop ik} \right\}$,

$$\left\{ {l \atop il} \right\} = \tfrac{1}{2}g^{ls}(g_{is,k} + g_{ks,i} - g_{ik,s}). \tag{5.90a}$$

The left-hand side of eq. (5.89) is called the Christoffel symbol of the first kind. It is denoted by the sign $[ik, s]$,

$$[ik, s] = \tfrac{1}{2}(g_{is,k} + g_{ks,i} - g_{ik,s}). \tag{5.89a}$$

In the case of Cartesian coördinates, both kinds of Christoffel three-index symbols vanish.

The concept of parallel displacement is independent of the existence of a metric tensor. We call a space with a law of parallel displacement an *affinely connected space* and the Γ^l_{ik} *the components of the affine connection*. When a metric is defined, covariant and contravariant vectors become equivalent, and the Γ^l_{ik} must take the values $\left\{ {l \atop ik} \right\}$ so that the parallel displacement of a vector does not depend on which of the two representations has been chosen.

We shall return now to our original program, the formation of new tensors by differentiation. Consider a "tensor field," that is, a tensor the components of which are functions of the coördinates. Now, we can take the value of this tensor at a point (ξ^s) and then displace it parallel to itself to the point $(\xi^s + \delta\xi^s)$. The value of the tensor field at the point $(\xi^s + \delta\xi^s)$, minus the value of the parallel displaced tensor, is itself a tensor. In the case of a mixed tensor of rank two, the value of the tensor at the point $(\xi^s + \delta\xi^s)$ is

$$t_{i\cdot}^{\cdot k} + t_{i\cdot,s}^{\cdot k}\delta\xi^s,$$

and the value of the parallel displaced tensor

$$t_{i\cdot}^{\cdot k} + \left(\left\{ {r \atop is} \right\} t_{r\cdot}^{\cdot k} - \left\{ {k \atop rs} \right\} t_{i\cdot}^{\cdot r}\right) \delta\xi^s;$$

the difference between these two expressions is

$$\left(t_{i\cdot,s}^{\cdot k} - \left\{ {r \atop is} \right\} t_{r\cdot}^{\cdot k} + \left\{ {k \atop rs} \right\} t_{i\cdot}^{\cdot r}\right) \delta\xi^s. \tag{5.91}$$

This expression is a tensor because of the way we have obtained it. As $\delta\xi^s$ is itself a vector and arbitrary, the expression in the parenthesis is a tensor. This tensor is called the covariant derivative of $t_{i\cdot}^{\cdot k}$ with respect to ξ^s. It is identical with the ordinary derivative when the $\left\{ {l \atop ik} \right\}$ vanish, and, therefore, particularly in the case of Cartesian coördinates. Two of the more usual notations of the covariant derivative are $\nabla_s t_{i\cdot}^{\cdot k}$ and $t_{i\cdot;s}^{\cdot k}$. We shall use the latter.

The covariant derivatives of an arbitrary tensor are formed by adding to the ordinary derivatives for each index of the differentiated tensor a further term, which for contravariant indices takes the form

$$t_{\cdots}{}^{i}{}_{\cdots;s} = \cdots + \begin{Bmatrix} i \\ rs \end{Bmatrix} t_{\cdots}{}^{r}{}_{\cdots} + \cdots ,$$

while for covariant indices it is

$$t_{\cdots k \cdots ;s} = \cdots - \begin{Bmatrix} r \\ ks \end{Bmatrix} t_{\cdots r \cdots} + \cdots .$$

This definition satisfies the rule for product differentiation,

$$(AB \cdots)_{;s} = A_{;s}B \cdots + AB_{;s} \cdots + \cdots ,$$

regardless of whether some of the indices of A, B, $\cdots$, are dummies.

The covariant derivatives of the metric tensor vanish, because of the vanishing parenthesis of eq. (5.88a). And since the covariant differentiation obeys the law of product differentiation, indices can be raised and lowered under the differentiation,

$$a_{i;s} = g_{ik}a^{k}{}_{;s} , \qquad \text{and so forth.} \tag{5.92}$$

Geodesic lines. The Christoffel symbols appear not only in connection with covariant differentiation and parallel displacement, but also in connection with a problem which is more directly related to the metric of a space, that is, the setting up of the differential equations of straight lines or shortest lines in terms of general coördinates. In a Euclidean space, the shortest line connecting two points is a straight line. In the case of general Riemannian spaces, there may not exist lines having all the properties of straight lines, but there is, in general, a uniquely determined shortest connecting line between two points. In the case of the surface of a sphere, for instance, these lines are great circles. Such shortest connecting lines are called *geodesics*. The length of an arbitrary line connecting two points P_1 and P_2 is

$$s_{12} = \int_{P_1}^{P_2} ds = \int_{P_1}^{P_2} \sqrt{g_{ik}\, d\xi^i\, d\xi^k} = \int_{P_1}^{P_2} \sqrt{g_{ik} \frac{d\xi^i}{dp} \frac{d\xi^k}{dp}}\, dp, \tag{5.93}$$

where p is an arbitrary parameter.

In order to find the minimum value of s_{12} with fixed end points of integration, one has to carry out the variation according to the Euler-Lagrange equations,

$$\delta \int_A^B L(y_\alpha , y'_\alpha)\, dx = \int_A^B \sum_\alpha \left[\frac{\partial L}{\partial y_\alpha} - \frac{d}{dx}\left(\frac{\partial L}{\partial y'_\alpha}\right)\right] \delta y_\alpha\, dx, \tag{5.94}$$

and the extremals are given by the equations

$$\frac{\partial L}{\partial y_\alpha} - \frac{d}{dx}\left(\frac{\partial L}{\partial y'_\alpha}\right) = 0. \tag{5.95}$$

In our case, the Lagrangian is the integrand of the last expression of eq. (5.93), while the variables y_α are the coördinates ξ^i. The derivatives $\frac{\partial L}{\partial \xi^l}$ and $\frac{\partial L}{\partial \xi^{l'}}$, where $\xi^{l'}$ stands for $\frac{d\xi^l}{dp}$, are given by

$$\frac{\partial L}{\partial \xi^l} = \frac{\partial L}{\partial g_{ik}} g_{ik,l} = \frac{1}{2}\frac{g_{ik,l}\xi^{i'}\xi^{k'}}{\sqrt{g_{rs}\xi^{r'}\xi^{s'}}} = \frac{1}{2} g_{ik,l}\xi^{i'}\xi^{k'}\frac{dp}{ds},$$

$$\frac{\partial L}{\partial \xi^{l'}} = \frac{g_{il}\xi^{i'}}{\sqrt{g_{rs}\xi^{r'}\xi^{s'}}} = g_{il}\xi^{i'}\frac{dp}{ds}.$$

$\frac{d}{dp}\left(\frac{\partial L}{\partial \xi^{l'}}\right)$ is

$$\frac{d}{dp}\left(\frac{\partial L}{\partial \xi^{l'}}\right) = \frac{dp}{ds}\left[g_{il,k}\xi^{i'}\xi^{k'} + g_{il}\xi^{i''} + g_{il}\xi^{i'}\cdot\frac{d^2p/ds^2}{(dp/ds)^2}\right].$$

We have, thus,

$$\delta s_{12} = \int_{P_1}^{P_2}\frac{dp}{ds}\left\{(\tfrac{1}{2}g_{ik,l} - g_{il,k})\xi^{i'}\xi^{k'} - g_{il}\xi^{i''} - g_{il}\xi^{i'}\frac{d^2p/ds^2}{(dp/ds)^2}\right\}\delta\xi^l\,dp. \tag{5.96}$$

Because the parenthesis is multiplied by an expression symmetric with respect to i and k, we symmetrize the parenthesis itself and write:

$$\delta s_{12} = -\int_{P_1}^{P_2}\frac{dp}{ds}\left\{\frac{1}{2}(g_{il,k} + g_{kl,i} - g_{ik,l})\xi^{i'}\xi^{k'} + g_{il}\xi^{i''} + g_{il}\xi^{i'}\frac{d^2p/ds^2}{(dp/ds)^2}\right\}\delta\xi^l\,dp. \tag{5.97}$$

The parenthesis is now an expression encountered before, eq. (5.89a). The differential equations for the geodesic lines are, thus,

$$[ik, l]\xi^{i'}\xi^{k'} + g_{il}\xi^{i''} + g_{il}\xi^{i'}\frac{d^2p/ds^2}{(dp/ds)^2} = 0,$$

or, multiplied by g^{ls},

$$\xi^{s''} + \left\{\begin{matrix} s \\ ik \end{matrix}\right\}\xi^{i'}\xi^{k'} + \xi^{s'}\frac{d^2p/ds^2}{(dp/ds)^2} = 0. \tag{5.98}$$

If we choose as the parameter the arc length s itself, the last term vanishes, and we have:

$$\frac{d^2\xi^l}{ds^2} + \begin{Bmatrix} l \\ ik \end{Bmatrix} \frac{d\xi^i}{ds}\frac{d\xi^k}{ds} = 0. \tag{5.99}$$

When the coördinate system is Cartesian, the second term vanishes identically, and eq. (5.99) simply states that the ξ^l must be linear functions of s.

Minkowski world and Lorentz transformations. We can now return to our starting point, Minkowski's treatment of the theory of relativity. He considered the ordinary, three dimensional space plus the time as a four dimensional continuum, the "world," with the invariant "length" or "interval" defined by eq. (5.1). A "world point" is an ordinary point at a certain time, its four coördinates x, y, z, and t, which we shall often denote from now on by x^1, x^2, x^3, and x^4. By introducing a "metric tensor" $\eta_{\mu\nu}$ with the components

$$\eta_{\mu\nu} = \begin{Bmatrix} -\frac{1}{c^2}, & 0, & 0, & 0 \\ 0, & -\frac{1}{c^2}, & 0, & 0 \\ 0, & 0, & -\frac{1}{c^2}, & 0 \\ 0, & 0, & 0, & +1 \end{Bmatrix}, \tag{5.100}$$

we can write eq. (5.1) in the form

$$\tau_{12}^{\ 2} = \eta_{\mu\nu}\Delta x^\mu \Delta x^\nu. \tag{5.101}$$

The Lorentz transformations are those linear coördinate transformations which carry the metric tensor $\eta_{\mu\nu}$ over into itself. The inertial systems of the special relativity theory are, thus, analogous to the Cartesian coördinate systems of ordinary three dimensional Euclidean geometry, and the Lorentz transformations correspond to the three dimensional orthogonal transformations. Their transformation coefficients are also subject to conditions similar to (5.10a).

When we carry out any linear transformation (not necessarily a Lorentz transformation), the transformation equations are of the form

$$x^{*\kappa} = \gamma^\kappa{}_\iota x^\iota + x_0^{*\kappa}, \tag{5.102}$$

and the coördinate differences transform as contravariant vectors,

$$\Delta x^{*\kappa} = \gamma^\kappa{}_\iota \Delta x^\iota. \tag{5.103}$$

The conditions for Lorentz transformations are that

$$\eta_{\mu\nu}\Delta x^{*\mu}\Delta x^{*\nu} = \eta_{\iota\kappa}\Delta x^{\iota}\Delta x^{\kappa} \tag{5.104}$$

for arbitrary Δx^{ι}. By substituting $\Delta x^{*\mu}$ from eq. (5.103), we get

$$\eta_{\mu\nu}\gamma^{\mu}{}_{\iota}\gamma^{\nu}{}_{\kappa}\Delta x^{\iota}\Delta x^{\kappa} = \eta_{\iota\kappa}\Delta x^{\iota}\Delta x^{\kappa}, \tag{5.105}$$

and, because the Δx^{ι} are arbitrary,

$$\eta_{\mu\nu}\gamma^{\mu}{}_{\iota}\gamma^{\nu}{}_{\kappa} = \eta_{\iota\kappa}\,. \tag{5.106}$$

These are the conditions for the transformation coefficients of Lorentz transformations, corresponding to the conditions (5.10a) for orthogonal transformations.†

The difference between a Euclidean four dimensional space and the Minkowski world is that in the latter the invariant $\tau_{12}^{\;2}$ is not positive definite. That is why no real coördinate transformation can carry eq. (5.101) over into eq. (5.2b), page 59. We have, therefore, to distinguish between contravariant and covariant indices.

In order to recognize coördinates and tensors of the Minkowski world as such, we shall adopt the convention of using in general the Latin alphabet for indices belonging to the ordinary space, running from 1 to 3, and of using Greek letters for Minkowski indices, running from 1 to 4. We shall call vectors and tensors of the Minkowski world "world vectors" and "world tensors."

The contravariant metric tensor has the components

$$\eta^{\mu\nu} = \begin{Bmatrix} -c^2, & 0\,, & 0\,, & 0 \\ 0\,, & -c^2, & 0\,, & 0 \\ 0\,, & 0\,, & -c^2, & 0 \\ 0\,, & 0\,, & 0\,, & +1 \end{Bmatrix}. \tag{5.107}$$

When contravariant tensors transform with the coefficients $\gamma^{\rho}{}_{\sigma}$ which satisfy the conditions (5.106), the coefficients of the covariant transformation law are the solutions of the equations

$$\gamma^{\rho}{}_{\sigma}\gamma_{\rho}{}^{\tau} = \delta^{\tau}_{\sigma}\,. \tag{5.108}$$

In order to find an explicit expression for these coefficients $\gamma_{\rho}{}^{\tau}$, one can multiply the transformation equations of a covariant vector

$$\overset{*}{v}_{\alpha} = \gamma_{\alpha}{}^{\beta}v_{\beta} \tag{5.109}$$

by $\eta^{\alpha\rho}$ and replace v_{β} by $\eta_{\beta\sigma}v^{\sigma}$. The result is

$$\eta^{\alpha\rho}\overset{*}{v}_{\alpha} = \eta^{\alpha\rho}\gamma_{\alpha}{}^{\beta}\eta_{\beta\sigma}v^{\sigma}$$

†See Appendix B.

Now, the left-hand side is equal to $v^{*\rho}$, and, therefore, to $\gamma^{\rho}{}_{\sigma}v^{\sigma}$,

$$\gamma^{\rho}{}_{\sigma}v^{\sigma} = \eta^{\rho\alpha}\gamma_{\alpha}{}^{\beta}\eta_{\beta\sigma}v^{\sigma}.$$

Further, since v^{σ} is arbitrary, the coefficients must be equal,

$$\gamma^{\rho}{}_{\sigma} = \eta^{\rho\alpha}\gamma_{\alpha}{}^{\beta}\eta_{\beta\sigma}\,. \tag{5.110}$$

Finally, by multiplying by $\eta_{\rho\mu}\eta^{\sigma\nu}$ and switching sides, we obtain

$$\gamma_{\mu}{}^{\nu} = \eta_{\mu\rho}\gamma^{\rho}{}_{\sigma}\eta^{\sigma\nu}. \tag{5.111}$$

All the algebraic operations of the general tensor calculus can be applied to world tensors. As in the case of orthogonal transformations, the determinant of the transformation coefficients takes only the values ± 1. Thus, the densities of even weight transform like tensors, and the densities of odd weight transform similarly to the "axial" vectors of the three dimensional orthogonal formalism.

The components of the metric tensor $\eta_{\mu\nu}$ are constant; therefore, the Christoffel symbols vanish, and the covariant derivatives are simply the ordinary derivatives. We shall denote such ordinary derivatives by the comma,

$$\frac{\partial t^{\cdots}_{\cdots}}{\partial x^{\alpha}} = t^{\cdots}_{\cdots,\alpha}. \tag{5.112}$$

We shall now demonstrate how the transformation coefficients of a Lorentz transformation are related to the relative velocity of the two coördinate systems S and S^*. When a point is in a state of straight-line, uniform motion, its velocity is described by the ratios of the coördinate differences of any two world points along its path,

$$u^{i} = \frac{\Delta x^{i}}{\Delta x^{4}}.$$

The velocity of the system S relative to the system S^* is the velocity of a particle P, which is at rest in S, relative to S^*. The first three coördinate differences of P with respect to S, Δx^{i}, vanish. The coördinate differences with respect to S^* are, therefore, given by

$$\left.\begin{aligned} \Delta x^{*i} &= \gamma^{i}{}_{4}\Delta x^{4}, \\ \Delta x^{*4} &= \gamma^{4}{}_{4}\Delta x^{4}. \end{aligned}\right\} \tag{5.113}$$

S has, therefore, relative to S^*, the velocity

$$v^{*i} = \frac{\gamma^{i}{}_{4}}{\gamma^{4}{}_{4}}. \tag{5.114}$$

Conversely, we can compute the velocity of S^* with respect to S by employing the "inverse" transformation coefficients $\gamma_\mu{}^\nu$, given by eq. (5.108). Because they are the transformation coefficients of the transformation $S^* \rightarrow S$, we can write, referring to eq. (5.114),

$$v^i = \frac{\gamma_4{}^i}{\gamma_4{}^4}. \tag{5.115}$$

By making use of eqs. (5.111), (5.100), and (5.107), we can write the right-hand side in the form

$$v^i = -c^2 \frac{\gamma^4{}_i}{\gamma^4{}_4}. \tag{5.116}$$

Now it is easy to show that in general v^{*2} is equal to v^2. Let us first form the three dimensional norm of v^{*i}, eq. (5.114). We have

$$v^{*2} = \sum_{i=1}^{3} \frac{(\gamma^i{}_4)^2}{(\gamma^4{}_4)^2}.$$

Because of eq. (5.106) the numerator can be rewritten,

$$\sum_{i=1}^{3} (\gamma^i{}_4)^2 = c^2[(\gamma^4{}_4)^2 - 1],$$

and we find, therefore,

$$v^{*2} = c^2 \frac{(\gamma^4{}_4)^2 - 1}{(\gamma^4{}_4)^2}. \tag{5.117}$$

Now we treat eq. (5.116) in an analogous way. It is

$$v^2 = c^4 \sum_{i=1}^{3} \frac{(\gamma^4{}_i)^2}{(\gamma^4{}_4)^2}.$$

Again we have

$$\sum_{i=1}^{3} (\gamma^4{}_i)^2 = \frac{1}{c^2} [(\gamma^4{}_4)^2 - 1],$$

and v^2 is

$$v^2 = c^2 \frac{(\gamma^4{}_4)^2 - 1}{(\gamma^4{}_4)^2}. \tag{5.118}$$

Incidentally, we find that $\gamma^4{}_4$ is always given by the expression

$$\gamma^4{}_4 = \frac{1}{\sqrt{1 - v^2/c^2}}. \tag{5.119}$$

Paths, world lines. Ordinarily, the motion of a particle along its path is described by stating the functional dependence of the three space coördinates on the time t,

$$x^i = f^i(t). \tag{5.120}$$

The components of the velocity are given as the derivatives,

$$u^i = \frac{df^i}{dt} = \dot{x}^i. \tag{5.121}$$

This kind of description is, of course, possible in the theory of relativity as well as in non-relativistic physics. However, it is often useful to choose a description in which the time is not set apart from the spatial coördinates as in eqs. (5.120) and (5.121).

The motion of a mass point is, in terms of the Minkowski world, represented by a line, a "world line," which we can advantageously describe by a parameter representation,

$$x^\mu = g^\mu(p), \tag{5.122}$$

where p is an arbitrary parameter defined along the world line. Such parameter descriptions are commonly used in analytical geometry.

In three dimensional geometry, the arc length is often chosen as the parameter. In the Minkowski world, we can use as the parameter the proper time τ along a world line. Just as the arc length S in ordinary geometry is defined as the line integral

$$s = \int_{P_1}^{P_2} \sqrt{dx^2 + dy^2 + dz^2},$$

τ is given as the integral of the differential $d\tau$ along the world line of the particle,

$$\int d\tau = \int \sqrt{dt^2 - \frac{1}{c^2}(dx^2 + dy^2 + dz^2)}$$

$$= \int \sqrt{\eta_{\mu\nu}\, dx^\mu\, dx^\nu} = \int \sqrt{\eta_{\mu\nu} \frac{dx^\mu}{d\tau} \frac{dx^\nu}{d\tau}}\, d\tau. \tag{5.123}$$

We can, therefore, describe the path by equations of the form

$$x^\mu = F^\mu(\tau). \tag{5.124}$$

τ is related to the x^μ by a differential equation. When we divide the integrand of eq. (5.123) by dt and take account of eq. (5.121), we obtain

$$\frac{d\tau}{dt} = \sqrt{1 - \frac{1}{c^2}\sum_{i=1}^{3}\left(\frac{dx^i}{dt}\right)^2} = \sqrt{1 - u^2/c^2}. \tag{5.125}$$

The velocity of a body in terms of the usual description (5.120), (5.121) is replaced by the direction of the world line in the four dimensional description. When the body is at rest, the line is parallel to the X^4-axis; and when the body is moving, the line will run at an angle relative to the X^4-axis. We can describe its direction in terms of its tangential vector,

$$U^\mu = \frac{dx^\mu}{d\tau} \tag{5.126}$$

The four quantities U^μ are the components of a contravariant unit vector,

$$\eta_{\mu\nu} U^\mu U^\nu = 1. \tag{5.127}$$

This can be verified easily by replacing $d\tau$ in eq. (5.126) by its definition,

$$d\tau = \sqrt{\eta_{\mu\nu}\, dx^\mu\, dx^\nu}.$$

The U^μ are related very simply to the velocity components u^i of eq. (5.121); making use of eq. (5.125), we have

$$\left.\begin{aligned} U^4 &= \frac{dt}{d\tau} = \frac{1}{\sqrt{1 - u^2/c^2}}, \\ U^i &= \frac{dx^i}{d\tau} = \frac{dx^i}{dt}\frac{dt}{d\tau} = u^i U^4. \end{aligned}\right\} \tag{5.128}$$

In the following chapters, we shall use u^i and U^μ consistently in the way they were introduced here. v^i shall be used exclusively to denote the relative motion of two coördinate systems.

When a body moves without being accelerated, its direction in the Minkowski world is constant and its world line is a straight line. The law of inertia takes a very simple form in our new description:

$$U^\mu = \text{const.} \tag{5.129}$$

PROBLEMS

1. Prove that the right-hand side of eq. (5.90a) transforms according to eq. (5.82).

2. Prove that the symmetry properties with respect to indices of the same transformation character are invariant.

3. For three dimensional Riemannian space, define the differential operations gradient (of a scalar), divergence, and curl, and prove that the relation holds:

$$\text{curl grad } V = 0.$$

Treating "axial" vectors as skewsymmetric tensors of rank 2, define also the divergence of an axial vector and prove the relation

$$\text{div curl } \mathbf{A} = 0,$$

where $\mathbf{A}$ is a polar vector.

4. Prove that the following relations hold in three dimensional space:

$$\delta_{ikl}\delta^{ikm} = 2\delta_l^m \,; \qquad \delta_{ikl}\delta^{imn} = \delta_k^m\delta_l^n - \delta_k^n\delta_l^m \,.$$

State similar relations for the Minkowski world.

5. With the help of Problem 4, prove that in Euclidean, three dimensional geometry the relation holds:

$$\text{curl curl } \mathbf{A} = \text{grad div } \mathbf{A} - \nabla^2\mathbf{A},$$

$\mathbf{A}$ being either a polar vector or an axial vector.

6. Compute in three dimensional space the two triple products of the three polar vectors:

$$[\mathbf{A} \times [\mathbf{B} \times \mathbf{C}]], \qquad (\mathbf{A} \cdot [\mathbf{B} \times \mathbf{C}]).$$

7. In a plane, introduce polar coördinates.
(*a*) Compute the components of the metric tensor.
(*b*) State the differential equations of the straight lines.

8. On the surface of a sphere of radius R, introduce the so-called Riemannian homogeneous coördinates, which are characterized by an expression for the infinitesimal length of the form

$$ds^2 = f(\xi^2 + \eta^2)(d\xi^2 + d\eta^2), \qquad \text{with } f(0) = 1.$$

(*a*) State the function $f(\xi^2 + \eta^2)$ and the transformation equations between Riemannian coördinates and the usual coördinates of longitude and latitude.

Answer:

$$f = \left[\frac{R^2}{R^2 + \frac{1}{4}(\xi^2 + \eta^2)}\right]^2,$$

$$\xi = \frac{2R\cos\varphi}{1 + \sin\varphi}\cos\vartheta,$$

$$\eta = \frac{2R\cos\varphi}{1 + \sin\varphi}\sin\vartheta.$$

(*b*) Compute the differential equations for the great circles in either coördinate system.

Remark: The Riemannian coördinates are obtained by a conformal transformation leading from a plane to a spherical surface, familiar from the theory of complex functions.

9. The Laplacian operator in n dimensions is defined as the divergence of the gradient of a scalar, in general coördinates:

$$\nabla^2 V = g^{rs} V_{;rs} .$$

(*a*) Using the $\left\{ {i \atop mn} \right\}$, write out the right-hand side.

(*b*) Introduce a coördinate system the coördinates of which are everywhere orthogonal to each other, in other words, where the line element takes the form

$$ds^2 = \sum_{i=1}^{n} \epsilon_i (h_i)^2 (d\xi^i)^2, \qquad \epsilon_i = \pm 1.$$

The h_i are functions of ξ^s.

Express the Laplacian in terms of the h_i .

Answer:

$$\nabla^2 V = \frac{1}{\Pi} \sum_{i=1}^{n} \left(\frac{\Pi}{\epsilon_i (h_i)^2} V_{,i} \right)_{,i}, \qquad \Pi = h_1 \cdot h_2 \cdot \cdots \cdot h_n .$$

Remark: This expression is frequently used in order to obtain the Schroedinger equation in other than Cartesian coördinates. The student can easily derive the expression for $\nabla^2 V$ in three dimensions for spherical coördinates, cylinder coördinates, and so forth.

10. (*a*) Show that this relation holds in n dimensions:

$$\left\{ {s \atop ms} \right\} = \frac{1}{\sqrt{g}} (\sqrt{g})_{,m}, \qquad g = | g_{ik} | .$$

(*b*) Show that the following expressions are a scalar and a vector, respectively, when V^i is a vector and F^{ik} a skewsymmetric, contravariant tensor:

$$\frac{1}{\sqrt{g}} (\sqrt{g} \cdot V^i)_{,i} ; \qquad \frac{1}{\sqrt{g}} (\sqrt{g} \cdot F^{ik})_{,k} .$$

(*c*) In general coördinates, bring $\nabla^2 V$ into a form which is a generalization of that given sub 9(*b*).

11. V_i is a vector, F_{ik} a skewsymmetric tensor. Show that the following expressions are tensors with respect to arbitrary transformations, even though the derivatives are *ordinary* derivatives,

$$V_{i,k} - V_{k,i} ; \qquad F_{ik,l} + F_{kl,i} + F_{li,k} .$$

12. Schwarz's inequality,

$$\sum_{i=1}^{n} (a_i)^2 \sum_{k=1}^{n} (b_k)^2 \geqslant \left(\sum_{l=1}^{n} a_l b_l\right)^2,$$

states in effect that, in an n dimensional Euclidean space, any side of a triangle is shorter than the sum of the two others. For the latter statement can be written in the form

$$|\mathbf{a}| + |\mathbf{b}| \geqslant |\mathbf{a} + \mathbf{b}|.$$

When both sides are squared,

$$\mathbf{a}^2 + \mathbf{b}^2 + 2|\mathbf{a}|\cdot|\mathbf{b}| \geqslant \mathbf{a}^2 + \mathbf{b}^2 + 2(\mathbf{a}\cdot\mathbf{b}),$$

the squares on either side cancel; and when the remaining terms are squared once more, Schwarz's inequality is obtained.

By introducing a suitable Cartesian coördinate system, prove Schwarz's inequality, and thereby that the above statement about the sides of triangles is true regardless of the number of dimensions.

Another method of proof uses the positive norm of the skewsymmetric product,

$$\tfrac{1}{2} \sum_{i,k} (a_i b_k - a_k b_i)^2 = [\mathbf{a} \times \mathbf{b}]^2.$$

13. In an n dimensional Euclidean space, m unit vectors may be defined, $\overset{1}{v}_i, \cdots \overset{m}{v}_i$, $m \leqslant n$, which are mutually orthogonal to each other,

$$\overset{k}{v}_i \overset{l}{v}_i = \delta_{kl}, \qquad k, l = 1 \cdots m, \qquad i = 1 \cdots n.$$

Show that for any vector f_i, Bessel's inequality holds:

$$\sum_{s=1}^{n} \left\{ f_s^2 - \left[\sum_{a=1}^{m} (f_i \overset{a}{v}_i) \overset{a}{v}_s \right]^2 \right\} \geqslant 0.$$

The inequality goes over into an equality if $m = n$.

14. Prove that a vector field V_i in an n dimensional space can be represented as the gradient field of a scalar function if and only if its skewsymmetric derivatives vanish,

$$V_{i,k} - V_{k,i} = 0.$$

CHAPTER VI

Relativistic Mechanics of Mass Points

Program for relativistic mechanics. In Chapter IV, we have laid the foundation for the special theory of relativity. But so far we have dealt only with uniform, straight-line motion. The clocks and scales which we used for the determination of coördinate values were not accelerated. We replaced the Galilean transformation equations connecting two inertial systems, (4.14), by the Lorentz equations (4.13). The Lorentz equations are linear transformation equations, just as the Galilean equations were; that is, the new coördinate values (space and time) are linear functions of the old ones. Therefore, an unaccelerated motion in one inertial system will remain unaccelerated when a Lorentz transformation is carried out.

The law of inertia (2.1) *is invariant with respect to Lorentz transformations.*

The remaining chapters of Part I will discuss accelerated motion, in other words, will develop a relativistic mechanics. This will be more involved than the development of classical mechanics. The difficulties are twofold: In the first place, the equations of classical mechanics are covariant with respect to Galilean transformations, but not with respect to Lorentz transformations. Therefore, we shall have to develop a Lorentz invariant formalism so that our statements may be independent of the coördinate system used. The second difficulty is more profound: In classical mechanics, the force which acts on a body at a given time is determined by the positions of the other interacting bodies at the same time. An "action-at-a-distance" force law can be formulated only if it is meaningful to speak of the "positions of the other interacting bodies at the same time"; that is, if the "same time" is independent of the frame of reference used. This condition, we know, runs counter to the theory of relativity.

It is, therefore, impossible to transform automatically every conceivable classical law of force into a Lorentz covariant law. We can treat only those theories from which the concept of action at a distance can be eliminated. This possibility exists in the theory of collisions, and in this chapter we shall concern ourselves principally with collision forces. Another theory without action at a distance concerns the motion of electric charges in an electromagnetic field. The original action at a distance law of Coulomb soon turned out to be valid only for electrostatic fields. In general, it is not the positions of the other (distant) charges which determine the force acting on a charged particle. However, the relativistic formulation of the electromagnetic ponderomotive law can be tackled only when we know the transformation laws of the electromagnetic field. We shall, therefore, treat the theory of collisions first.

The theory of collisions is an abstraction from the ordinary mechanics of mass points, which applies when the range of the forces is short compared with the dimensions of the mechanical system.

One assumes that interaction takes place only during the instant that the distance between two bodies or mass points is negligible. Before and after this infinitesimally short time interval, the motion of the bodies is not accelerated. During the short period of interaction, the conservation laws of momentum and energy are fulfilled. If the kinetic energy remains unaltered by the collision, we speak of an *elastic* collision; if, on the other hand, part or all of the kinetic energy is transformed into other forms of energy (heat, and so forth), we speak of an *inelastic* collision. Naturally, the mechanism which underlies elastic collisions is less involved.

The form of the conservation laws. There are four conservation laws in classical mechanics, three concerning the three components of the momentum of an isolated system, and one concerning its energy. With respect to spatial transformations, the three momentum conservation laws transform together as the components of a three dimensional vector, while the energy conservation law is invariant.

With respect to Galilean transformations, the momentum laws are invariant, while the energy law holds in the new system only because the energy *and* momentum laws are satisfied in the first system. The classical conservation laws are not covariant with respect to Lorentz transformations which involve time. They must, therefore, be modified so that they are Lorentz covariant, but also go over into the classical laws for moderate velocities.

The relativistic conservation laws should have the same transformation properties with respect to spatial coördinate transformations as the classical laws; in other words, there should again be a vector law (with three components) and a scalar law. This will determine, to a large extent, the form of the relativistic laws.

The state of a mass point is completely characterized by its mass m and its velocity $\mathbf{u}$. If its "relativistic momentum" is to transform as a vector with respect to spatial transformations, and if this momentum is to depend only on the state of the mass point, the momentum vector must be parallel to the velocity; that is, the momentum of a mass point must take the form

$$p_s = \mu(m, u)\cdot u^s, \tag{6.1}$$

where u is the amount of the velocity and μ is a function of m and u which remains to be determined.

Likewise, the energy must be some function of m and u. This function is tied up with μ in that the change of energy is the product of the change of momentum with time and the distance traveled, that is, the scalar product of the change of momentum and the velocity,

$$dE = d\mathbf{p}\cdot\mathbf{u} = d(\mu\mathbf{u})\cdot\mathbf{u} = d(\mu\cdot u^2) - \mu\mathbf{u}\cdot d\mathbf{u}$$
$$= d(\mu u^2) - \mu u\, du = u\cdot d(\mu u);$$

or we can write:

$$\frac{dE}{dt} = u\cdot\frac{d}{dt}(\mu u). \tag{6.2}$$

Once we know the function μ, the function E can be obtained by solving eq. (6.2).

A model example. To obtain the function μ, we shall consider an example of an elastic collision which is chosen so that with respect to one coördinate system the speeds of the colliding particles are the same before and after the collision, which means that the factors μ remain constant. Thus, the conservation laws can be satisfied by choosing only the appropriate *directions* for the velocities before and after the collision. Then, when we go over to another coördinate system, we shall find that only one choice of μ is compatible with the conservation laws in the new system.

Two mass points of equal mass m approach the origin of the coördinate system S from opposite directions and reach it at the time $t = 0$. The components of their respective velocities shall be

$$\left.\begin{aligned} \underset{1}{u_x} &= a = -\underset{2}{u_x}, \\ \underset{1}{u_y} &= b = -\underset{2}{u_y}, \\ \underset{1}{u_z} &= 0 = \underset{2}{u_z}. \end{aligned}\right\} \tag{6.3}$$

Their equations of motion before the collision are, thus,

$$\left.\begin{aligned} \underset{1}{x} &= at = -\underset{2}{x}, \\ \underset{1}{y} &= bt = -\underset{2}{y}, \\ \underset{1}{z} &= \underset{2}{z} = 0. \end{aligned}\right\} \tag{6.4}$$

Upon colliding, we assume, they continue their respective motions in the x-direction and exchange their velocities in the y-direction. Their velocities after the collision are, thus,

$$\left.\begin{aligned} \underset{1}{\bar{u}_x} &= -\underset{2}{\bar{u}_x} = a, \\ \underset{1}{\bar{u}_y} &= -\underset{2}{\bar{u}_y} = -b, \\ \underset{1}{\bar{u}_z} &= \underset{2}{\bar{u}_z} = 0, \end{aligned}\right\} \tag{6.5}$$

and their equations of motion

$$\left.\begin{aligned} \underset{1}{\bar{x}} &= -\underset{2}{\bar{x}} = at, \\ \underset{1}{\bar{y}} &= -\underset{2}{\bar{y}} = -bt, \\ \underset{1}{\bar{z}} &= \underset{2}{\bar{z}} = 0. \end{aligned}\right\} \tag{6.6}$$

The motions of the two particles are illustrated in Fig. 7. The speeds of the two particles are the same before and after the collision and equal to each other.

Even though we have not yet made any assumptions about the functional dependence of μ on m and u, we are assured that our example is consistent with the relativistic conservation laws. The behavior of the particles in our example is, therefore, correctly described, regardless of the modifications which the condition of relativistic invariance may force upon the laws of classical mechanics.

In classical mechanics, μ is assumed to be equal to m (and thus independent of u). The classical laws are at least approximately valid for small velocities (compared with c); therefore, μ must assume the value m for vanishing u. In order to find its dependence on u, we shall represent the behavior of our two particles in terms of a new coördinate system, S^*. S^* is connected with the original system S by eqs. (4.13) or (4.15). To avoid unnecessary complications, we shall choose v, the relative velocity of S^* and S, to be equal to the constant a of eqs (6.3) ff.

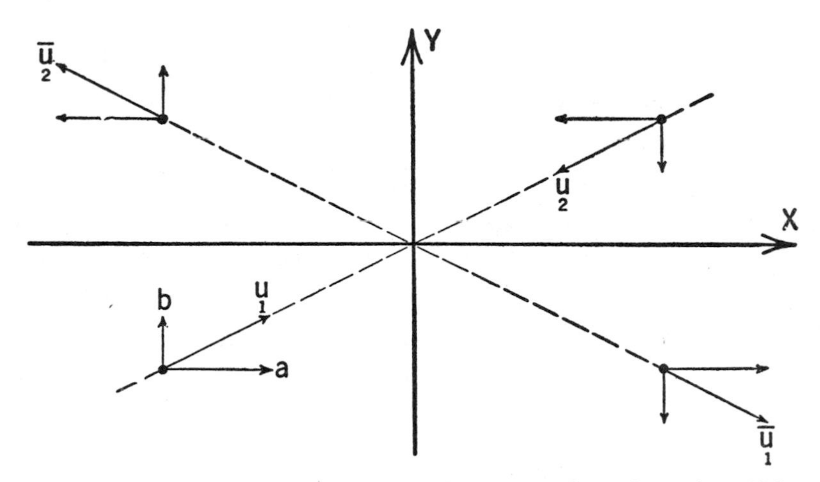

Fig. 7. Two particles, which have equal masses and equal speeds, collide.

Upon transforming eqs. (6.4) and (6.6), we obtain the equations of motion, which are, before the collision,

$$\left.\begin{aligned} x^*_1 &= 0, & x^*_2 &= -\frac{2a}{1 + a^2/c^2}\, t^*, \\ y^*_1 &= \frac{bt^*}{\sqrt{1 - a^2/c^2}}, & y^*_2 &= -\frac{\sqrt{1 - a^2/c^2}}{1 + a^2/c^2}\, bt^*, \end{aligned}\right\} \tag{6.4a}$$

and, after the collision,

$$\left.\begin{aligned} \bar{x}^*_1 &= 0, & \bar{x}^*_2 &= -\frac{2a}{1 + a^2/c^2}\, t^*, \\ \bar{y}^*_1 &= -\frac{bt^*}{\sqrt{1 - a^2/c^2}}, & \bar{y}^*_2 &= +\frac{\sqrt{1 - a^2/c^2}}{1 + a^2/c^2}\, bt^*. \end{aligned}\right\} \tag{6.6a}$$

The velocities are, before the collision,

$$\left.\begin{aligned} u^*_{x\,1} &= 0, & u^*_{x\,2} &= -\frac{2a}{1 + a^2/c^2}, \\ u^*_{y\,1} &= \frac{b}{\sqrt{1 - a^2/c^2}}, & u^*_{y\,2} &= -\frac{\sqrt{1 - a^2/c^2}}{1 + a^2/c^2}\, b, \\ u^*_1 &= \frac{b}{\sqrt{1 - a^2/c^2}}, & u^*_2 &= \frac{\sqrt{4a^2 + b^2(1 - a^2/c^2)}}{1 + a^2/c^2}, \end{aligned}\right\} \tag{6.7}$$

and, after the collision,

$$\left.\begin{aligned} &\underset{1}{\bar{u}}{}^*_x = 0, && \underset{2}{\bar{u}}{}^*_x = -\frac{2a}{1 + a^2/c^2}, \\ &\underset{1}{\bar{u}}{}^*_y = -\frac{b}{\sqrt{1 - a^2/c^2}}, && \underset{2}{\bar{u}}{}^*_y = \frac{\sqrt{1 - a^2/c^2}}{1 + a^2/c^2}\, b, \\ &\underset{1}{\bar{u}}{}^* = \frac{b}{\sqrt{1 - a^2/c^2}}, && \underset{2}{\bar{u}}{}^* = \frac{\sqrt{4a^2 + b^2(1 - a^2/c^2)}}{1 + a^2/c^2}. \end{aligned}\right\} \quad (6.8)$$

Using the expressions (6.7) and (6.8), we can set up the equations for the "momenta," containing the unknown function μ. We shall call the "momenta" of the individual particles $\underset{1}{\mathbf{p}}$ and $\underset{2}{\mathbf{p}}$ and their sum, the "total momentum," $\mathbf{p}$. We have, before the collision,

$$\left.\begin{aligned} p^*_x &= \underset{1}{p}{}^*_x + \underset{2}{p}{}^*_x = 0 - \mu(m, \underset{2}{u}{}^*)\cdot\frac{2a}{1 + a^2/c^2}, \\ p^*_y &= \underset{1}{p}{}^*_y + \underset{2}{p}{}^*_y = \mu(m, \underset{1}{u}{}^*)\,\frac{b}{\sqrt{1 - a^2/c^2}} \\ &\qquad\qquad - \mu(m, \underset{2}{u}{}^*)\,\frac{\sqrt{1 - a^2/c^2}}{1 + a^2/c^2}\, b, \end{aligned}\right\} \quad (6.9)$$

and, after the collision,

$$\left.\begin{aligned} \bar{p}^*_x &= 0 - \mu(m, \underset{2}{\bar{u}}{}^*)\cdot\frac{2a}{1 + a^2/c^2}, \\ \bar{p}^*_y &= -\mu(m, \underset{1}{\bar{u}}{}^*)\,\frac{b}{\sqrt{1 - a^2/c^2}} + \mu(m, \underset{2}{\bar{u}}{}^*)\,\frac{\sqrt{1 - a^2/c^2}}{1 + a^2/c^2}\, b. \end{aligned}\right\} \quad (6.10)$$

The conservation law for p^*_x is satisfied identically, as $\underset{2}{u}{}^*$ equals $\underset{2}{\bar{u}}{}^*$, and thus $\mu(m, \underset{2}{u}{}^*)$ is equal to $\mu(m, \underset{2}{\bar{u}}{}^*)$. The expressions for p^*_y and $\bar{p}^*_y$, on the other hand, differ only with respect to their signs, as $\underset{1}{u}{}^*$ also equals $\underset{1}{\bar{u}}{}^*$. Thus, the conservation law for p^*_y requires that p^*_y vanish. We obtain, thus, a functional equation for μ:

$$\left.\begin{aligned} &\mu(m, \underset{1}{u}{}^*) - \frac{1 - a^2/c^2}{1 + a^2/c^2}\,\mu(m, \underset{2}{u}{}^*) = 0, \\ &\qquad\qquad \underset{1}{u}{}^* = \frac{b}{\sqrt{1 - a^2/c^2}}, \\ &\qquad\qquad \underset{2}{u}{}^* = \frac{\sqrt{4a^2 + b^2(1 - a^2/c^2)}}{1 + a^2/c^2}. \end{aligned}\right\} \quad (6.11)$$

By going over to the limit $b \to 0$, we get the simpler equation

$$\mu(m, 0) = \frac{1 - a^2/c^2}{1 + a^2/c^2} \cdot \mu\left(m, \frac{2a}{1 + a^2/c^2}\right). \tag{6.12}$$

We mentioned before that $\mu(m, 0)$ equals m. In order to obtain the explicit function μ, we introduce u as the argument of the other μ,

$$\left.\begin{aligned} &u = \frac{2a}{1 + a^2/c^2}, \qquad a = \frac{c^2}{u}[1 \pm \sqrt{1 - u^2/c^2}], \\ &\frac{1 - a^2/c^2}{1 + a^2/c^2} = \sqrt{1 - u^2/c^2}, \end{aligned}\right\} \tag{6.13}$$

so that eq. (6.12) becomes

$$\mu(m, u) = \frac{m}{\sqrt{1 - u^2/c^2}}. \tag{6.14}$$

In other words, if Lorentz covariant conservation laws exist at all, the vector quantities occurring in them take the form

$$p_s = \sum_a \frac{m_a u^s_a}{\sqrt{1 - u_a^2/c^2}}. \tag{6.15}$$

We call this expression the "relativistic" momentum, in order to distinguish it from the analogous classical vector.

The relativistic energy of one mass point can be computed with the help of eq. (6.2),

$$\frac{dE}{dt} = u\frac{d}{dt}\left(\frac{um}{\sqrt{1 - u^2/c^2}}\right) = \frac{d}{dt}\left(\frac{mc^2}{\sqrt{1 - u^2/c^2}}\right); \tag{6.16}$$

in other words, the relativistic kinetic energy equals

$$E = \frac{mc^2}{\sqrt{1 - u^2/c^2}} + E_0, \tag{6.17}$$

where E_0 is the constant of integration.

Lorentz covariance of the new conservation laws. It would be quite difficult to show the Lorentz covariant character of the expressions (6.15) and (6.17) without the help of the formalism developed in the preceding chapter. Upon comparing eq. (6.15) with eq. (5.128), we find that the components of the relativistic momentum of the a-th particle are the first three components of the contravariant world vector $\underset{a}{m}\,\underset{a}{U}^{\iota}$, the fourth component of which is

$$\underset{a}{m}\underset{a}{U}^4 = \frac{\underset{a}{m}}{\sqrt{1 - \underset{a}{u}^2/c^2}}. \tag{6.18}$$

This expression is $(1/c^2)$ times the non-constant term of the relativistic energy, eq. (6.17). The four equations

$$\left.\begin{aligned} \sum_a \frac{\underset{a}{m}\underset{a}{u}^s}{\sqrt{1 - \underset{a}{u}^2/c^2}} &= \text{const.}, \\ \sum_a \frac{\underset{a}{m}}{\sqrt{1 - \underset{a}{u}^2/c^2}} &= \text{const.}, \end{aligned}\right\} \tag{6.19}$$

are, therefore, Lorentz covariant. The first term of eq. (6.17) is called the total (relativistic) energy of a particle. On account of its transformation property, it is to be considered as the fundamental expression for the energy. As it does not vanish when $\mathbf{u}$ becomes zero, it is often separated into two expressions, mc^2 and

$$T = mc^2[(1 - u^2/c^2)^{-1/2} - 1]. \tag{6.20}$$

mc^2 is called the "rest energy" of a particle, while T is its "relativistic kinetic energy."

Relation between energy and mass. The ratio between the momentum and the mass, the quantity μ, is often called "the relativistic mass" of a particle, and m is referred to as "the rest mass." The relativistic mass is equal to the total energy divided by c^2, and likewise the rest mass is $(1/c^2)$ times the rest energy. There exists, thus, a very close correlation between mass and energy which has no parallel in classical physics.

This correlation is further emphasized by the consideration of inelastic collisions. We know from classical physics that in the case of inelastic collisions, the momentum conservation laws remain unchanged, while the kinetic energy is partly transformed into other forms of energy. By considering Lorentz transformations, we find that the momentum conservation laws can hold only if the energy conservation law in the form (6.19) also holds; and this means that if the kinetic energy (6.20) of a mechanical system decreases, its rest energy increases, and therefore at least some of the rest masses of the constituent system must increase. We are thus led to the conclusion that all forms of energy contribute to the mass at the rate

$$\Delta E = c^2 \Delta m. \tag{6.21}$$

The recognition that energy and mass are equivalent is probably the most important accomplishment of relativistic mechanics. This equivalence accounts, for instance, for the mass defects occurring in atomic nuclei.

In classical physics, there are separate conservation laws for mass and energy. In relativistic mechanics, there is a conservation law for the total energy of an isolated system, but the rest masses of the constituent mass points change whenever kinetic energy is changed into other forms of energy, and vice versa. The mass of a material body is approximately constant as long as the bulk of its energy, the rest energy, is not involved in such changes. But rest masses change appreciably when interaction energies have the order of magnitude of the rest energies. This is the case in many nuclear reactions and particularly in the formation and annihilation of electron-positron pairs.

For small values of u/c, the relativistic kinetic energy (6.20) goes over into the classical expression, as shown by the power expansion

$$mc^2[(1 - u^2/c^2)^{-1/2} - 1] = \frac{m}{2} u^2 \left(1 + \frac{3}{4}\frac{u^2}{c^2} + \cdots\right). \tag{6.20a}$$

The four-vector

$$p^\rho = mU^\rho \tag{6.22}$$

is often referred to as the energy-momentum vector.

The Compton effect. An interesting application of the relativistic conservation laws is the theory of the Compton effect. It is assumed that a γ-quantum and a free electron "collide." Without making any assumptions about the nature of the interaction forces, we can determine the wave length of the γ-quantum and the velocity of the electron after the collision.

As for the γ-quantum, eqs. (6.15) and (6.17) are no longer applicable, because the denominator $\sqrt{1 - u^2/c^2}$ vanishes when u equals c. Furthermore, it is meaningless to speak of the "rest mass" of a photon, as there is no frame of reference with respect to which it is at rest. We replace (6.17) by the quantum relation between frequency and energy,

$$E = h\nu. \tag{6.23}$$

The components of the momentum form a contravariant world vector together with the energy, divided by c^2. On the other hand, ν transforms so that the expression

$$\nu t - \frac{\nu}{c} k_s x^s \tag{6.24}$$

is invariant ($\mathbf{k}$ is the three dimensional unit vector perpendicular to the wave front). It follows that the quantities $(-\frac{\nu}{c}\mathbf{k},\ \nu)$ are the components of a covariant world vector, and, therefore, that $(+c\mathbf{k}\nu,\ \nu)$ are the contravariant components of the same world vector. The transformation equations connecting $\mathbf{p}$, E/c^2 suggest, therefore, that we ascribe to the momentum of a γ-quantum the expression

$$\mathbf{p}_\gamma = h \cdot \frac{\nu}{c} \cdot \mathbf{k}, \tag{6.25}$$

and the magnitude of the momentum becomes

$$p_\gamma = h\,\frac{\nu}{c}. \tag{6.26}$$

We are now prepared to state all the conditions determining the collision of γ-quantum and electron. We assume that the γ-quantum approaches the electron along the X-axis, and that the electron is initially at rest. (Fig. 8.)

The rest mass of the electron is denoted by m. After the collision, the electron will move along a path at an angle α to the X-axis, while a γ-quantum of a different frequency $\bar{\nu}$ leaves the scene of the collision in some other direction, characterized by the angle β. We shall introduce our coördinates so that everything takes place in the X, Y-plane.

Before the collision takes place, the total energy E and the total momentum P are given by

$$\left.\begin{aligned} E &= mc^2 + h\nu, \\ P &= \frac{h\nu}{c}. \end{aligned}\right\} \tag{6.27}$$

After the collision, the total energy is

$$\bar{E} = \frac{mc^2}{\sqrt{1 - u^2/c^2}} + h\bar{\nu}, \tag{6.28}$$

and the total momenta in the x-direction and the y-direction are

$$\left.\begin{aligned} \bar{P}_1 &= \frac{h\bar{\nu}}{c}\cdot\cos\beta + \frac{mu}{\sqrt{1-u^2/c^2}}\cos\alpha, \\ \bar{P}_2 &= -\frac{h\bar{\nu}}{c}\cdot\sin\beta + \frac{mu}{\sqrt{1-u^2/c^2}}\sin\alpha. \end{aligned}\right\} \tag{6.29}$$

On account of the conservation laws, we have, therefore,

$$\left.\begin{aligned} mc^2 + h\nu &= \frac{mc^2}{\sqrt{1-u^2/c^2}} + h\bar{\nu}, \\ \frac{h\nu}{c} &= \frac{h\bar{\nu}}{c}\cos\beta + \frac{mu}{\sqrt{1-u^2/c^2}}\cos\alpha, \\ 0 &= -\frac{h\bar{\nu}}{c}\sin\beta + \frac{mu}{\sqrt{1-u^2/c^2}}\sin\alpha. \end{aligned}\right\} \tag{6.30}$$

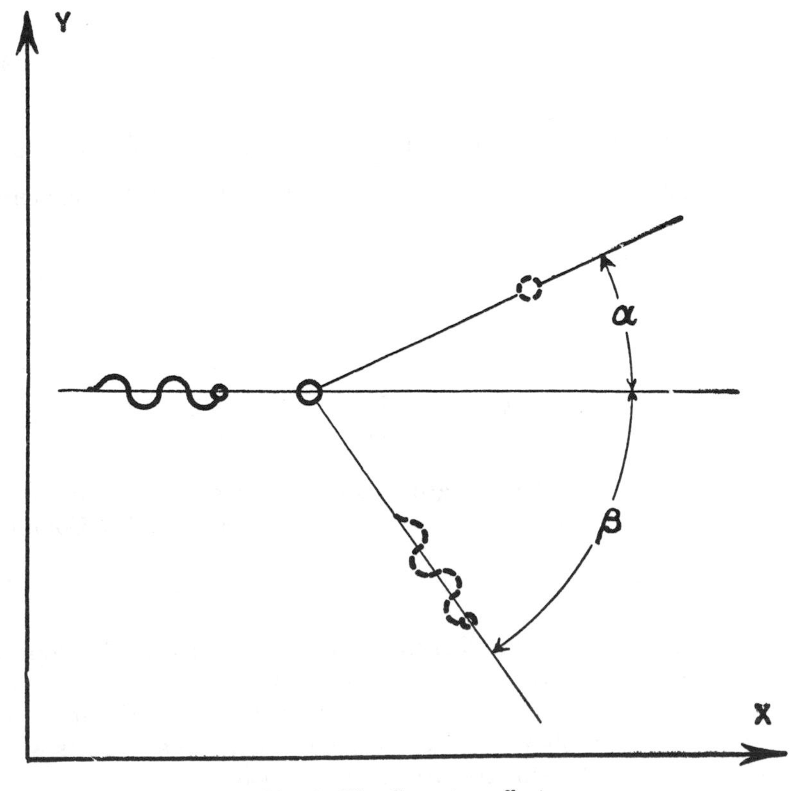

Fig. 8. The Compton effect.

We are interested in $\bar{\nu}$ as a function of the scattering angle β. Therefore, we eliminate from the equation first α and then u. In order to eliminate α, we first separate in the two momentum equations the one term containing α, and then square and add them. The result of these operations is the equation

$$\frac{h^2}{c^2}(\nu^2 - 2\nu\bar{\nu}\cos\beta + \bar{\nu}^2) = \frac{m^2u^2}{1 - u^2/c^2}, \tag{6.31}$$

which replaces the two last equations (6.30). As for the first equation (6.30), we remove the one square root by separating the one term containing it from the other terms, and squaring. We obtain:

$$m^2c^2 + 2mh(\nu - \bar{\nu}) + \frac{h^2}{c^2}(\nu^2 - 2\nu\bar{\nu} + \bar{\nu}^2) = \frac{m^2c^2}{1 - u^2/c^2}. \tag{6.32}$$

Now we eliminate u from eqs. (6.31) and (6.32) by subtracting them from each other, and we get one equation,

$$m(\nu - \bar{\nu}) - \frac{h}{c^2}(1 - \cos\beta)\nu\bar{\nu} = 0. \tag{6.33}$$

We express this relation in its usual form by writing $2\sin^2\frac{\beta}{2}$ instead of $(1 - \cos\beta)$ and by introducing the wave length instead of the frequency,

$$\nu = \frac{c}{\lambda}, \qquad \bar{\nu} = \frac{c}{\bar{\lambda}}. \tag{6.34}$$

We obtain, finally,

$$\bar{\lambda} - \lambda = \frac{2h}{mc}\sin^2\frac{\beta}{2}. \tag{6.35}$$

The greatest shift in wave length takes place when the γ-quantum is scattered straight back. It is equal to twice the so-called Compton wave length $\frac{h}{mc}$.

Relativistic analytical mechanics. Now we are able to develop the fundamentals of relativistic analytical mechanics. It is assumed that the student is fairly familiar with the rudiments of classical analytical mechanics, and the following review will stress only those aspects which have a bearing on the considerations of this book.

Whenever a (classical) mechanical system is subject to conservative forces, the differential equations can be written in the form

$$\frac{\partial L}{\partial x^s} - \frac{d}{dt}\left(\frac{\partial L}{\partial \dot{x}^s}\right) = 0, \qquad L = T - V, \tag{6.36}$$

where T is the kinetic and V the potential energy. If the equations of motion are written in this form, they are covariant not only with respect to spatial orthogonal coördinate transformations, but with respect to arbitrary transformations of the n coördinates which represent the degrees of freedom of the system.

These equations are the so-called Euler-Lagrange equations of a variational problem: A variational problem is the problem of finding those curves connecting two fixed points along which a given line integral assumes stationary values. Consider an $(n + 1)$ dimensional space with the coördinates x^s and t, and a line integral,

$$I = \int_{P_1}^{P_2} L\,dt, \tag{6.37}$$

the integrand of which depends on the coördinates x^s (and possibly t) and on the ratios between the coördinate differentials, $\frac{dx^s}{dt}$, along the path of integration. If we carry out an infinitesimal variation of the path of integration, but so that the end points remain fixed, we obtain

$$\delta I = \int_{P_1}^{P_2} \left(\frac{\partial L}{\partial x^s}\,\delta x^s + \frac{\partial L}{\partial \dot{x}^s}\,\delta \dot{x}^s\right) dt = \int_{P_1}^{P_2} \left[\frac{\partial L}{\partial x^s} - \frac{d}{dt}\left(\frac{\partial L}{\partial \dot{x}^s}\right)\right] \delta x^s\,dt, \tag{6.38}$$

where $\dot{x}^s$ stands for $\frac{dx^s}{dt}$. The paths of integration for which I is stationary are those along which eqs. (6.36) are satisfied.

The momenta are defined by the equations

$$p_s = \frac{\partial L}{\partial \dot{x}^s}. \tag{6.39}$$

With their help, the n differential equations of the second order (6.36) may be transformed into $2n$ equations of the first order. One solves the n equations (6.39) with respect to $\dot{x}^s$ and forms the so-called Hamiltonian function,

$$H = -L + p_s \dot{x}^s, \tag{6.40}$$

with all the $\dot{x}_s$ replaced by the solutions of eqs. (6.39). The equations of motion take the "*canonical*" form,

$$\left.\begin{aligned} \dot{x}^s &= \frac{\partial H}{\partial p_s}, \\ \mathring{p}_s &= -\frac{\partial H}{\partial x^s}. \end{aligned}\right\} \tag{6.41}$$

The whole formalism is covariant with respect to general three dimensional coördinate transformations (and, in fact, to even more general transformations). One feature of particular interest to us is the possibility of introducing another parameter instead of the time t. We shall call this parameter θ and define it by the equation

$$dt = \frac{dt}{d\theta}\,d\theta, \tag{6.42}$$

$\frac{dt}{d\theta}$ being given. All ordinary derivatives with respect to t are denoted by dots, and those with respect to θ by primes. We have, then,

$$\left.\begin{aligned} I &= \int L^*(x^s, x^{s\prime}, t, t')\,d\theta, \\ L^* &= t' \cdot L\left(x^s, \frac{x^{s\prime}}{t'}, t\right). \end{aligned}\right\} \tag{6.43}$$

We can now write the Euler-Lagrange equations in terms of the new Lagrangian L^*. The derivatives of L^* with respect to its arguments are as follows:

$$\left.\begin{aligned} \frac{\partial L^*}{\partial x^s} &= t'\,\frac{\partial L}{\partial x^s}, & \frac{\partial L^*}{\partial t} &= t'\,\frac{\partial L}{\partial t}, \\ \frac{\partial L^*}{\partial x^{s\prime}} &= \frac{\partial L}{\partial \dot{x}^s}, & \frac{\partial L^*}{\partial t'} &= L - \dot{x}^s\,\frac{\partial L}{\partial \dot{x}^s} = -H. \end{aligned}\right\} \tag{6.44}$$

Inasmuch as we can write

$$\frac{d}{d\theta} = t' \cdot \frac{d}{dt},$$

the Euler-Lagrange equations take the form

$$\left.\begin{aligned} \frac{\partial L}{\partial x^s} - \frac{d}{dt}\left(\frac{\partial L}{\partial \dot{x}^s}\right) &= 0, \\ \frac{\partial L}{\partial t} + \frac{dH}{dt} &= 0. \end{aligned}\right\} \tag{6.45}$$

Note that the definition of the momenta (6.39) is invariant with respect to this parameter transformation. But an additional component which is canonically conjugate to t has been added to the momenta. Its value is $-H$. We shall denote this new component by $\mathfrak{E}$.

Again the $(n + 1)$ equations defining the momenta can be solved for the $(n + 1)$ quantities $x^{s\prime}$ and t', and the new Hamiltonian may be defined

$$H^* = -L^* + p_s x^{s\prime} + \mathfrak{E} \cdot t' = t'(-L + p_s \dot{x}^s + \mathfrak{E}). \tag{6.46}$$

This new Hamiltonian vanishes.† But since $\mathfrak{E}$ is to be considered as an independent variable, the various partial derivatives do not vanish. They are related to the derivatives of the Hamiltonian (6.40) as follows:

$$\left.\begin{aligned} \frac{\partial H^*}{\partial x^s} &= t' \frac{\partial H}{\partial x^s}, & \frac{\partial H^*}{\partial t} &= t' \frac{\partial H}{\partial t}, \\ \frac{\partial H^*}{\partial p_s} &= t' \frac{\partial H}{\partial p_s}, & \frac{\partial H^*}{\partial \mathfrak{E}} &= t'. \end{aligned}\right\} \tag{6.47}$$

We get, therefore, in addition to eqs. (6.41), the following:

$$\frac{dH}{dt} = \frac{\partial H}{\partial t}. \tag{6.48}$$

The value of the Hamiltonian (6.40) is the sum of the kinetic and potential energy. This total energy changes with time at the rate $\frac{\partial H}{\partial t}$. Eq. (6.48) is not independent of the equations (6.41). It can be derived directly from them, without using the method of parameter transformation. The total differential of H is

$$dH = \frac{\partial H}{\partial x^s} dx^s + \frac{\partial H}{\partial p_s} dp_s + \frac{\partial H}{\partial t} dt;$$

and, because of eqs. (6.41),

$$dH = -\frac{dp_s}{dt} dx^s + \frac{dx^s}{dt} dp_s + \frac{\partial H}{\partial t} dt.$$

Dividing this equation by dt, we obtain eq. (6.48).

The parameter transformation is useful insofar as it permits the switching from one parameter, t, to another parameter, θ, and we know that both representations are equivalent.

We shall now go on to a treatment of relativistic mechanics, and, for

†See Appendix B.

the time being, we shall discuss the motion of particles not subject to forces. The Lagrangian is some function of the $\frac{dx^s}{dt}$ (if t is chosen as parameter) or of the $\frac{dx^s}{d\tau}$ (if τ is the parameter). For the purpose of the following discussion, dots represent differentiation with respect to t, primes differentiation with respect to τ, the proper time of the particle. Its derivatives with respect to these arguments must be equal to the momenta. The momenta canonically conjugate to the derivatives of the x^s are given by eq. (6.15). The fourth momentum (the canonical conjugate of the time), according to (6.44), is

$$p_4 = L^{(t)} - \dot{x}^s p_s \,, \tag{6.49}$$

where $L^{(t)}$ is the Lagrangian belonging to the parameter t.

Let us start with the parameter t. The solution of the differential equations

$$\frac{\partial L^{(t)}}{\partial \dot{x}^s} = \frac{m\dot{x}^s}{\sqrt{1 - u^2/c^2}}, \qquad u^2 = \dot{x}^s \dot{x}^s, \tag{6.50}$$

for $L^{(t)}$ is

$$L^{(t)} = mc^2(k - \sqrt{1 - u^2/c^2}). \tag{6.51}$$

k is the integration constant. The integral I has the value

$$I = \int_{t_1}^{t_2} L^{(t)}\, dt = mc^2[k(t_2 - t_1) - \tau(P_2\,, P_1)]. \tag{6.52}$$

$\tau(P_2\,, P_1)$ denotes the proper time along the path of integration between the two end points P_1 and P_2. The integral is a Lorentz invariant when k is zero. The expression

$$mc^2k(t_2 - t_1)$$

is independent of the path of integration. Therefore, it does not contribute to the variation δI as long as the bounds of the integral are not being varied. We shall find that the value of k determines the value of the integration constant E_0 of eq. (6.17). As this term is not invariant, we shall drop it before going over to the four dimensional representation.

The equations

$$p_s = \frac{m\dot{x}^s}{\sqrt{1 - u^2/c^2}} \tag{6.15a}$$

can be solved with respect to the $\dot{x}^s$. The expressions are:

$$\dot{x}^s = \frac{p_s/m}{\sqrt{1 + \mathbf{p}^2/m^2 c^2}}. \tag{6.15b}$$

The Hamiltonian is defined by eq. (6.40). In this case it turns out to be

$$H^{(t)} = mc^2[\sqrt{1 + p^2/m^2 c^2} - k] = mc^2\left(\frac{1}{\sqrt{1 - u^2/c^2}} - k\right). \tag{6.53}$$

When k is chosen to be 1, $H^{(t)}$ is equal to the relativistic kinetic energy; when it is chosen to be 0, $H^{(t)}$ is the total energy.

The Hamiltonian equations are

$$\left.\begin{aligned} \dot{x}^s &= \frac{\partial H}{\partial p_s} = \frac{p_s/m}{\sqrt{1 + p^2/m^2 c^2}}, \\ \dot{p}_s &= -\frac{\partial H}{\partial x^s} = 0. \end{aligned}\right\} \tag{6.54}$$

Let us now go over to the parameter τ. According to eq. (6.43), $L^{(\tau)}$ is given by

$$L^{(\tau)} = t'\cdot L^{(t)} = -mc^2\sqrt{t'^2 - \frac{1}{c^2}x^{s\prime 2}} = -mc^2\sqrt{\eta_{\iota\kappa}x^{\iota\prime}x^{\kappa\prime}}. \tag{6.55}$$

The momenta are

$$\left.\begin{aligned} p_i &= \frac{\partial L^{(\tau)}}{\partial x^{i\prime}} = \frac{mx^{i\prime}}{\sqrt{\eta_{\iota\kappa}x^{\iota\prime}x^{\kappa\prime}}}, \\ p_4 &= \frac{\partial L^{(\tau)}}{\partial x^{4\prime}} = -\frac{mc^2 t'}{\sqrt{\eta_{\iota\kappa}x^{\iota\prime}x^{\kappa\prime}}}, \qquad p_\rho = -\frac{mc^2\eta_{\rho\sigma}x^{\sigma\prime}}{\sqrt{\eta_{\iota\kappa}x^{\iota\prime}x^{\kappa\prime}}}. \end{aligned}\right\} \tag{6.56}$$

The root appearing in eqs. (6.55) and (6.56) has the value 1. Unless this is kept in mind, it is impossible to solve the equations (6.56) with respect to the $x^{\rho\prime}$. As a matter of fact, we could have chosen some other parameter instead of τ, let us say θ.† If we had, eqs. (6.56) would have taken exactly the same form, except that the primes would signify differentiation with respect to θ. Then it would not be possible to express the $x^{\rho\prime}$ by the p_ρ alone. As it is, we make use of the fact that the $x^{\rho\prime}$ are the contravariant components of a unit vector, and that the $-\frac{p_\rho}{mc^2}$ are the covariant components of the same unit vector. The $x^{\rho\prime}$ are identical with U^ρ. The solution may be written in the form

$$x^{\rho\prime} = -\frac{\eta^{\rho\sigma}p_\sigma}{\sqrt{\eta^{\iota\kappa}p_\iota p_\kappa}}, \tag{6.57}$$

†See Appendix B.

so that it resembles eq. (6.15b). This way of writing the solution is really quite arbitrary. The value of the square root in the denominator is constant.

$$\sqrt{\eta^{\iota\kappa}p_\iota p_\kappa} = mc^2. \tag{6.58}$$

Therefore, the solution can be written in the general form

$$x^{\rho\prime} = -\frac{\eta^{\rho\sigma}p_\sigma}{mc^2}\cdot\varphi(\eta^{\iota\kappa}p_\iota p_\kappa/m^2c^4), \qquad \varphi(1) = 1, \tag{6.57a}$$

where φ is a completely arbitrary function of its argument except that it is 1 when the argument is 1. Using the expression (6.57a), we have for $L^{(\tau)}$, $x^\rho p_\rho'$, and $H^{(\tau)}$,

$$\left.\begin{aligned} L^{(\tau)} &= -\sqrt{\eta^{\iota\kappa}p_\iota p_\kappa}\cdot\varphi, \\ x^{\rho\prime}p_\rho &= -\frac{\eta^{\rho\sigma}p_\rho p_\sigma}{mc^2}\cdot\varphi, \\ H^{(\tau)} &= \left[\sqrt{\eta^{\rho\sigma}p_\rho p_\sigma} - \frac{\eta^{\rho\sigma}p_\rho p_\sigma}{mc^2}\right]\cdot\varphi. \end{aligned}\right\} \tag{6.59}$$

As an abbreviation, we write p instead of the square root. The Hamiltonian equations are

$$\left.\begin{aligned} p_\rho' &= -\frac{\partial H^{(\tau)}}{\partial x^\rho} = 0, \\ x^{\rho\prime} &= \frac{\partial H^{(\tau)}}{\partial p_\rho} = \frac{2\eta^{\rho\sigma}p_\sigma}{m^2c^4}\left[p - \frac{p^2}{mc^2}\right]\cdot\varphi' + \left(\frac{\eta^{\rho\sigma}p_\sigma}{p} - 2\,\frac{\eta^{\rho\sigma}p_\sigma}{mc^2}\right)\cdot\varphi. \end{aligned}\right\} \tag{6.60}$$

φ' denotes the derivative of φ with respect to its argument. The first term on the right-hand side vanishes because the square bracket is zero. The second term has the value

$$x^{\rho\prime} = -\frac{\eta^{\rho\sigma}p_\sigma}{mc^2} = -\frac{p^\rho}{mc^2}. \tag{6.60a}$$

The factor φ may be omitted, since its value is one.

The appearance of the arbitrary function φ is not accidental. The integral (6.52), with $k = 0$, is equal to the arc length of the world line in the Minkowski world. The Euler-Lagrange equations are the differential equations of the geodesic line, corresponding to eq. (5.99).

The variational principle (5.93) is not the only one leading to the differential equations (5.99). Any variational principle

$$I = \int_{P_1}^{P_2}\psi\left(g_{ik}\frac{d\xi^i}{ds}\frac{d\xi^k}{ds}\right)ds \tag{6.61}$$

has the variation

$$\delta I = -2\int_{P_1}^{P_2}\left(g_{rn}\frac{d^2\xi^r}{ds^2} + [rt, n]\frac{d\xi^r}{ds}\frac{d\xi^t}{ds}\right)\cdot\psi'\cdot\delta\xi^n\cdot ds.$$

In our case, any function

$$\Lambda^{(\tau)} = \Phi(\eta_{\iota\kappa}x^{\iota\prime}x^{\kappa\prime}) \tag{6.62}$$

would be suitable as a Lagrangian, and the momenta would be defined as

$$\pi_\rho = 2\eta_{\rho\mu}x^{\mu\prime}\cdot\Phi'(\eta_{\iota\kappa}x^{\iota\prime}x^{\kappa\prime}), \tag{6.63}$$

where Φ' is the derivative of Φ with respect to its argument. These π_ρ are identical with the momenta (6.56) if

$$\Phi'(1) = -\frac{m}{2}c^2. \tag{6.64}$$

In the same way, any function of the momenta π_ρ having the form

$$\mathrm{H}^{(\tau)} = \mathrm{X}(\eta^{\iota\kappa}\pi_\iota\pi_\kappa) \tag{6.65}$$

would be a suitable Hamiltonian. The equations connecting the momenta with the velocities are

$$x^{\rho\prime} = \frac{\partial \mathrm{H}^{(\tau)}}{\partial \pi_\rho} = 2\eta^{\rho\mu}\pi_\mu\cdot\mathrm{X}'. \tag{6.66}$$

The π_ρ are identical with the p_ρ if

$$\mathrm{X}'(m^2c^4) = -\frac{1}{2mc^2}. \tag{6.67}$$

Relativistic force. The force acting on an accelerated mass point may be defined in different ways. One can closely adhere to the classical definition and define it as the change of momentum with time,

$$f_s = \frac{d}{dt}\left(\frac{m\dot{x}^s}{\sqrt{1 - u^2/c^2}}\right), \qquad u^2 = \dot{x}^s\dot{x}^s. \tag{6.68}$$

If we carry out the differentiation on the right-hand side, we get a rather unwieldy expression,

$$f_s = (1 - u^2/c^2)^{-3/2}\cdot m\cdot\left[\delta_{st}(1 - u^2/c^2) + \frac{\dot{x}^s\dot{x}^t}{c^2}\right]\ddot{x}^t. \tag{6.69}$$

In general, the force thus defined is not parallel to the acceleration. It is parallel only when the acceleration is either parallel or perpendicular to the velocity. When it is parallel, eq. (6.69) takes the form

$$f_s = (1 - u^2/c^2)^{-3/2}m\ddot{x}^s. \tag{6.70}$$

When force and velocity are orthogonal, eq. (6.69) becomes

$$f_s = (1 - u^2/c^2)^{-1/2} m\ddot{x}^s. \tag{6.71}$$

The coefficients of the acceleration on the right-hand sides of eqs. (6.70) and (6.71) are occasionally referred to as "longitudinal mass" and "transversal mass," respectively.

The f_s defined by eq. (6.68) have no simple transformation properties. There exists, however, a world vector which is closely analogous to the three dimensional force. We define the "world force" as the change of the momentum with *proper* time,

$$F_\rho = \frac{dp_\rho}{d\tau}. \tag{6.72}$$

The first three components are equal to the expressions (6.68), multiplied by $\frac{dt}{d\tau}$,

$$F_s = \frac{m\ddot{x}^s}{1 - u^2/c^2} + \frac{mu\dot{u}\dot{x}^s}{c^2(1 - u^2/c^2)^2}. \tag{6.73}$$

For small values of u/c, these expressions approach $m\ddot{x}^s$. The fourth component is

$$F_4 = -\frac{mu\dot{u}}{(1 - u^2/c^2)^2}. \tag{6.74}$$

For small u/c, this component approaches the negative work performed per unit time. It is equal to the negative product of the first three components multiplied by $\dot{x}^s$.

PROBLEMS

1. Starting with a Lagrangian (6.62), determine the corresponding Hamiltonian. Show that eq. (6.67) is satisfied by the Hamiltonian if the Lagrangian satisfies eq. (6.64).

2. An excited atom of total mass M, at rest with respect to the inertial system chosen, goes over into a lower state with an energy smaller by ΔW. It emits a photon, and thereby undergoes a recoil. The frequency of the photon will, therefore, not be exactly $\nu = \Delta W/h$, but smaller. Compute this frequency.

Answer:

$$\nu = \frac{\Delta W}{h} \cdot \left(1 - \frac{1}{2}\frac{\Delta W}{Mc^2}\right).$$

3. Show that the integral of the "relativistic force" (6.68) over the three dimensional path is equal to the change of the relativistic kinetic energy (6.20), and that the integral of the world vector (6.72) over the world line dx^ρ vanishes if the rest mass is assumed not to change. What happens if the rest mass changes during the motion?

CHAPTER VII

Relativistic Electrodynamics

When Einstein derived the transformation laws of Maxwell's equations, Minkowski's four dimensional formalism was not yet known. He proceeded by expressing the derivatives appearing in Maxwell's equations in terms of derivatives with respect to a Lorentz transformed set of coördinates. Applying the principle of relativity, he demanded that the electrodynamic equations take the same form in the new coördinate system as in the first system. He identified, therefore, the linear combinations of field intensities which appeared in the transformed Maxwell equations, with the field intensities with respect to the new coordinates. He showed that the transformation laws so obtained have the required properties, namely, that two successive transformations are equivalent to one transformation of the same type, and that the inverse transformation is obtained by changing the sign of the relative velocity of the two coördinate systems.

This computation is straightforward, but somewhat lengthy. We shall follow here a different procedure, which takes advantage of the four dimensional formalism developed in Chapter V.

Maxwell's field equations. We shall use electrostatic units throughout. Furthermore, we shall accept Lorentz' electron theory, according to which the equations for empty space (with the dielectric constant and the permeability equal to unity) are also valid in the interior of matter, and according to which the polarization and magnetization of a material medium are the results of the displacement of actual charges and the partial ordering of elementary magnets. Maxwell's equations take the form

$$\operatorname{div} \mathbf{E} = 4\pi\sigma, \tag{7.1}$$

$$\operatorname{curl} \mathbf{E} + \frac{1}{c}\frac{\partial \mathbf{H}}{\partial t} = 0, \tag{7.2}$$

$$\operatorname{div} \mathbf{H} = 0, \tag{7.3}$$

$$\operatorname{curl} \mathbf{H} - \frac{1}{c}\frac{\partial \mathbf{E}}{\partial t} = \frac{4\pi}{c}\mathbf{I}, \tag{7.4}$$

where **E** is the electric field strength, **H** the magnetic field strength, σ the charge density, and **I** the current density. In empty space, the charge and current densities vanish.

From eqs. (7.2) and (7.3), it can be concluded that the field intensities can be represented as derivatives of a scalar φ and a vector **A**, as follows:

$$\mathbf{H} = \text{curl } \mathbf{A}, \tag{7.5}$$

$$\mathbf{E} = -\text{ grad } \varphi - \frac{1}{c}\frac{\partial \mathbf{A}}{\partial t}. \tag{7.6}$$

By differentiating eq. (7.1) with respect to the time, and by forming the divergence of eq. (7.4), we obtain the conservation laws for charge and current densities,

$$\text{div } \mathbf{I} + \frac{\partial \sigma}{\partial t} = 0. \tag{7.7}$$

Preliminary remarks on transformation properties. With respect to spatial orthogonal transformations, the quantities φ, **A**, **E**, **H**, σ, and **I** transform independently of each other. But not all of the vectors are of the same type. We found, in Chapter V, that there exist two types of vectors in three dimensional vector calculus, the "polar" and the "axial" vectors. They differ in their behavior with respect to reflections of the coördinate system in that the direction of an "axial vector" is reversed upon reflection. The curl of a "polar vector" is an axial vector, and vice versa.

In a linear equation, all terms must, of course, transform according to the same law, otherwise the equation would not be covariant with respect to reflections. Eqs. (7.2) and (7.4) show that **E** and **H** cannot be vectors of the same type.

We are accustomed to considering the sign of a charge as independent of the screw sense of the coördinate system. The electric field intensity is defined as the force acting upon the unit charge, and its direction remains, therefore, unchanged by a reflection of the coördinate system. **E** is a polar vector. It follows that **H** is an axial vector, and **I** and **A** are polar vectors.

"Axial vectors," we found, are equivalent—in three dimensions—to skewsymmetric tensors of rank 2. This is why they can appear together with "polar" quantities in linear, covariant equations. From the point of view of covariance, it is often preferable to express this equivalence explicitly and to treat **H** as a skewsymmetric tensor with the components

$$\left.\begin{aligned} H_{12} &= -H_{21} = H_z\,, \\ H_{23} &= -H_{32} = H_x\,, \\ H_{31} &= -H_{13} = H_y\,. \end{aligned}\right\} \tag{7.8}$$

If we write Maxwell's equations as vector-tensor equations, they take the form:

$$E_{s,s} = 4\pi\sigma, \tag{7.1a}$$

$$E_{s,r} - E_{r,s} + \frac{1}{c}\frac{\partial H_{rs}}{\partial t} = 0, \tag{7.2a}$$

$$H_{rs,t} + H_{st,r} + H_{tr,s} = 0, \tag{7.3a}$$

$$H_{rs,s} - \frac{1}{c}\frac{\partial E_r}{\partial t} = \frac{4\pi}{c} I_r, \tag{7.4a}$$

and the equations (7.5) and (7.6) take the form

$$H_{rs} = A_{s,r} - A_{r,s}\,, \tag{7.5a}$$

$$E_r = -\frac{1}{c}\frac{\partial A_r}{\partial t} - \varphi_{,r}\,. \tag{7.6a}$$

Eq. (7.7) becomes

$$I_{s,s} + \frac{\partial \sigma}{\partial t} = 0. \tag{7.7a}$$

The covariant character of these equations with respect to all spatial, orthogonal coördinate transformations (including reflections) is assured. Had we expressed **H** as a vector density, eqs. (7.4), for instance, would have taken the form

$$\mathfrak{H}_{2,1} - \mathfrak{H}_{1,2} - \frac{1}{c}\frac{\partial E_3}{\partial t} = \frac{4\pi}{c} I_3,$$

and so forth, in which the covariance is much less obvious. Note that the set (7.3a) contains only one significant equation; those equations in which not all three indices are different are satisfied identically.

When we carry out a coördinate transformation involving the transition to a new frame of reference, we can no longer expect all the quantities **E**, **H**, and so forth, to transform independently of each other. Let us consider, for instance, an electrostatic field in which the charges are at rest with respect to a given frame of reference. With respect to another frame of reference, the charges move; and we find, therefore, a nonvanishing current even though there is no current with respect to the original coördinate system. The current induces, in turn, a magnetic field, even though there is no magnetic field with respect to the original frame of reference. Finally, although the vector potential **A** vanishes in the original coördinate system, **A** cannot vanish in the new coördinate system with respect to which we encounter a magnetic field strength. This example shows that, with respect to Lorentz transformations, the scalar and vector potentials transform as the components of one quantity, and that the same is true of electric and magnetic field strengths and of the charge and current densities. Therefore, we must try to represent the electric and magnetic field strength, for instance, by one world tensor which decomposes into two three dimensional vectors when we consider purely spatial coördinate transformations.

The representation of four dimensional tensors in three plus one dimensions. In the preceding chapter, we found that different three dimensional quantities are frequently part of one four dimensional quantity. For instance, the "relativistic mass," a scalar with respect to spatial orthogonal transformations, and the "relativistic momentum," a spatial vector, are *together* the components of one world vector, the energy-momentum vector.

Conversely, if we consider only the group of spatial transformations instead of the group of all Lorentz transformations, a world vector or tensor can be decomposed into several parts which transform independently of each other. In the case of purely spatial transformations, the four-rowed form of the transformation coefficients, $\{\gamma^{\iota}{}_{\kappa}\}$, assumes the form

$$\begin{Bmatrix} \gamma^{i}{}_{k}, & \gamma^{i}{}_{4} \\ \gamma^{4}{}_{k}, & \gamma^{4}{}_{4} \end{Bmatrix} \longrightarrow \begin{Bmatrix} c_{ik}, & 0 \\ 0, & 1 \end{Bmatrix}, \tag{7.9}$$

where c_{ik} are the coefficients of an orthogonal transformation. A contravariant world vector V^{ι} transforms as follows:

$$V'^{i} = c_{ik}V^{k}, \qquad V'^{4} = V^{4}. \tag{7.10}$$

A world tensor of rank 2, $V^{\iota\kappa}$ can be similarly decomposed into a three dimensional tensor, two three dimensional vectors. and a three dimensional scalar:

$$V'^{ik} = c_{im}c_{kn}V^{mn}, \quad V'^{i4} = c_{im}V^{m4}, \quad V'^{4k} = c_{kn}V^{4n}, \quad V'^{44} = V^{44}. \tag{7.11}$$

If the world tensor $V^{\iota\kappa}$ is symmetric, the three dimensional tensor is also symmetric, and the two vectors are identical. If $V^{\iota\kappa}$ is skewsymmetric, the three dimensional tensor is also skewsymmetric (and, therefore, equivalent to an "axial vector"), the two vectors are identical except for the sign, and the scalar vanishes. Covariant vectors and tensors can be decomposed in the same manner.

Just as world vectors and tensors can be decomposed into three dimensional tensors, vectors, and scalars, four dimensional covariant operations and equations can be decomposed. Let us denote the two parts of a world vector V^ι by v_i and v. The divergence of V^ι is

$$W = V^\iota{}_{,\iota} = v_{i,i} + v_{,4} = \operatorname{div} \mathbf{v} + \frac{\partial v}{\partial t}. \tag{7.12}$$

Likewise, we can denote the parts of a world tensor $V^{\iota\kappa}$ by v_{ik}, $\underset{1}{v_i}$, $\underset{2}{v_k}$ and v,

$$V^{\iota\kappa} = \begin{Bmatrix} v_{ik}, & \underset{1}{v_i} \\ \underset{2}{v_k}, & v \end{Bmatrix}.$$

The divergence of such a tensor, W^ι, may be denoted by w_i and w. The set of equations

$$V^{\iota\kappa}{}_{,\kappa} = W^\iota \tag{7.13}$$

is decomposed into two sets:

$$\left.\begin{aligned} V^{i\kappa}{}_{,\kappa} &= v_{ik,k} + \frac{\partial}{\partial t}(\underset{1}{v_i}) = w_i \\ V^{4\kappa}{}_{,\kappa} &= \underset{2}{v_{k,k}} + \frac{\partial v}{\partial t} = w \end{aligned}\right\}. \tag{7.13a}$$

If the world tensor $V^{\iota\kappa}$ is skewsymmetric, these equations can be written so that they contain only vectors. Instead of the three dimensional skewsymmetric tensor v_{ik}, we can introduce the "axial vector" $\mathbf{t}_s$ by the equations

$$v_{ik} = \delta_{iks}\mathbf{t}_s, \qquad \mathbf{t}_s = \tfrac{1}{2}\delta_{iks}v_{ik}. \tag{7.14}$$

The tensor divergence $v_{ik,k}$ then goes over into

$$v_{ik,k} = \delta_{iks}\mathbf{t}_{s,k}, \tag{7.15}$$

which is, according to eq. (5.38b), the ith component of the curl of $\mathbf{t}$. The first set of eqs. (7.13a) becomes, therefore:

$$\operatorname{curl} \mathbf{t} + \frac{\partial}{\partial t}\underset{1}{\mathbf{v}} = \mathbf{w}. \tag{7.13b}$$

As mentioned before, $\underset{2}{\mathbf{v}}$ equals $(-\underset{1}{\mathbf{v}})$ if $V^{\iota\kappa}$ is skewsymmetric. The last equation (7.13a) is, therefore,

$$-\mathrm{div}\,\underset{1}{\mathbf{v}} = w. \tag{7.13c}$$

Skewsymmetric derivatives can be decomposed similarly: Consider a covariant world vector B_ι, with the three dimensional parts b_i, b. Its four dimensional curl, $C_{\iota\kappa}$, can be decomposed as follows:

$$\left.\begin{aligned} C_{ik} &= b_{k,i} - b_{i,k}, \\ C_{i4} &= b_{,i} - \frac{\partial b_i}{\partial t} = \underset{1}{c_i}. \end{aligned}\right\} \tag{7.16}$$

If we introduce the "axial vector" $\mathfrak{D}_s$ by the equations

$$\tfrac{1}{2} C_{ik}\delta_{iks} = \mathfrak{D}_s,$$

the equations become

$$\left.\begin{aligned} \boldsymbol{\mathfrak{D}} &= \mathrm{curl}\,\mathbf{b}, \\ \underset{1}{\mathbf{c}} &= \mathrm{grad}\, b - \frac{\partial \mathbf{b}}{\partial t}. \end{aligned}\right\} \tag{7.16a}$$

These instances illustrate how typical four dimensional relations are split up into several apparently unrelated three dimensional vector relations. We are now ready to attempt the four dimensional reformulation of Maxwell's field equations.

The Lorentz invariance of Maxwell's field equations. Eqs. (7.16a) are the same as the eqs. (7.5) and (7.6) if $\mathbf{H}$ is identified with $\boldsymbol{\mathfrak{D}}$, $\mathbf{A}$ with $\mathbf{b}$, $\mathbf{E}$ with $\frac{1}{c}\underset{1}{\mathbf{c}}$, and φ with $-\frac{1}{c}b$. This means that eqs. (7.5) and (7.6) are Lorentz covariant if we assume that $\mathbf{A}$ and $(-c\varphi)$ transform as the components of a covariant world vector, and that $\mathbf{H}$ and $c\cdot\mathbf{E}$ are the six components of a covariant skewsymmetric world tensor.

The fusion of $\mathbf{H}$ and $\mathbf{E}$ into one skewsymmetric world tensor of rank 2 furnishes the basis for relativistic electrodynamics. We shall find that the remaining field equations are also compatible with this transformation law.

We shall denote the covariant world vector with the components

$(A_s, -c\varphi)$ by φ_ι, and the covariant electromagnetic field tensor with the components

$$\begin{Bmatrix} 0, & -H_3, & +H_2, & -cE_1 \\ H_3, & 0, & -H_1, & -cE_2 \\ -H_2, & +H_1, & 0, & -cE_3 \\ +cE_1, & +cE_2, & +cE_3, & 0 \end{Bmatrix} \tag{7.17}$$

by $\varphi_{\iota\kappa}$. The equations (7.5) and (7.6) are represented by the set

$$\varphi_{\iota\kappa} = \varphi_{\iota,\kappa} - \varphi_{\kappa,\iota}. \tag{7.18}$$

A skewsymmetric expression, $\varphi_{\iota\kappa}$, which is the curl of a vector, satisfies the equations

$$\varphi_{\iota\kappa,\lambda} + \varphi_{\kappa\lambda,\iota} + \varphi_{\lambda\iota,\kappa} = 0. \tag{7.19}$$

This can be verified if we substitute the expressions (7.18) in the left-hand sides of eqs. (7.19). Of the equations (7.19), all those are *identically* satisfied which have at least two indices equal, whether the skew-symmetric $\varphi_{\iota\kappa}$ satisfy eq. (7.18) or not. For instance,

$$\varphi_{12,2} + \varphi_{22,1} + \varphi_{21,2} \equiv 0,$$

simply because of the skewsymmetry of $\varphi_{\iota\kappa}$. In a four dimensional world, there remain only four significant equations, those with the index combinations (2, 3, 4), (1, 3, 4), (1, 2, 4), (1, 2, 3). If we substitute the expressions (7.17) into eq. (7.19), we obtain the equations (7.2) and (7.3). These two field equations are, therefore, shown to be equivalent to the world tensor relation (7.19).

There remain eqs. (7.1) and (7.4). These two equations are similar in form to the equations (7.13b) and (7.13c). Eqs. (7.13b, c) were obtained as the three dimensional representation of the divergence of a contravariant skewsymmetric world tensor of rank 2. If we raise both indices of the covariant tensor $\varphi_{\iota\kappa}$, we obtain the contravariant tensor

$$\varphi^{\iota k} = \begin{Bmatrix} 0, & -c^4H_3, & +c^4H_2, & +c^3E_1 \\ +c^4H_3, & 0, & -c^4H_1, & +c^3E_2 \\ -c^4H_2, & +c^4H_1, & 0, & +c^3E_3 \\ -c^3E_1, & -c^3E_2, & -c^3E_3, & 0 \end{Bmatrix}. \tag{7.17a}$$

The divergence of this tensor has the components

$$\varphi^{i\kappa}{}_{,\kappa} = -\,c^4\,(\text{curl}\,\mathbf{H})_i + c^3\,\frac{\partial E_i}{\partial t},$$

$$\varphi^{4\kappa}{}_{,\kappa} = -\,c^3\,\text{div}\,\mathbf{E}.$$

The right-hand sides are equal to $-4\pi c^3 I_i$ and to $-4\pi c^3\sigma$. We conclude, therefore, that **I** and σ together form a contravariant world vector, *the world current density*, I^ρ, which is connected with the field tensor $\varphi^{\iota\kappa}$ by the equation

$$\varphi^{\iota\kappa}{}_{,\kappa} = -4\pi c^3 I^\iota. \tag{7.20}$$

This equation is equivalent to Maxwell's equations (7.1) and (7.4).

If we form the divergence of eq. (7.20), the left-hand side vanishes identically, because of the skewsymmetry of $\varphi^{\iota\kappa}$. We obtain, therefore, one equation which contains only the components of I^ι,

$$I^\iota{}_{,\iota} = 0. \tag{7.21}$$

This equation is identical with eq. (7.7).

Maxwell's field equations are Lorentz covariant if the quantities appearing in them are identified with components of world vectors and world tensors in the manner indicated above. Before going on to a discussion of the ponderomotive law, we shall briefly consider the physical significance of the transformation laws.

The physical significance of the transformation laws. Let us write out explicitly the transformation law of the tensor $\varphi^{\iota\kappa}$,

$$\varphi^{*\mu\nu} = \gamma^\mu{}_\iota \gamma^\nu{}_\kappa \varphi^{\iota\kappa}. \tag{7.22}$$

By separating the components **H** and **E**, we obtain these laws:

$$\left.\begin{aligned} \varphi^{*mn} &= \gamma^m{}_i \gamma^n{}_k \varphi^{ik} + (\gamma^m{}_i \gamma^n{}_4 - \gamma^m{}_4 \gamma^n{}_i)\varphi^{i4}, \\ \varphi^{*m4} &= \gamma^m{}_i \gamma^4{}_k \varphi^{ik} + (\gamma^m{}_i \gamma^4{}_4 - \gamma^m{}_4 \gamma^4{}_i)\varphi^{i4}. \end{aligned}\right\} \tag{7.22a}$$

Instead of considering the most general Lorentz transformations (which may involve spatial orthogonal transformations), we shall choose a particular set of $\gamma^\mu{}_\iota$, the transformation coefficients of eqs. (4.13). Then eqs. (7.22a) go over into the transformation equations

$$\left.\begin{aligned} \varphi^{*12} &= (1 - v^2/c^2)^{-1/2}(\varphi^{12} + v\varphi^{24}), \\ \varphi^{*13} &= (1 - v^2/c^2)^{-1/2}(\varphi^{13} + v\varphi^{34}), \\ \varphi^{*23} &= \varphi^{23}, \\ \varphi^{*14} &= \varphi^{14}, \\ \varphi^{*24} &= (1 - v^2/c^2)^{-1/2}\left(\varphi^{24} - \frac{v}{c^2}\varphi^{21}\right), \\ \varphi^{*34} &= (1 - v^2/c^2)^{-1/2}\left(\varphi^{34} - \frac{v}{c^2}\varphi^{31}\right). \end{aligned}\right\} \tag{7.22b}$$

When we introduce the quantities E_s and H_s according to eq. (7.17a), we obtain

$$\left.\begin{aligned} H^*_3 &= (1 - v^2/c^2)^{-1/2}\left(H_3 - \frac{v}{c}E_2\right), \\ H^*_2 &= (1 - v^2/c^2)^{-1/2}\left(H_2 + \frac{v}{c}E_3\right), \\ H^*_1 &= H_1, \\ E^*_1 &= E_1, \\ E^*_2 &= (1 - v^2/c^2)^{-1/2}\left(E_2 - \frac{v}{c}H_3\right), \\ E^*_3 &= (1 - v^2/c^2)^{-1/2}\left(E_3 + \frac{v}{c}H_2\right). \end{aligned}\right\} \tag{7.22c}$$

The contributions of **H** to the transformed **E**, and vice versa, are proportional to $\frac{v}{c}$. The components of **H** and **E** in the direction of v do not transform at all.

These transformation laws are closely related to the laws of Ampère and Faraday. Let us consider a point charge which is at rest in a coördinate system S. In a system S^*, which is moving relatively to S, we find a magnetic field strength which is $\frac{1}{c}$ times the vector product of the electric field strength and the velocity of the particle in S^*.

Consider, on the other hand, a magnetic field, such as one produced by a permanent magnet, which is at rest with respect to S. When we go over to a new system S^*, we shall encounter an electric field strength which is $\frac{1}{c}$ times the vector product of the velocity of the magnet (with respect to S^*) and the magnetic field strength. The integral of **E** over a closed curve will, in general, not vanish.

As for the transformation law of **I** and σ: Let us consider a finite material body with a uniform distribution of charge throughout, which is at rest with respect to S. If the volume of the body is V_0 and the charge density σ_0, the total charge C is

$$C = V_0\sigma_0\,. \tag{7.23}$$

Let us now carry out a Lorentz transformation (4.13). Because of the Lorentz contraction in the x-direction, the total volume of the body in S^* will be

$$V^* = \sqrt{1 - v^2/c^2}\cdot V_0\,. \tag{7.24}$$

The charge density, on the other hand, undergoes a change according to the transformation law for contravariant vectors. With respect to S, the current vanishes; and we have, therefore,

$$\sigma^* = \gamma^4_{\ 4}\cdot\sigma_0 = (1 - v^2/c^2)^{-1/2}\cdot\sigma_0 . \tag{7.25}$$

The product of V and σ remains unchanged; in other words, the charge C is an invariant.

The current density has one component in S^*, I^{*1} which is given by

$$I^{*1} = \gamma^1_{\ 4}\sigma_0 = -(1 - v^2/c^2)^{-1/2}\cdot v\sigma_0 = -v\sigma^*. \tag{7.26}$$

The current density is the product of charge density and velocity. The relativistic transformation law for σ, $\mathbf{I}$, therefore, is in agreement with Lorentz' electron theory, according to which all currents are moving charges.

Gauge transformations. The so-called second set of Maxwell's equations, the four equations (7.19), implies the existence of a world vector potential φ_ι, of which the field tensor $\varphi_{\iota\kappa}$ is the curl. *However, the world vector potential is not uniquely determined by a given electromagnetic field.* To a world vector φ_ι which satisfies eqs. (7.18), we may add an arbitrary gradient field $\Phi_{,\iota}$, and the sum

$$\bar{\varphi}_\iota = \varphi_\iota + \Phi_{,\iota} \tag{7.27}$$

will again satisfy eqs. (7.18), the left-hand sides of which remain unchanged. This transformation of the world vector potential, the addition of a gradient, is called a *gauge transformation.* Gauge transformations, of course, have nothing to do with coördinate transformations. Neither the field tensor $\varphi_{\iota\kappa}$ nor the current density world vector I^ι is affected by gauge transformations. These quantities and Maxwell's equations, (7.19) and (7.20), are said to be *gauge invariant.*

The ponderomotive equations. We cannot measure electromagnetic fields directly. We observe only the forces (that is, accelerations) to which charged particles are subjected. In order to relate the field equations to actual physical experience, we have, therefore, to know the laws governing the behavior of charged particles. The classical law (the Lorentz force) states that the force acting on a particle is given by

$$m\dot{\mathbf{u}} = e\left[\mathbf{E} + \frac{\mathbf{u}}{c}\times\mathbf{H}\right], \tag{7.28}$$

where $\mathbf{u}$ is the velocity of the particle and $\times$ signifies the vector product. $\mathbf{E}$ and $\mathbf{H}$ are those parts of the electric and magnetic field strength which are independent of the presence of the particle itself. The contributions of the particle itself to the field are, of course, singular.[1]

In (three dimensional) components, the equations take the form

$$m\dot{u}^s = e\left(E_s + \frac{1}{c}u^r H_{sr}\right), \qquad u^s = \dot{x}^s. \tag{7.29}$$

Eqs. (7.28) and (7.29) are the Euler-Lagrange equations of the variational principle

$$\left.\begin{aligned} &\delta \int L^{(c)}\, dt = 0, \\ &L^{(c)} = \frac{m}{2}u^2 - e\varphi + \frac{e}{c}u^s A_s. \end{aligned}\right\} \tag{7.30}$$

The momenta which are canonically conjugate to the coördinates are given by

$$p_s^{(c)} = \frac{\partial L^{(c)}}{\partial u^s} = mu^s + \frac{e}{c}A_s, \tag{7.31}$$

and the Hamiltonian is

$$\left.\begin{aligned} H^{(c)} &= -L^{(c)} + u^s p_s^{(c)} = \frac{1}{2m}\left(p_s^{(c)} - \frac{e}{c}A_s\right)^2 + e\varphi \\ &= \frac{m}{2}u^2 + e\varphi. \end{aligned}\right\} \tag{7.32}$$

Let us consider again the Lagrangian $L^{(c)}$. The expression

$$\varphi - \frac{1}{c}u^s A_s \tag{7.33}$$

differs from a Lorentz invariant scalar only by a factor. For, if we multiply it by $(1 - u^2/c^2)^{-1/2}$, it becomes

$$\left.\begin{aligned} (1 - u^2/c^2)^{-1/2}\left(\varphi - \frac{1}{c}u^s A_s\right) &= \frac{dt}{d\tau}\left(-\frac{1}{c}\varphi_4 - \frac{1}{c}u^s \varphi_s\right) \\ &= -\frac{1}{c}U^\rho \varphi_\rho. \end{aligned}\right\} \tag{7.34}$$

---

[1] In all classical and special relativistic field theories, only that part of the field which is not created by the particle under consideration appears in the ponderomotive laws. This splitting up of the field is unsatisfactory, particularly because it is not a uniquely determined mathematical operation. Only the general theory of relativity furnishes us with an approach to the ponderomotive laws which considers the field as a whole (cf. Chapter XV).

As for the first term of $L^{(c)}$, we know that in relativistic mechanics it has to be replaced by the expression (6.51). This expression (with $k = 0$),

$$-mc^2\sqrt{1 - u^2/c^2} \tag{7.35}$$

also differs from a scalar only by the same factor, $\sqrt{1 - u^2/c^2}$. Therefore, the expressions (7.35) and (7.33) transform in the same way. Multiplied by dt, they are scalars. We can, therefore, form the integral

$$I = \int_{P_1}^{P_2} \left\{-mc^2\sqrt{1 - u^2/c^2} - e\varphi + \frac{e}{c}u^s A_s\right\} dt, \tag{7.36}$$

which is Lorentz invariant. Its Euler-Lagrange equations must also be covariant equations. If a parameter transformation is carried out so that τ is chosen as parameter, eq. (7.36) goes over into

$$I = \int_{P_1}^{P_2} \left\{-mc^2\sqrt{\eta_{\iota\kappa} U^\iota U^\kappa} + \frac{e}{c} U^\rho \varphi_\rho\right\} d\tau, \qquad U^\rho = \frac{dx^\rho}{d\tau}. \tag{7.37}$$

This integral is not only invariant with respect to Lorentz coördinate transformations, but it has one further important property. We have seen that the φ_ρ are determined by the $\varphi_{\rho\sigma}$ except for an arbitrary gradient. The ponderomotive equations belonging to the Lagrangian, therefore, must not be affected by the addition of such a gradient to φ_ρ. Now, when we replace φ_ρ in eq. (7.37) by $\bar{\varphi}_\rho$ of eq. (7.27), the value of the integral is changed as follows:

$$\begin{aligned} \bar{I} = I + \int_{P_1}^{P_2} \frac{e}{c} U^\rho \Phi_{,\rho}\, d\tau &= I + \frac{e}{c}\int_{P_1}^{P_2} \Phi_{,\rho}\, dx^\rho \\ &= I + \frac{e}{c}[\Phi(P_2) - \Phi(P_1)]. \end{aligned} \tag{7.38}$$

The value of the integral I is changed by the gauge transformation, but the change depends on the values of Φ only at the end points, not along the path. The *variation* of I with fixed end points of integration is, therefore, not affected by a gauge transformation; in other words, the Euler-Lagrange equations which belong to the variational problem

$$\delta I = 0$$

are gauge invariant.

The first term of the integrand of eq. (7.37) may be replaced by any scalar function satisfying eqs. (6.62) and (6.64). However, with t as parameter, eq. (7.36) is the only Lorentz invariant Lagrangian which produces gauge invariant equations.

Working first with the Lagrangian of eq. (7.36), we obtain the following momenta:

$$p_s = \frac{\partial L^{(t)}}{\partial u^s} = \frac{mu^s}{\sqrt{1 - u^2/c^2}} + \frac{e}{c} A_s, \tag{7.39}$$

with u^s given by

$$u^s = \frac{1}{m}\left(p_s - \frac{e}{c} A_s\right)\left[1 + \frac{\left(p_k - \frac{e}{c} A_k\right)^2}{m^2 c^2}\right]^{-1/2}. \tag{7.40}$$

Substituting these expressions in $L^{(t)}$, we get

$$L^{(t)} = \left[1 + \frac{\left(p_k - \frac{e}{c} A_k\right)^2}{m^2 c^2}\right]^{-1/2} \left(-mc^2 + \frac{e}{c} A_s \frac{p_s - \frac{e}{c} A_s}{m}\right) - e\varphi, \tag{7.41}$$

and, for $p_s u^s$,

$$p_s u^s = \left[1 + \frac{\left(p_k - \frac{e}{c} A_k\right)^2}{m^2 c^2}\right]^{-1/2} \frac{1}{m} p_s \left(p_s - \frac{e}{c} A_s\right); \tag{7.42}$$

therefore, for $H^{(t)}$,

$$H^{(t)} = mc^2 \left[1 + \frac{\left(p_k - \frac{e}{c} A_k\right)^2}{m^2 c^2}\right]^{1/2} + e\varphi. \tag{7.43}$$

The Euler-Lagrange equations belonging to eq. (7.36) are

$$\frac{d}{dt}\left(\frac{mu^s}{\sqrt{1 - u^2/c^2}}\right) = eE_s + \frac{e}{c} u^r H_{sr}. \tag{7.44}$$

The Hamiltonian equations belonging to eq. (7.43) are

$$\left.\begin{aligned}
u^s &= \frac{\partial H^{(t)}}{\partial p_s} = \left[1 + \frac{\left(p_k - \frac{e}{c} A_k\right)^2}{m^2 c^2}\right]^{-1/2} \cdot \frac{1}{m}\left(p_s - \frac{e}{c} A_s\right), \\
\dot{p}_s &= -\frac{\partial H^{(t)}}{\partial x^s} \\
&= \frac{e}{mc}\left[1 + \frac{\left(p_l - \frac{e}{c} A_l\right)^2}{m^2 c^2}\right]^{-1/2} \left(p_k - \frac{e}{c} A_k\right) A_{k,s} - e\varphi_{,s} \\
&= \frac{e}{c} u^k A_{k,s} - e\varphi_{,s},
\end{aligned}\right\} \tag{7.45}$$

with

$$\dot{p}_s = \frac{d}{dt}\left(\frac{mu^s}{\sqrt{1 - u^2/c^2}}\right) + \frac{e}{c}(A_{s,k} u^k + A_{s,4}). \tag{7.46}$$

Of course, eqs. (7.45) are equivalent to eq. (7.44). Eq. (7.44) is the expression most suitable for practical applications.

We can give the formulas a more symmetric appearance by using τ as a parameter. We shall replace the first term in eq. (7.37) by the expression

$$-\tfrac{1}{2}mc^2\eta_{\iota\kappa}U^\iota U^\kappa, \qquad U^\iota = \frac{dx^\iota}{d\tau},$$

which satisfies eqs. (6.62) and (6.64). This choice of the function Φ of eq. (6.62) is, of course, arbitrary, but with its help we obtain very simple equations. The Lagrangian is

$$L^{(\tau)} = -\tfrac{1}{2}mc^2\eta_{\iota\kappa}U^\iota U^\kappa + \frac{e}{c}U^\iota\varphi_\iota. \tag{7.47}$$

The momenta are

$$p_\rho = \frac{\partial L^{(\tau)}}{\partial U^\rho} = -mc^2 U_\rho + \frac{e}{c}\varphi_\rho, \tag{7.48}$$

and the Euler-Lagrange equations take the form

$$+mc^2\frac{dU_\rho}{d\tau} - \frac{e}{c}\varphi_{\rho,\sigma}U^\sigma + \frac{e}{c}U^\iota\varphi_{\iota,\rho} = 0,$$

or

$$-mc^2\eta_{\rho\sigma}\frac{dU^\sigma}{d\tau} = \frac{e}{c}\varphi_{\sigma\rho}U^\sigma. \tag{7.49}$$

The first three equations (7.49) are identical with eq. (7.44), multiplied by $\dfrac{dt}{d\tau}$. The fourth equation is not independent of the three others. For the product of eq. (7.49) by U^ρ yields an identity,

$$-mc^2\eta_{\rho\sigma}U^\rho\frac{dU^\sigma}{d\tau} - \frac{e}{c}\varphi_{\sigma\rho}U^\sigma U^\rho \equiv 0. \tag{7.50}$$

The second term vanishes identically because $\varphi_{\sigma\rho}$ is skewsymmetric. The first term vanishes, too, for this reason: The U^ρ are the components of a unit vector; the infinitesimal change of a vector of constant length is always orthogonal to the vector itself. Or, to put it in a slightly different way,

$$\eta_{\rho\sigma}U^\rho\frac{dU^\sigma}{d\tau} = \frac{1}{2}\frac{d}{d\tau}(\eta_{\rho\sigma}U^\rho U^\sigma) = \frac{1}{2}\frac{d}{d\tau}(1) \equiv 0. \tag{7.51}$$

It follows that eq. (7.49) contains only three really independent equations.

The U^ι can be expressed by the momenta through the equations

$$U^\iota = -\frac{1}{mc^2}\left(p^\iota - \frac{e}{c}\varphi^\iota\right). \tag{7.52}$$

We can now compute $H^{(\tau)}$ and obtain

$$H^{(\tau)} = -\frac{1}{2mc^2}\eta_{\iota\kappa}\left(p^\iota - \frac{e}{c}\varphi^\iota\right)\left(p^\kappa - \frac{e}{c}\varphi^\kappa\right). \tag{7.53}$$

The Hamiltonian equations are

$$\left.\begin{aligned} U^\rho &= \frac{\partial H^{(\tau)}}{\partial p_\rho} = -\frac{1}{mc^2}\left(p^\rho - \frac{e}{c}\varphi^\rho\right), \\ \frac{dp_\rho}{d\tau} &= -\frac{\partial H^{(\tau)}}{\partial x^\rho} = -\frac{1}{mc^2}\left(p^\sigma - \frac{e}{c}\varphi^\sigma\right)\cdot\frac{e}{c}\varphi_{\sigma,\rho}. \end{aligned}\right\} \tag{7.54}$$

They are equivalent to eqs. (7.49).

CHAPTER VIII

The Mechanics of Continuous Matter

Introductory remarks. The field equations of Maxwell contain charge density and current density. We know that in reality the charge, for instance, is not at all continuously distributed through space; it vanishes everywhere except in the interior of the elementary particles, electrons, protons, and so forth. Whether this interior has a finite volume or is a "point," and whether Maxwell's equations hold in these regions, are questions still unanswered. It is only for a "macroscopic" treatment that we may, under certain circumstances, spread the highly concentrated charges evenly over an extended volume and, thereby, obtain a mean electromagnetic field.

From a "macroscopic" point of view, we are interested in the behavior of matter as a whole rather than in the equations of motion of individual mass points and charged particles. This chapter is devoted to the treatment of "continuous matter" in that sense.

If the mechanics of continuous matter is only an approximation, why is it important for the development of relativity theory? From the point of view of relativity theory, the most promising parts of classical physics are field theories, for action-at-a-distance theories are not capable of relativistic modification. In a field theory, we consider continuous field variables not as averages of random distributions of mass points, but as the basic physical quantities. The treatment of the mechanics of continuous matter will show us how to introduce mechanical concepts, energy and momentum, into the field theories, particularly into the theory of the electromagnetic field.

Nonrelativistic treatment. When we describe the behavior of so-called continuous media—elastic bodies, liquids, gases—we fix our attention not on the individual particle, but on the local volume element. In the course of time, particles enter and others leave such a volume element. The motion of each molecule is subject to the general laws of mechanics, but it is important to reformulate these laws in a language that does not refer to mass points and their positions, but which speaks of local mass density, momentum density, and so forth. A "volume element" must contain a sufficient number of individual particles to give these averages a meaning and make them reasonably continuous functions of the four coördinates x, y, z, and t. We shall assume that we can define a local average density and a local average velocity. First let us proceed along prerelativistic lines. We shall call the density ρ, the average velocity $\mathbf{u}$. The change of the density in a volume element is given by the influx of matter into that volume element,

$$\frac{\partial \rho}{\partial t} + \operatorname{div}(\rho \mathbf{u}) = 0. \tag{8.1}$$

This equation, the "continuity equation," is the expression of the mass conservation law.

In order to express the equations of motion, we shall introduce the concept of "local" and of "material" differentiation with respect to time. Let us consider some local property q of the continuum (a scalar, a component of a vector, and so forth) and describe its change with time. When we describe its change with time at a fixed point (x^i), its derivative, the "*local differential quotient*," is $\frac{\partial q}{\partial t}$. On the other hand, we may refer its change to a coördinate system which is locally connected with the average state of motion of the matter; in other words, we can describe the change of q as it would present itself to an observer who participates in the local average motion of the matter. This derivative, the "*material differential quotient*," is usually denoted by $\frac{Dq}{Dt}$. It is connected with the local differential quotient by the relation

$$\frac{Dq}{Dt} = \frac{\partial q}{\partial t} + \mathbf{u} \cdot \operatorname{grad} q = \frac{\partial q}{\partial t} + u^s \cdot q_{,s}. \tag{8.2}$$

Let us denote the force which acts on a unit volume of matter by g_i. Then the equations of force are

$$\rho \frac{Du^i}{Dt} = g^i, \tag{8.3}$$

or, by making use of eq. (8.2),

$$\rho \left(\frac{\partial u^i}{\partial t} + u^s u^i_{,s} \right) = g^i. \tag{8.3a}$$

The left-hand side can be transformed by partial differentiation:

$$\left.\begin{aligned}\rho\left(\frac{\partial u^i}{\partial t} + u^s u^i_{,s}\right) &= \frac{\partial}{\partial t}(\rho u^i) - u^i\frac{\partial \rho}{\partial t} + (\rho u^s u^i)_{,s} - u^i(\rho u^s)_{,s}\\ &= \frac{\partial}{\partial t}(\rho u^i) + (\rho u^s u^i)_{,s} - u^i\left[\frac{\partial \rho}{\partial t} + (\rho u^s)_{,s}\right].\end{aligned}\right\} \tag{8.4}$$

The contents of the square bracket vanish because of the continuity equation (8.1), and we have:

$$g^i = \frac{\partial}{\partial t}(\rho u^i) + (\rho u^i u^s)_{,s}. \tag{8.5}$$

The forces themselves can be of two different types. Our continuum may be subject to a gravitational or electric field or to a similar field which exerts a force on the matter in a given infinitesimal volume element proportional to the size of the volume element. Such forces are called "volume forces." We shall denote them by f^i. The other type of force which we have to take into account is the stress within the material itself. The particles contained in two adjoining volume elements may exert mutual forces on each other which are proportional to the area of the face shared by the two volume elements. These forces are, therefore, called "area forces." The components of the area force, dq^i, depend on the components of the oriented face dA_k linearly,

$$dq^i = t^{ik} dA_k\,, \tag{8.6}$$

where the vector dA_k is perpendicular to the face and has a length equal to its area. The dq^i must be linear in the dA_k, because the total force acting on an infinitesimal volume element depends on its size, but not on its shape. The sign of the form t^{ik} is chosen so that we obtain the force exerted by the matter in a given volume element if we choose as the direction of dA_k the *outward* normal. This is the usual convention, and it is chosen so that the stress tensor which corresponds to ordinary isotropic pressure has positive components.

t^{ik} must be symmetric in its indices,

$$t^{ik} = t^{ki}. \tag{8.7}$$

Otherwise, an isolated body ($\mathbf{f} = 0$) would change its total angular momentum at the rate

$$\frac{dM_i}{dt} = \int_V \delta_{ikl} t^{kl}\, dV,$$

integrated over the whole body.[1]

---

[1] δ_{ikl} is the Levi-Civita tensor density.

The force which acts on the matter contained in a (finite) volume is given as

$$G^i = \int_V f^i \, dV - \oint_s t^{ik} \, dA_k, \tag{8.8}$$

where f^i are the volume forces per unit volume. With the help of Gauss's theorem, the second integral, over the closed surface, can be transformed into a volume integral,

$$G^i = \int_V (f^i - t^{ik}{}_{,k}) \, dV. \tag{8.9}$$

The force per unit volume becomes, thus,

$$g^i = f^i - t^{ik}{}_{,k} \,. \tag{8.10}$$

By comparing the last equation with eq. (8.5), we obtain the non-relativistic equations of motion,

$$\frac{\partial(\rho u^i)}{\partial t} + (\rho u^i u^k + t^{ik})_{,k} - f^i = 0. \tag{8.11}$$

These three equations, together with the continuity equation (8.1), determine the behavior of continuous matter under the influence of stresses and other forces.

In addition, there is an energy conservation law, which states that the energy contained in a volume element changes as energy flows into and out of it.

The mechanical system is, of course, completely determined only when we know the t^{ik} and f^i. The volume forces are given by the exterior conditions, such as the presence of a gravitational field, while the stresses depend on the internal deformation or on the flow of the matter. In the case of a perfect fluid, for instance, t^{ik} is equal to the pressure p, multiplied by the Kronecker tensor δ_{ik}. p itself is a function of the density ρ and the temperature, which is given by the equation of state of the fluid. It is not necessary for us to determine the stress components; we are interested only in their transformation properties.

A special coördinate system. In order to obtain Lorentz covariant laws, we shall first introduce a special coördinate system $\overset{0}{S}$ with respect to which the matter is at rest at a world point $\overset{0}{P}$. At this world point, the formulation of the equations is greatly simplified, because all classical terms containing undifferentiated velocity components vanish.

The first derivatives of u^i are equal to the first derivatives of U^i, and the first derivatives of U^4 vanish. U^4 itself equals unity. As the matter does not move, its total energy density consists of rest energy density and is equal to c^2 times its mass density ρ. The only changes in the classical laws are caused by our recognition that the flow of energy contributes to the momentum and must, therefore, be considered in all conservation laws.

Once we have formulated the laws with respect to the special coordinate system, we shall give them the form of tensor equations.

For the time being, we shall not consider forms of energy transfer other than those caused by the motion of matter under the action of mechanical stress. Later, we shall extend our formalism to electromagnetic interaction as well. With this restriction in mind, we shall first formulate the law of the conservation of energy.

The total energy contained in a volume is changed at the rate at which energy flows into or out of that volume. The energy flux, because of the relativistic relationship between energy and mass, is nothing but c^2 times the momentum density. The latter consists of two parts: In the first place, we have again the mass density, multiplied by the velocity, $\rho\mathbf{u}$. But, in addition, we have to consider the flux of energy on account of the stress. Let us consider an oriented infinitesimal face $d\mathbf{A}$. The matter on either side experiences a force, which is $t^{rs}dA_s$ on the side to which the normal points, and $(-t^{rs}dA_s)$ on the other side. If the matter in the neighborhood of this face moves with the velocity $\mathbf{u}$, the work done on one side will be $u^r t^{rs} dA_s$,[2] the work done on the other side $(-u^r t^{rs} dA_s)$. An amount of energy is gained on one side equal to that lost on the other side; in other words, a flux of of energy takes place through the face, and the vector of energy flux has the components $u^r t^{rs}$. The corresponding momentum density is smaller by the factor $1/c^2$. The total momentum density is, thus,

$$P^s = \rho u^s + \frac{1}{c^2} u^r t^{rs}. \tag{8.12}$$

At the point $\overset{0}{P}$, both terms vanish, but not their derivatives. For the conservation of energy, we obtain the equation

$$\frac{\partial \rho}{\partial t} + P^s{}_{,s} = 0,$$

or

$$\rho_{,4} + \rho u^s{}_{,s} + \frac{1}{c^2} u^r{}_{,s} t^{rs} = 0. \tag{8.13}$$

[2] In three dimensional, orthogonal coordinates, there is no difference between covariant and contravariant transformation laws. The positions of the indices are, therefore, meaningless.

Likewise, we can write down the equation connecting the change of momentum density with time and the force per unit volume. Because of our choice of coördinate system, there is no distinction between "material" and "local" differential quotients at the point $\overset{0}{P}$. We write, therefore,

$$\rho u^{i}{}_{,4} + \frac{1}{c^2} u^{r}{}_{,4} t^{ri} + t^{ir}{}_{,r} = f^{i}. \tag{8.14}$$

The equations (8.13) and (8.14) replace the nonrelativistic equations (8.1) and (8.11).

Tensor form of the equations. Our previous experience has taught us that relativistic laws frequently differ from classical laws only by a suitable extension of three dimensional vectors and tensors into four dimensions. The three dimensional momentum $m\mathbf{u}$ has to be replaced by the world vector mU^{ρ}, the three dimensional skewsymmetric tensor H_{mn} by the skewsymmetric world tensor $\varphi_{\mu\nu}$, and so forth. The form of the nonrelativistic equations (8.11) suggests that a symmetric tensor of rank 2 plays an important role in the equations, which is the sum of two terms, one corresponding to the nonrelativistic $\rho u^{i} u^{k}$, the other to the t^{ik}. Accordingly, we shall first introduce a symmetric world tensor $t^{\iota\kappa}$, which in the special coördinate system $\overset{0}{S}$ and at the world point $\overset{0}{P}$ shall have the components

$$t^{\iota\kappa} = \begin{Bmatrix} t^{ik}, & 0 \\ 0, & 0 \end{Bmatrix}. \tag{8.15}$$

This means that it satisfies the covariant equations

$$t^{\iota\kappa} U_{\kappa} = 0. \tag{8.16}$$

With the help of this world tensor, we can re-express the term $\frac{1}{c^2} u^{r}{}_{,4} t^{ri}$ which appears in eq. (8.14). It is:

$$\frac{1}{c^2} u^{r}{}_{,4} t^{ri} = \frac{1}{c^2} (u^{r} t^{ri})_{,4} = -\ (U_{r} t^{ri})_{,4} = (-\ U_{\rho} t^{\rho i} + U_{4} t^{4i})_{,4}. \tag{8.17}$$

Because of eq. (8.16), the first term in the parenthesis vanishes; and in the second term, U^4 equals unity, so that we can write:

$$\frac{1}{c^2} u^{r}{}_{,4} t^{ri} = t^{i4}{}_{,4}. \tag{8.18}$$

In the same manner, we rewrite the term $\frac{1}{c^2} u^r{}_{,s} t^{rs}$ which appears in eq. (8.13). It is

$$\underline{\frac{1}{c^2} u^r{}_{,s} t^{rs}} = \frac{1}{c^2} (u^r t^{rs})_{,s} = -(U_r t^{rs})_{,s} = +(U_4 t^{4s})_{,s} = \underline{t^{4s}{}_{,s}}. \tag{8.19}$$

Thus eqs. (8.13) and (8.14) take the form

$$\rho_{,4} + (\rho u^s + t^{4s})_{,s} = 0, \tag{8.13a}$$

$$(\rho u^i + t^{4i})_{,4} + t^{ir}{}_{,r} = f^i. \tag{8.14a}$$

As mentioned, U^4 equals unity, and its first derivatives vanish. We change only the form of the equations by replacing u^i by U^i and by adding a few factors U^4,

$$(\rho U^4 U^4)_{,4} + (\rho U^s U^4 + t^{s4})_{,s} = 0, \tag{8.13b}$$

$$(\rho U^i U^4 + t^{i4})_{,4} + t^{is}{}_{,s} = f^i. \tag{8.14b}$$

All that remains to be done is to show that we may add t^{44} in the first parenthesis of eq. (8.13b), and to replace f^i by a world vector f^ι.

According to eq. (8.16), $t^{44}{}_{,4}$ can be expressed by other components,

$$t^{44}{}_{,4} = (U_4 t^{44})_{,4} = -(U_s t^{s4})_{,4} = \frac{1}{c^2} (u^s t^{s4})_{,4}. \tag{8.20}$$

Both u^s and t^{s4} vanish at $\overset{0}{P}$; therefore the derivative of $u^s t^{s4}$ must vanish. We may, therefore, add $t^{44}{}_{,4}$ to the expression $(\rho U^4 U^4)_{,4}$ in eq. (8.13b) without changing its value.

On the right-hand side of the equations (8.13b) and (8.14b) we define the world vector f^ρ, the components of which at $\overset{0}{P}$ and with respect to $\overset{0}{S}$ are $(f^i, 0)$. It satisfies the covariant equation

$$f^\rho U_\rho = 0. \tag{8.21}$$

With these two changes, we can combine (8.13b) and (8.14b) into the four components of a covariant law,

$$P^{\mu\nu}{}_{,\nu} = f^\mu, \qquad P^{\mu\nu} = \rho U^\mu U^\nu + t^{\mu\nu}. \tag{8.22}$$

$P^{\mu\nu}$ transforms as a contravariant tensor, and its divergence is, therefore, a contravariant vector.

Let us shortly consider the physical significance of the tensor $P^{\mu\nu}$ in a coördinate system other than $\overset{0}{S}$. As the simplest procedure, let us carry out a Lorentz transformation (4.13), that is, go over to a coördinate system in which matter moves along the X-axis at the rate of speed v. In that coördinate system, the components of the tensor $P^{\mu\nu}$ take the form

$$\left.\begin{aligned}
P^{*11} &= \frac{t^{11} + \rho v^2}{1 - v^2/c^2}, \\
P^{*12} &= \frac{t^{12}}{\sqrt{1 - v^2/c^2}}, \qquad P^{*13} = \frac{t^{13}}{\sqrt{1 - v^2/c^2}}, \\
P^{*22} &= t^{22}, \qquad P^{*23} = t^{23}, \qquad P^{*33} = t^{33}, \\
P^{*14} &= -\frac{\rho v + \dfrac{v}{c^2}\, t^{11}}{1 - v^2/c^2}, \\
P^{*24} &= -\frac{v}{c^2}\,\frac{t^{12}}{\sqrt{1 - v^2/c^2}}, \qquad P^{*34} = -\frac{v}{c^2}\,\frac{t^{13}}{\sqrt{1 - v^2/c^2}}, \\
P^{*44} &= \frac{\rho + \dfrac{v^2}{c^4}\, t^{11}}{1 - v^2/c^2}.
\end{aligned}\right\} \tag{8.23}$$

The total energy of a free-moving particle is increased over the rest energy by the factor $(1 - u^2/c^2)^{-1/2}$. If we consider energy density rather than energy, we have to take into account the fact that the volume containing a given number of molecules is Lorentz-contracted by another factor $(1 - u^2/c^2)^{-1/2}$. These two effects together bring it about that the total energy density (and, therefore, also mass density) is enlarged over the rest energy density (or rest mass density) by a factor $(1 - u^2/c^2)^{-1}$. [In our case, u equals $(-v)$.] This does not take into account the stress under the influence of which the matter moves, but explains merely why ρ is multiplied everywhere in eq. (8.23) by the factor $(1 - u^2/c^2)^{-1}$.

As for the components P^{rs} (with no index 4), they contain the t^{rs} multiplied by varying powers of $(1 - u^2/c^2)^{-1/2}$, and the "relativistic mass density," $(1 - u^2/c^2)^{-1}\cdot\rho$, multiplied by $u^r u^s$.

The components P^{r4} retain their character as components of the momentum density; as was mentioned before, the stress makes a contribution to the momentum density. P^{44} remains the mass density. The stress contributes a term which is quadratic with respect to the velocity of the matter.

Just as in classical hydrodynamics, the system is completely determined only when the tensor $P^{\mu\nu}$ is given. Let us consider again the case of a perfect fluid. In the special coördinate system $\overset{0}{S}$, the tensor $P^{\mu\nu}$ has the components

$$\begin{Bmatrix} p\delta_{ik}, & 0 \\ 0\ , & \rho \end{Bmatrix};$$

that is, there are no shearing stresses, the pressure is isotropic. When we carry out a Lorentz transformation, $P^{\mu\nu}$ goes over into the expression

$$\left.\begin{aligned} P^{\mu\nu} &= \rho U^\mu U^\nu - \frac{1}{c^2}(\eta^{\mu\nu} - U^\mu U^\nu)p \\ &= \left(\rho + \frac{1}{c^2}p\right)U^\mu U^\nu - \frac{1}{c^2}p\eta^{\mu\nu}. \end{aligned}\right\} \tag{8.24}$$

ρ and p are connected with each other and with the temperature by the equation of state of the fluid.

$P^{\mu\nu}$ is usually called the stress-energy tensor. In our case, it has a special character in that the P^{r4} vanish in the special coördinate system. To borrow an expression from the algebra of quadratic forms: U^ι is an eigenvector of the matrix $P^\iota_{\cdot\kappa}$, and ρ is the corresponding eigenvalue. This is not a general property of the stress-energy tensor; U^ι is an eigenvector of $P^\iota_{\cdot\kappa}$ as long as we consider only energy transfer by means of mechanical interaction. Our next step will be to consider electromagnetic interaction as well, and U^ι will no longer be an eigenvector of $P^\iota_{\cdot\kappa}$.

The stress-energy tensor of electrodynamics. In the preceding chapter, we found that the world force acting on a charged mass point, $m\,\dfrac{dU^\rho}{d\tau}$, according to eq. (7.49), is $\dfrac{e}{c^3}\varphi^\rho_{\cdot\sigma}U^\sigma$, where $\varphi^\rho_{\cdot\sigma}$ is the (mixed) field tensor. Let us assume now that the continuous matter considered in this chapter contains charged particles, for instance, electrons, and in such numbers that the relative contribution of every particle to the total field is negligible. We may then consider $\varphi^\rho_{\cdot\sigma}$ as the total average field in the neighborhood of the particle under consideration. Let us consider a volume element $d\overset{0}{V}$ at rest in $\overset{0}{S}$. Let us call σ the proper charge density of the charged particles, that is, the charge density which would be measured in the system $\overset{1}{S}$ in which the charged particles are at rest. Let us denote the state of motion of the charged particles from now on by W^ι, and the state of motion of $\overset{0}{S}$ by U^ι. The volume of $d\overset{0}{V}$ with respect to the system $\overset{1}{S}$ is contracted by a factor $\sqrt{1 - \overset{1}{u}{}^2/c^2}$, where $\overset{1}{\mathbf{u}}$ is the relative velocity of $\overset{0}{S}$ and $\overset{1}{S}$. The total charge contained in $d\overset{0}{V}$ is, therefore,

$$e = \sigma \cdot \sqrt{1 - \overset{1}{u}{}^2/c^2}\, d\overset{0}{V}. \tag{8.25}$$

According to eq. (7.49), the force exerted on this charge is

$$\frac{dp^\kappa}{d\overset{1}{\tau}} = \sigma \sqrt{1 - \overset{1}{u}{}^2/c^2} \cdot d\overset{0}{V} \cdot \frac{1}{c^3}\, \varphi^{\kappa}_{\cdot\cdot\sigma} W^\sigma, \tag{8.26}$$

where p^κ is the energy-momentum vector [see eq. (6.22)], and $\overset{1}{\tau}$ is the time in the system $\overset{1}{S}$. According to Lorentz' electron theory, the proper charge density σ, multiplied by the four velocity W^ι, is the world current density I^ι; so that we may write:

$$\frac{dp^\kappa}{d\overset{1}{\tau}} = \sqrt{1 - \overset{1}{u}{}^2/c^2}\, d\overset{0}{V}\, \frac{1}{c^3}\, \varphi^{\kappa}_{\cdot\cdot\sigma} I^\sigma. \tag{8.26a}$$

On the other hand, it is desirable to introduce $\overset{0}{S}$-time, instead of proper time $\overset{1}{\tau}$. Because of the time dilatation, the change of momentum per unit $\overset{0}{S}$-time is slower than the change per unit $\overset{1}{S}$-time by the factor $\sqrt{1 - \overset{1}{u}{}^2/c^2}$. This factor cancels the identical factor on the right-hand side of eq. (8.26a), and we get

$$\frac{dp^\kappa}{d\overset{0}{t}} = \frac{1}{c^3}\, \varphi^{\kappa}_{\cdot\cdot\sigma} I^\sigma\, d\overset{0}{V}. \tag{8.26b}$$

This last equation no longer contains quantities referring to the system $\overset{1}{S}$. It holds, therefore, even if the matter contains various kinds of charged particles (electrons, atomic nuclei, ions with different masses and charges) with different average velocities. For the total current density is the sum of the partial current densities of these different types of particle, and, likewise, the change of the total momentum density per unit $\overset{0}{S}$-time is the sum of the changes of the momentum densities of the various types of particle.

Eq. (8.26b) is, therefore, correct whenever it is at all possible to treat matter as if it were continuous. We may, therefore, consider $\frac{1}{c^3}\varphi^{\kappa}_{\cdot\sigma}I^{\sigma}$ as the world force per unit proper volume, and substitute it in the right-hand side of eq. (8.22). We obtain

$$P^{\mu\nu}{}_{,\nu} = \frac{1}{c^3}\varphi^{\mu}_{\cdot\sigma}I^{\sigma}. \tag{8.27}$$

It is possible to transform the right-hand side so that it becomes the divergence of a symmetric world tensor. First of all, I^{ι} is determined by the field intensities, and may be taken from eq. (7.20). Then we obtain:

$$P^{\mu\nu}{}_{,\nu} = -\frac{c^{-6}}{4\pi}\varphi^{\mu}_{\cdot\sigma}\varphi^{\sigma\nu}{}_{,\nu}. \tag{8.27a}$$

Next, we integrate by parts,

$$\varphi^{\mu}_{\cdot\sigma}\varphi^{\sigma\nu}{}_{,\nu} = -(\varphi^{\mu}_{\cdot\sigma}\varphi^{\nu\sigma})_{,\nu} + \varphi^{\nu\sigma}\varphi^{\mu}_{\cdot\sigma,\nu}. \tag{8.28}$$

In the last term, we exchange the indices ν and σ in the second factor. Because of the skewsymmetry of $\varphi^{\nu\sigma}$, we obtain

$$\left.\begin{aligned}\varphi^{\nu\sigma}\varphi^{\mu}_{\cdot\sigma,\nu} &= \tfrac{1}{2}\varphi^{\nu\sigma}(\varphi^{\mu}_{\cdot\sigma,\nu} - \varphi^{\mu}_{\cdot\nu,\sigma}) = \tfrac{1}{2}\eta^{\mu\rho}\varphi^{\nu\sigma}(\varphi_{\rho\sigma,\nu} - \varphi_{\rho\nu,\sigma})\\ &= \tfrac{1}{2}\eta^{\mu\rho}\varphi^{\nu\sigma}(\varphi_{\rho\sigma,\nu} + \varphi_{\nu\rho,\sigma}).\end{aligned}\right\} \tag{8.29}$$

Considering the equations (7.19), this last expression is equal to

$$\varphi^{\nu\sigma}\varphi^{\mu}_{\cdot\sigma,\nu} = -\tfrac{1}{2}\eta^{\mu\rho}\varphi^{\nu\sigma}\varphi_{\sigma\nu,\rho} = +\tfrac{1}{4}\eta^{\mu\rho}(\varphi^{\nu\sigma}\varphi_{\nu\sigma})_{,\rho}. \tag{8.29a}$$

Changing a few dummy indices, we get for eq. (8.28)

$$\varphi^{\mu}_{\cdot\sigma}\varphi^{\sigma\nu}{}_{,\nu} = -(\varphi^{\mu}_{\cdot\sigma}\varphi^{\nu\sigma} - \tfrac{1}{4}\eta^{\mu\nu}\varphi^{\rho\sigma}\varphi_{\rho\sigma})_{,\nu}\,; \tag{8.28a}$$

and substituting that expression in eq. (8.27a), we obtain finally the expression

$$\left[P^{\mu\nu} + \frac{c^{-6}}{4\pi}(\tfrac{1}{4}\eta^{\mu\nu}\varphi^{\rho\sigma}\varphi_{\rho\sigma} - \varphi^{\mu}_{\cdot\sigma}\varphi^{\nu\sigma})\right]_{,\nu} = 0. \tag{8.30}$$

Again, we shall consider the components of this new four dimensional tensor one by one. We set

$$\frac{c^{-6}}{4\pi}(\tfrac{1}{4}\eta^{\mu\nu}\varphi^{\rho\sigma}\varphi_{\rho\sigma} - \varphi^{\mu}_{\cdot\sigma}\varphi^{\nu\sigma}) \equiv M^{\mu\nu}, \tag{8.31}$$

and replace the $\varphi_{\mu\nu}$ and $\varphi^{\mu\nu}$ by the expressions (7.17) and (7.17a). The various components of $M^{\mu\nu}$ are given by the expressions

$$\left.\begin{aligned}
M^{44} &= \frac{1}{8\pi c^2}(\mathbf{E}^2 + \mathbf{H}^2), \\
M^{4s} &= M^{s4} = \frac{1}{4\pi c}\,\delta_{sik}E_iH_k, \\
M^{rs} &= \frac{1}{4\pi}\,[\tfrac{1}{2}\delta_{rs}(\mathbf{E}^2 + \mathbf{H}^2) - H_rH_s - E_rE_s].
\end{aligned}\right\} \quad (8.32)$$

The M^{rs} are the components of the stress tensor of the electromagnetic field, and were already known to Maxwell; the M^{4s} are the components of Poynting's vector, divided by c^2; and M^{44} is the energy density of the electromagnetic field, divided by c^2, which also occurs in the classical theory of electromagnetism.

A plane electromagnetic wave has a certain energy density, energy flux, and produces a certain stress. A plane wave the direction of propagation of which is the X-axis has a four dimensional energy-stress tensor with the components

$$\left.\begin{aligned}
M^{44} &= \frac{1}{4\pi c^2}\,A^2, \\
M^{14} &= \frac{1}{4\pi c}\,A^2, \\
M^{11} &= \frac{1}{4\pi}\,A^2,
\end{aligned}\right\} \quad (8.33)$$

where A is the common magnitude of $\mathbf{E}$ and $\mathbf{H}$. All other components vanish. The stress in the direction of propagation, which is positive, is called the "pressure of radiation."

PROBLEM

(*a*) How do the radiation pressure, the momentum density, and the energy density of a plane electromagnetic wave transform when a Lorentz transformation (4.13) is applied to eq. (8.33)?

(*b*) Transform the frequency ν of the same plane wave. Assuming that the energy of one photon is equal to $h\nu$, state the transformation law of photon density.

CHAPTER IX

Applications of the Special Theory of Relativity

Experimental verifications of the special theory of relativity. We found, in Chapter IV, that it is only the theory of relativity which explains all three of these experimental effects: The Michelson-Morley experiment, Fizeau's experiment, and the aberration.

The experiment of Michelson and Morley has been repeated many times under varying conditions.[1] All these new experiments have confirmed the original results, with the exception of those which were carried out by D. C. Miller. It is very difficult to decide why Miller's experiments show an effect which appears to indicate an "ether drift" of about 10 km/sec. However, since all the evidence of other experiments points to the accuracy of the Lorentz transformation equations, it is reasonable to assume that Miller's results were caused by a systematic experimental error which has not yet been discovered.†

Recently the Lorentz transformation equations have been confirmed again in an entirely new manner by H. E. Ives.[2] Ives measured the so-called relativistic Doppler effect. The frequency of a light ray is not invariant with respect to Lorentz transformations. (See Problem 3, Chapter IV, p. 45.) The transformation law is

$$\nu^* = \nu \frac{1 - \cos \alpha \cdot v/c}{\sqrt{1 - v^2/c^2}}, \tag{9.1}$$

where α is the angle between the relative velocity of the two coördinate systems and the direction of the propagation of light in the original coördinate system. This relativistic law differs from the classical law only by the denominator. Ordinarily, the relativistic, second-order effect is overshadowed by the classical Doppler effect (the term in the

---

[1] R. J. Kennedy, *Proc. Nat. Acad. Sci.*, **12**, 621 (1926); *Astrophys. J.*, **68**, 367 (1928).

Piccard and Stahel, *Comptes Rendus*, **183**, 420 (1926); **185**, 1198 (1928); *Naturwissenschaften*, **14**, 935 (1926), **16**, 25 (1928).

D. C. Miller, *Rev. Mod. Phys.*, **5**, 203 (1933), where further references are listed.

G. Joos, *Ann. d. Physik*, **7**, 385 (1930).

[2] *J. Opt. Soc. Am.*, **28**, 215 (1938).

†See Appendix B.

numerator which depends on the angle), which is of the first order in v/c.

Ives measured the change of frequency of the H_β line emitted by hydrogen *canal rays*, using accelerating voltages up to 18,000 volts or velocities up to about 1.8×10^8 cm/sec., that is, $v/c \sim 6 \times 10^{-3}$. To separate the small second-order effect from the much greater first-order effect, Ives measured the wave length of H_β emitted by the canal rays in two directions, with and against their motion. With the help of a mirror, both these lines and the undisplaced line of hydrogen atoms at rest were photographed simultaneously on the same photographic plate. According to the relativistic transformation law (9.1), the two frequencies of the displaced lines are

$$\nu_1^* = \nu \frac{1 - v/c}{\sqrt{1 - v^2/c^2}}, \qquad \nu_2^* = \nu \frac{1 + v/c}{\sqrt{1 - v^2/c^2}},$$

and their average is

$$\bar{\nu} = \frac{\nu}{\sqrt{1 - v^2/c^2}}. \tag{9.2}$$

Ives measured the displacement of $\bar{\nu}$ with respect to the frequency of the undisplaced line, ν. He confirmed the relativistic transformation law.

This experiment is important because it measures directly an effect of the "time dilatation," the apparent slowing down of moving clocks. The outcome of the Michelson-Morley experiment can be and, in fact, was first explained by the "Lorentz contraction" alone, that is, by the shortening of the distance between the half-silvered glass plate and the mirror in the direction of "motion through the ether."

In Ives' experiment, on the other hand, no moving scales are involved, but it is the motion of the "atomic clocks," the canal rays, relative to the observer, and the slowing down of these "clocks," which produces the "relativistic Doppler effect."

All these tests of the theory of relativity concern only the Lorentz transformation equations applied to light rays. We shall now turn to other modifications of classical theories, which were considered in Chapters VI and VII.

Many facts of nuclear physics show that inertial mass is equivalent to energy. The mass of a nucleus is generally less than the sum of the masses of the constituent protons and neutrons. In many cases, it has been possible to confirm that the mass defect is $1/c^2$ times the energy gained by the synthesis of the nucleus.[3] A very striking example of the

[3] Cf. Rasetti, Franco, *Elements of Nuclear Physics*, Prentice-Hall, Inc., New York, 1936, p. 165.

destruction of mass is the annihilation of electron-positron pairs. The masses of the electron and the positron are wholly transformed into electromagnetic radiation. If no third particle (nucleus) is sufficiently near to interact with the system, the total momentum must be the same before and after the annihilation. As the velocities of positron and electron are usually small immediately before the annihilation, energy and momentum are conserved only if two γ-quanta are emitted in opposite directions so that their momenta cancel each other, while the energy of each corresponds to the mass of one electron. Radiation with this energy—about 5×10^5 EV—is actually observed wherever positrons are annihilated.

Charged particles in electromagnetic fields. Further tests of the special theory of relativity are based on particles under the influence of forces. We have already treated the "collision" of γ-quanta with electrons, the Compton effect, in Chapter VI, p. 93. We shall now consider the action of stationary electromagnetic fields on charged particles. The equations of motion have been stated in Chapter VII, eq. (7.44),

$$\frac{d}{dt}\left(\frac{m\mathbf{u}}{\sqrt{1 - u^2/c^2}}\right) = e\left(\mathbf{E} + \frac{\mathbf{u}}{c} \times \mathbf{H}\right). \tag{9.3}$$

To obtain the energy conservation law, let us form the inner product of this equation by $\mathbf{u}$ and carry out an integration by parts on both sides. On the left-hand side we get

$$\left.\begin{aligned}\underline{\mathbf{u}\frac{d}{dt}\left(\frac{m\mathbf{u}}{\sqrt{1 - u^2/c^2}}\right)} &= \frac{d}{dt}\left(\frac{mu^2}{\sqrt{1 - u^2/c^2}}\right) - \frac{m\mathbf{u}\dot{\mathbf{u}}}{\sqrt{1 - u^2/c^2}} \\ &= \frac{d}{dt}\left(\frac{mu^2}{\sqrt{1 - u^2/c^2}}\right) - \frac{mu\dot{u}}{\sqrt{1 - u^2/c^2}} \\ &= \frac{d}{dt}\left(\frac{mu^2}{\sqrt{1 - u^2/c^2}}\right) + mc^2\frac{d}{dt}(\sqrt{1 - u^2/c^2}) \\ &= \underline{mc^2\frac{d}{dt}\left(\frac{1}{\sqrt{1 - u^2/c^2}}\right)},\end{aligned}\right\} \tag{9.4}$$

while the right-hand side becomes

$$\underline{e\mathbf{u}\mathbf{E} + \mathbf{u}\cdot\left(\frac{\mathbf{u}}{c} \times \mathbf{H}\right)} = e\mathbf{u}\mathbf{E} = \underline{e\mathbf{u}\left(-\operatorname{grad}\varphi - \frac{1}{c}\frac{\partial\mathbf{A}}{\partial t}\right)}.$$

We find, therefore, that the change of energy along the path of the particle is given by the expression

$$\frac{d}{dt}\left[\frac{mc^2}{\sqrt{1 - u^2/c^2}}\right] = -e\mathbf{u}\left(\text{grad } \varphi + \frac{1}{c}\frac{\partial \mathbf{A}}{\partial t}\right). \tag{9.5}$$

From the expression $\mathbf{u}$ grad φ, we can split off an exact differential, if we consider the change of the potential φ along the path,

$$\frac{d\varphi}{dt} = \varphi_{,s}\frac{dx^s}{dt} + \varphi_{,4} = \mathbf{u}\cdot\text{grad } \varphi + \frac{\partial \varphi}{\partial t}. \tag{9.6}$$

Eq. (9.5) then takes the form

$$\frac{d}{dt}\left[\frac{mc^2}{\sqrt{1 - u^2/c^2}} + e\varphi\right] = +e\left(\frac{\partial \varphi}{\partial t} - \frac{1}{c}\mathbf{u}\frac{\partial \mathbf{A}}{\partial t}\right). \tag{9.7}$$

If the field is static, so that φ and $\mathbf{A}$ do not change in the course of time, the expression in the square bracket on the left-hand side is constant.

Eq. (9.7) enables us to determine the velocity of particles which have entered a strong electric field with negligible velocity. After they have passed through a potential drop V, their kinetic energy is

$$mc^2\left\{\frac{1}{\sqrt{1 - u^2/c^2}} - 1\right\} = eV, \tag{9.8}$$

and their velocity is

$$u = \sqrt{\frac{2eV}{m}\cdot\frac{1 + \frac{e}{m}\frac{V}{2c^2}}{\left(1 + \frac{e}{m}\frac{V}{c^2}\right)^2}}. \tag{9.9}$$

The classical formula,

$$u_{\text{class}} = \sqrt{2\frac{e}{m}V},$$

is a good approximation when $\frac{e}{m}\frac{V}{c^2}$ is small compared with unity.

Eq. (9.8) shows that the change of energy of a particle is equal to the potential difference multiplied by the charge. It applies in all cases where the energy of particles is experimentally determined by the accelerating voltage, such as in the Van de Graaff electrostatic generator.

In a cloud chamber, the velocity of a charged particle is usually determined by measuring the radius of its path in a magnetic field perpendicular to its velocity. Let us compute this radius.

The acceleration of a charged particle in a magnetic field is given by eq. (9.3),

$$\frac{d}{dt}\left(\frac{m\mathbf{u}}{\sqrt{1-u^2/c^2}}\right) = \frac{e}{c}\,\mathbf{u}\times\mathbf{H}. \tag{9.10}$$

Eq. (9.7) shows that the speed u is constant when $\mathbf{H}$ is stationary. We may, therefore, replace the vector $\mathbf{u}$ by the product of the constant speed u and a unit vector $\mathbf{s}$ which is parallel to $\mathbf{u}$ and changes its direction in the course of time. The left-hand side of eq. (9.10) becomes

$$\frac{mu}{\sqrt{1-u^2/c^2}}\frac{d\mathbf{s}}{dt},$$

and the right-hand side is

$$\frac{e}{c}\,u\mathbf{s}\times\mathbf{H},$$

so that we obtain eq. (9.10) in the form

$$\frac{m}{\sqrt{1-u^2/c^2}}\frac{d\mathbf{s}}{dt} = \frac{e}{c}\,\mathbf{s}\times\mathbf{H}. \tag{9.10a}$$

We have yet to replace the differentiation with respect to t by the differentiation with respect to the arc length l. As $\frac{dl}{dt}$ is the speed u we get, instead of (9.10a),

$$\frac{mu}{\sqrt{1-u^2/c^2}}\frac{d\mathbf{s}}{dl} = \frac{e}{c}\,\mathbf{s}\times\mathbf{H}. \tag{9.11}$$

$\mathbf{s}$ is the tangential unit vector of the path, $\frac{d\mathbf{s}}{dl}$ is, therefore, the curvature, the magnitude of which is the reciprocal radius of curvature, R. We obtain, therefore,

$$\frac{mu}{\sqrt{1-u^2/c^2}} = R\cdot\frac{e}{c}\cdot H\cdot\sin(\mathbf{s},\mathbf{H}). \tag{9.12}$$

The angle $(\mathbf{s},\mathbf{H})$ remains constant along the path, for, if we form the scalar product of eq. (9.10a) by $\mathbf{H}$, the right-hand side vanishes, and the left-hand side contains as a factor the derivative of $(\mathbf{s}\cdot\mathbf{H})$ with respect to t. The path of the particle is, therefore, a spiral. When $\mathbf{s}$ and $\mathbf{H}$ are perpendicular to each other, the path is a circle. *The product RH is, then, a measure of the magnitude of the relativistic momentum of the particle,*

$$RH = \frac{c}{e}\cdot\frac{mu}{\sqrt{1-u^2/c^2}} = \frac{c}{e}\,p. \tag{9.13}$$

If e and m are known, velocity and energy may be computed.

Occasionally one determines the deflection of a charged particle by an electrostatic field which is perpendicular to the path of the particle. The acceleration is again given by eq. (9.3),

$$\frac{d}{dt}\left\{\frac{m\mathbf{u}}{\sqrt{1 - u^2/c^2}}\right\} = e\mathbf{E}. \tag{9.14}$$

As long as $\mathbf{E}$ and $\mathbf{u}$ are perpendicular to each other, the speed u remains constant. If we introduce again the unit vector $\mathbf{s}$, eq. (9.14) becomes

$$\frac{mu}{\sqrt{1 - u^2/c^2}}\frac{d\mathbf{s}}{dt} = e\mathbf{E}. \tag{9.14a}$$

The introduction of the arc length l leads to the equation

$$\frac{mu^2}{\sqrt{1 - u^2/c^2}}\frac{d\mathbf{s}}{dl} = e\mathbf{E}; \tag{9.14b}$$

or, if we use the radius of curvature, R,

$$\frac{mu^2}{\sqrt{1 - u^2/c^2}} = eRE. \tag{9.15}$$

In the case of an electric field, the paths are, of course, not circles. Eq. (9.15) refers only to the radius of curvature of that part of the path in which the direction of motion is perpendicular to the lines of force.

To determine both the rest mass and the velocity of a particle, one must observe its behavior in both electric and magnetic fields. One can, for instance, give a particle a known energy by accelerating it with a measured potential drop, eq. (9.8); if, then, its momentum is determined in a magnetic field, eq. (9.13), both its mass and momentum can be computed.

This and other, similar arrangements have, at the same time, been used to test the relativistic laws of motion; the determinations were carried out on many particles of the same kind, at varying velocities. All these experiments have uniformly confirmed the correctness of the relativistic laws. A comprehensive account of these experiments is to be found in the article of W. Gerlach in the *Handbuch der Physik*.[4]

The field of a rapidly moving particle. The particles of cosmic radiation usually move with velocities approximating the speed of light. Their electromagnetic field is similar to that of an electromagnetic wave.

To compute the field of such a rapidly moving particle, we shall

[4] Vol. XXII, Berlin, 1926, pp. 61–82.

first consider its field with respect to a frame of reference in which the particle is at rest. Let us choose the coördinate system S^* so that the particle is always at the point of origin, $x^* = y^* = z^* = 0$. The field produced by the particle is purely electric and is given by the equations

$$E_s^* = ex_s^*/r^{*3}. \tag{9.16}$$

We shall carry out the transformation to another coördinate system S in two steps: First, we shall apply the transformation equations (7.22c) to the components of the electromagnetic field. Then we shall express the starred coördinates in terms of the unstarred coördinates.

The transformation equations which are inverse to eqs. (7.22c) are obtained by changing the sign of v everywhere. All the components of $\mathbf{H}^*$ vanish, and we have:

$$\left.\begin{aligned}
H_3 &= (1 - v^2/c^2)^{-1/2}\cdot v/c\cdot E_2^*,\\
H_2 &= -(1 - v^2/c^2)^{-1/2}\cdot v/c\cdot E_3^*,\\
H_1 &= 0,\\
E_1 &= E_1^*,\\
E_2 &= (1 - v^2/c^2)^{-1/2}\cdot E_2^*,\\
E_3 &= (1 - v^2/c^2)^{-1/2}\cdot E_3^*.
\end{aligned}\right\} \tag{9.17}$$

The vectors $\mathbf{H}$ and $\mathbf{E}$ are perpendicular to each other.

Let us now compute E_s^* as functions of the S-coördinates. r^* is given by the expression

$$r^{*2} = x^{*2} + y^{*2} + z^{*2};$$

and, if we replace these coördinates by the expressions (4.13),

$$r^{*2} = \frac{(x - vt)^2}{1 - v^2/c^2} + y^2 + z^2. \tag{9.18}$$

For the three components E_s^*, we obtain

$$\left.\begin{aligned}
E_1^* &= e\left[\frac{(x - vt)^2}{1 - v^2/c^2} + y^2 + z^2\right]^{-3/2}\cdot\frac{x - vt}{\sqrt{1 - v^2/c^2}},\\
E_2^* &= e\left[\frac{(x - vt)^2}{1 - v^2/c^2} + y^2 + z^2\right]^{-3/2}\cdot y,\\
E_3^* &= \left[\frac{(x - vt)^2}{1 - v^2/c^2} + y^2 + z^2\right]^{-3/2}\cdot z.
\end{aligned}\right\} \tag{9.19}$$

The three components E_s are

$$\left.\begin{aligned} E_1 &= e\,\frac{x - vt}{N}, \\ E_2 &= e\,\frac{y}{N}, \qquad E_3 = e\,\frac{z}{N}, \\ N &= \sqrt{1 - v^2/c^2}\cdot\left[\frac{(x - vt)^2}{1 - v^2/c^2} + y^2 + z^2\right]^{3/2}, \end{aligned}\right\} \tag{9.20}$$

and the components H_s are

$$\left.\begin{aligned} H_1 &= 0, \\ H_2 &= -\frac{v}{c}\,e\,\frac{z}{N}, \\ H_3 &= +\frac{v}{c}\,e\,\frac{y}{N}. \end{aligned}\right\} \tag{9.21}$$

Let us assume that v, the speed of the cosmic ray particle, is almost equal to the speed of light; and let us consider the field at any point (x, y, z) at rest in the S-system. The amplitude of the electric field,

$$E = \sqrt{E_1^2 + E_2^2 + E_3^2},$$

reaches a maximum at the time $t_0 = \dfrac{x}{v}$; this maximum is more pronounced the more closely v approaches c. At this time, t_0, the electric field is perpendicular to the X-axis and equal in magnitude to the magnetic field strength. The field of the moving particle at a fixed point (x, y, z) off the path is approximately the same as the field of a short, plane wave train.

Sommerfeld's theory of the hydrogen fine structure. Shortly after Bohr and Sommerfeld had succeeded in explaining the spectrum of the hydrogen atom by the quantization of phase integrals, Sommerfeld attempted to apply relativistic corrections to the underlying mechanical model. He thought that relativistic corrections were responsible for the splitting up of the degenerate terms of the nonrelativistic theory, and that he would obtain a theory of the fine structure.

His attempt appeared successful in the case of the hydrogen atom, but his theory failed completely when applied to other spectra. We know today that Sommerfeld's explanation of the fine structure was erroneous, even in the case of the hydrogen atom; it does not take into account the spin of the electron, nor does it lead to the same result as a consistent relativistic wave-mechanical treatment with spin 0. But it so happens that the two errors—the neglect of wave mechanics and the neglect of spin—cancel each other in the case of the hydrogen atom, and so Sommerfeld's treatment, which was based on two errors, led to the correct result. Sommerfeld's work on the fine structure is interesting from an historical point of view because it brought to the light for the first time the significance of the "fine structure constant" α. In this section, we shall give an account of this theory.

A single electron moves in the field of a proton, which, because of its greater mass, may be assumed to be at rest. The equations of motion are eqs. (9.14), with **E** given by the equations

$$\mathbf{E} = -\operatorname{grad} \varphi, \qquad \varphi = \frac{e}{r}. \tag{9.22}$$

One integral of these equations is the energy integral (9.8),

$$mc^2[(1 - u^2/c^2)^{-1/2} - 1] - e\varphi(r) = W. \tag{9.23}$$

We obtain another integral, the angular momentum integral, when we form the vector product of the equation

$$\frac{d}{dt}\left\{\frac{m\mathbf{u}}{\sqrt{1 - u^2/c^2}}\right\} = +e \operatorname{grad} \varphi \tag{9.24}$$

by the radius vector **r**. As the gradient of φ and the radius vector **r** are everywhere parallel, the vector product on the right-hand side vanishes, and we obtain

$$0 = \mathbf{r} \times \frac{d}{dt}\left(\frac{m\mathbf{u}}{\sqrt{1 - u^2/c^2}}\right)$$

$$= \frac{d}{dt}\left(\frac{m\mathbf{r} \times \mathbf{u}}{\sqrt{1 - u^2/c^2}}\right) - \frac{d\mathbf{r}}{dt} \times \frac{m\mathbf{u}}{\sqrt{1 - u^2/c^2}}. \tag{9.25}$$

$\frac{d\mathbf{r}}{dt}$ in the last term is simply **u**, and the vector product of **u** by itself vanishes. We find, therefore, that the expression

$$\frac{m(\mathbf{r} \times \mathbf{u})}{\sqrt{1 - u^2/c^2}} = p_\theta \tag{9.26}$$

is another constant of the equation of motion.

If we introduce polar coördinates r, θ in the plane of motion, the two integrals take the form

$$\left.\begin{aligned} \frac{mc^2}{\sqrt{1 - u^2/c^2}} &= mc^2 + \frac{e^2}{r} + W, \\ \frac{mr^2\,\dot\theta}{\sqrt{1 - u^2/c^2}} &= I_\theta\,, \\ u^2 &= \dot r^2 + r^2\dot\theta^2. \end{aligned}\right\} \tag{9.27}$$

According to Sommerfeld, the integral

$$\oint I_\theta\, d\theta,$$

extended over a full period of θ, must be an integral multiple of h,

$$mr^2\dot\theta = n_\theta\cdot\hbar\cdot\sqrt{1 - u^2/c^2}, \qquad \hbar = \frac{h}{2\pi}. \tag{9.28}$$

To formulate the second quantum condition, we must compute the radial component of the momentum,

$$p_r = \frac{m\dot r}{\sqrt{1 - u^2/c^2}}. \tag{9.29}$$

With the help of the two integrals of motion, (9.27), we can express p_r as a function of r alone. We have

$$\left.\begin{aligned} p_r^2 &= m^2\frac{\dot r^2}{1 - u^2/c^2} = m^2\frac{u^2 - r^2\dot\theta^2}{1 - u^2/c^2} \\ &= m^2c^2\left(\frac{1}{1 - u^2/c^2} - 1\right) - \frac{m^2r^2\dot\theta^2}{1 - u^2/c^2}. \end{aligned}\right\} \tag{9.30}$$

First, we compute the expression $\dfrac{1}{1 - u^2/c^2}$ from the energy integral, the first equation (9.27). It is

$$\frac{1}{1 - u^2/c^2} = \left(\frac{mc^2 + \dfrac{e^2}{r} + W}{mc^2}\right)^2. \tag{9.31}$$

Then we compute the last term of eq. (9.30) from the quantum condition (9.28). We have:

$$\frac{m^2r^2\dot\theta^2}{1 - u^2/c^2} = \left(\frac{n_\theta\cdot\hbar}{r}\right)^2. \tag{9.32}$$

If we substitute the two expressions (9.31), (9.32) in eq. (9.30), we get

$$p_r^2 = m^2c^2\left[\left(\frac{mc^2 + e^2/r + W}{mc^2}\right)^2 - 1\right] - \left(\frac{n_\theta\hbar}{r}\right)^2. \tag{9.33}$$

The second quantum condition states that the integral

$$\oint p_r \, dr,$$

extended over a full period of the variable r, must be an integral multiple of h,

$$\oint \sqrt{\frac{1}{c^2}\left(W + \frac{e^2}{r}\right)\left(W + \frac{e^2}{r} + 2mc^2\right) - \left(\frac{n_\theta \hbar}{r}\right)^2} \, dr = n_r h. \tag{9.34}$$

The integrand

$$\left.\begin{aligned} &2mW\left(1 + \frac{W}{2mc^2}\right) + 2e^2 m\left(1 + \frac{W}{mc^2}\right)\frac{1}{r} - \frac{n_\theta^2 \hbar^2}{r^2}\left(1 - \frac{\alpha^2}{n_\theta^2}\right), \\ &\alpha = \frac{e^2}{\hbar c}, \end{aligned}\right\} \tag{9.35}$$

differs from the analogous nonrelativistic expression in that the second term in each parenthesis represents the relativistic correction.

The expression (9.35) has, for negative values of W, two zeros for positive values of r, which correspond to the perihelion and the aphelion positions of the electron. The integral (9.34) is to be carried out from one zero to the other zero, and back, with the reverse sign of the root. This may be done either by elementary methods or by the so-called complex integration.[5] The result is

$$\oint \sqrt{-A + 2B/r - C/r^2} \, dr = 2\pi\left(\frac{B}{\sqrt{A}} - \sqrt{C}\right), \tag{9.36}$$

where A, B, and C are positive constants. If we substitute their values, as given by eqs. (9.34) and (9.35), we obtain

$$2\pi\left\{\frac{me^2(1 + W/mc^2)}{\sqrt{-2mW(1 + W/2mc^2)}} - n_\theta \hbar \sqrt{1 - \frac{\alpha^2}{n_\theta^2}}\right\} = n_r h. \tag{9.37}$$

By solving this equation with respect to W, we obtain the fine-structure formula of Sommerfeld,

$$\left.\begin{aligned} W &= -mc^2\left\{1 + \frac{\alpha^2}{[n_r + \sqrt{n_\theta^2 - \alpha^2}]^2}\right\}^{-1/2} + mc^2 \\ &= \frac{me^4}{2(n_r + n_\theta)^2 \hbar^2}\left\{1 + \frac{\alpha^2}{(n_r + n_\theta)^2}\left(\frac{n_r}{n_\theta} + \frac{1}{4}\right) + \cdots\right\}. \end{aligned}\right\} \tag{9.38}$$

De Broglie waves. After the quantum theory of Bohr and Sommerfeld, the next step toward the development of a consistent quantum

[5] See, for instance, A. Sommerfeld, *Atomic Structure and Spectral Lines*, Annex.

mechanics was taken by de Broglie. He assumed that along the path of a moving particle a wave motion took place. If the path of the particle was closed, as in the case of the hydrogen atoms, the waves would interfere with themselves. If the length of the path was equal to an integral number of wave lengths, the waves would reinforce themselves. In every other case, they would cancel. Those paths along which the waves reinforce each other were the "permitted" paths of the Bohr-Sommerfeld theory.

Let us consider a free particle which is at rest in some coördinate system S. Its rest mass is m, its rest energy mc^2. We cannot associate a propagating wave with this particle, for there is nothing about the particle which might suggest a particular direction of propagation. But we may associate with the particle a frequency ν, according to Einstein's quantum law,

$$E = h\nu, \tag{9.39}$$

so that the frequency of the particle at rest is

$$\nu = \frac{mc^2}{h}. \tag{9.40}$$

The de Broglie wave can be represented by a "wave function,"

$$\psi = \psi_0 e^{2\pi i \nu t}. \tag{9.41}$$

Let us now carry out a Lorentz transformation (4.13), (4.15). With respect to the system S^*, the momentum and energy of the particle are given by the expressions

$$\left.\begin{aligned} p_x^* &= -\frac{mv}{\sqrt{1 - v^2/c^2}}, \qquad p_y^* = p_z^* = 0; \\ E^* &= \frac{mc^2}{\sqrt{1 - v^2/c^2}}. \end{aligned}\right\} \tag{9.42}$$

Let us now transform the wave function ψ. We shall assume that ψ is a scalar. Its dependence on the coördinates is then expressed by the equation

$$\psi = \psi_0 \cdot \exp\left\{2\pi i \nu \frac{t^* + \frac{v}{c^2} \cdot x^*}{\sqrt{1 - v^2/c^2}}\right\} \tag{9.43}$$

Its frequency and wave length have the values

$$\nu^* = \frac{\nu}{\sqrt{1 - v^2/c^2}} = \frac{mc^2/h}{\sqrt{1 - v^2/c^2}} = \frac{E^*}{h}, \tag{9.44}$$

$$\lambda^* = \frac{c^2}{v\nu}\sqrt{1 - v^2/c^2} = \frac{h}{mv}\sqrt{1 - v^2/c^2} = \frac{h}{p^*}. \tag{9.45}$$

Let us go on to compute the velocity with which a plane harmonic de Broglie wave propagates through space. This velocity is the product of frequency and wave length,

$$w = \nu^*\lambda^* = \frac{E^*}{p^*} = \frac{c^2}{v}. \tag{9.46}$$

This velocity, the so-called phase velocity, is greater than c: its product by v equals c^2. As the de Broglie wave phases cannot be used for the transmission of signals, eq. (9.46) is not in contradiction to the foundations of the theory of relativity.

Let us consider, on the other hand, a de Broglie wave which is not quite harmonic, but composed of two harmonic waves of nearly identical frequency. The amplitude of the combined wave will not be constant, and the maxima and minima of this amplitude will move through space with a certain velocity, which is called the "*group velocity.*" Let us compute this group velocity, w_g. Let us take as the two constituent waves:

$$\overset{1}{\psi} = \psi_0 \exp\left\{2\pi i\left(\overset{1}{\nu}t - \frac{x}{\overset{1}{\lambda}}\right)\right\}$$

and

$$\overset{2}{\psi} = \psi_0 \exp\left\{2\pi i\left(\overset{2}{\nu}t - \frac{x}{\overset{2}{\lambda}}\right)\right\}, \qquad \overset{2}{\nu} = \overset{1}{\nu} + 2\delta\nu, \qquad \overset{2}{\lambda} = \overset{1}{\lambda} + 2\delta\lambda.$$

The combined wave will be:

$$\psi = \psi_0\left[e^{2\pi i\left(\overset{1}{\nu}t - \frac{x}{\overset{1}{\lambda}}\right)} + e^{2\pi i\left(\overset{2}{\nu}t - \frac{x}{\overset{2}{\lambda}}\right)}\right]. \tag{9.47}$$

The square bracket can be expressed in a different form, with the help of the relation

$$e^{i\alpha} + e^{i\beta} = e^{i\frac{\alpha+\beta}{2}}\left(e^{i\frac{\alpha-\beta}{2}} + e^{-i\frac{\alpha-\beta}{2}}\right) = 2\cos\frac{\alpha-\beta}{2}\cdot e^{i\frac{\alpha+\beta}{2}}.$$

We have, therefore,

$$\psi = 2\psi_0 \cos\left[2\pi\left(\delta\nu\cdot t + \frac{\delta\lambda}{\overset{1}{\lambda}\overset{2}{\lambda}}\cdot x\right)\right] e^{2\pi i\left[(\overset{1}{\nu}+\delta\nu)t - \frac{\overset{1}{\lambda}+\delta\lambda}{\overset{1}{\lambda}\overset{2}{\lambda}}x\right]}. \tag{9.48}$$

The speed of propagation of the amplitude is, therefore,

$$w_g = -\frac{\lambda_1 \lambda_2 \delta\nu}{\delta\lambda};$$

or, when we consider $\delta\nu$ and $\delta\lambda$ as differentials,

$$w_g = \frac{d\nu}{d\left(\frac{1}{\lambda}\right)}. \tag{9.49}$$

In de Broglie waves, $\frac{1}{\lambda}$ equals $\frac{p}{h}$, while ν is $\frac{E}{h}$. The group velocity is, therefore,

$$w_g = \frac{dE}{dp}. \tag{9.50}$$

Now E and p are connected by the equation

$$E^2 - c^2 p^2 = m^2 c^4, \tag{9.51}$$

which states that the magnitude of the energy-momentum vector is the rest energy. From

$$E = \sqrt{m^2 c^4 + c^2 p^2}$$

we obtain

$$\frac{dE}{dp} = \frac{c^2 p}{E} = u, \tag{9.52}$$

the speed of the particle. The group velocity of a de Broglie wave equals the velocity of the particle.

PROBLEMS

1. Compute the velocity of: (*a*) electrons, (*b*) protons which have been accelerated by a voltage of 3×10^6 volts.

2. (*a*) When the accelerating potential V in eq. (9.9) is so small that u/c is small compared with unity, develop an approximate equation which shows the second order relativistic effect to be added to the classical equation.

(*b*) When V is so great that u approaches c, develop an approximative equation which shows the deviation of u from c as a function of the potential drop V.

3. From the radius of curvature in a cloud chamber, compute the velocity and the energy of a particle the mass and charge of which are known.

4. State an approximate relationship between the energy and the momentum of a particle which is moving almost with the speed of light. This condition is satisfied by the particles in the cosmic radiation.

PART II

The General Theory of Relativity

CHAPTER X

The Principle of Equivalence

Introduction. The special theory of relativity had its origin in the development of electrodynamics. The general theory of relativity is the relativistic theory of gravitation.

Once before the problem of gravitation gave rise to a new era in physics, the era of Newton's classical mechanics. The three fathers of scientific physics, Galileo, Kepler, and Newton, studied gravitation: Galileo the quasi-homogeneous gravitational field on the surface of the earth, Kepler and Newton the action of one mass point of great mass on another of much smaller mass. Galileo formulated the law of inertia and established the idea that the force acting on a body was measured by its acceleration (and not by its velocity, as had been assumed before). Newton determined the amount of the gravitational action of one mass point on another. This gravitational force is always an attraction, and its magnitude is

$$f = \kappa \frac{mM}{r^2}, \tag{10.1}$$

where m and M are the masses of the two mass points, r is the distance between them, and κ is a universal constant, which has the value

$$\kappa = 6.66 \times 10^{-8} \text{ dyn cm}^2 \text{ g}^{-2}. \tag{10.2}$$

The force which acts on the body with the mass m is the negative gradient of the gravitational potential G_M, multiplied by its mass, m, where

$$G_M = -\kappa \frac{M}{r} \tag{10.3}$$

at the distance r from the mass point with the mass M. The potential energy of the two-body system m, M is

$$U = mG_M = -\kappa \frac{mM}{r}. \tag{10.4}$$

This theory of gravitation is a typical example of a mechanical theory, in fact, the most important example. The force which acts on a mass point P_m at the time t is completely determined by the distances of all other mass points from P_m at the time t, their masses, and the mass m of P_m itself. It is, therefore, essential that the simultaneity of distant events and the distance of the two mass points have an invariant significance. Newton's theory of gravitation is covariant with respect to Galilean transformations, but, of course, not with respect to Lorentz transformations.

The work of Faraday, Maxwell, and Hertz in the field of electrodynamics brought about new concepts which differed sharply from those of classical mechanics. The action at a distance of one mass point on another, which is typical for mechanics, was replaced in electrodynamics by the action of the field on a mass point and the dependence of the field on the positions and the velocities of the mass points. In other words, interaction does not take place directly between distant mass points, but between points of the field which are separated by infinitesimal distances.

In prerelativistic physics, the mechanical theory of gravitation and the field theory of electrodynamics were based on the same concepts of space and time; and, therefore, though the two theories were fundamentally different, they did not contradict each other. They were no longer compatible, however, when the analysis of the transformation properties of Maxwell's equations led to the development of the special theory of relativity. While Maxwell's theory merely eliminated action at a distance from the realm of electrodynamics, the Lorentz transformation equations ruled out action at a distance from the whole of physics by depriving time and space of their absolute character. If the theory of gravitation was to be at all consistent with the other fields of physics, it had to be changed into a relativistic field theory. However, an examination of the fundamental assumptions of Newton's theory from the point of view of field physics revealed that the "relativization" of the theory of gravitation necessitated an expansion of the special theory of relativity into what is known today as the general theory of relativity. We shall now retrace this analysis.

The principle of equivalence. The gravitational force differs from all other forces in one respect: It is proportional to the *mass* of the body on which it acts. In the ponderomotive law of classical mechanics, eqs. (2.13), the components of the force acting on a body are proportional to the mass of that body. The constant factor, m_i, cancels on both sides of eqs. (2.13), and thus, *the acceleration of a body in a gravitational field is independent of its mass.*

Newton's theory of gravitation accepts this fact, but does not explain it. Within the framework of classical physics, an "explanation" was hardly called for. Other force laws, Coulomb's law of electrostatic forces, the nature of Van der Waals' forces, had not been "explained," either. Nevertheless, Newton's law is in a class by itself. The mass of a body is a constant which is characteristic for its behavior under the influence of any force, it is the ratio of force to acceleration. In this connection, we may call the mass of a body its "*inertial mass*," because it is a measure of its "inertial resistance to acceleration." The electrostatic force acting on a particle is the product of the electric field strength, which is *independent* of the particle, and the charge of the particle, which is *characteristic* of the particle. Likewise, the gravitational force is the product of the "gravitational field strength," the negative gradient of the gravitational potential, (10.3), and the mass of the particle. In its role as a "gravitational charge," we shall call the mass the "*gravitational mass*" of the particle According to Newton's theory of gravitation, the inertial mass and the gravitational mass of the same body are always equal. This proposition is called *the principle of equivalence* for reasons which will become apparent later.

Now, it might be that the "inertial mass" and the "gravitational mass" are approximately equal for most bodies, but that this approximate equality is accidental, and that an accurate determination would reveal that the two kinds of mass of a body are really different. Fortunately, it is possible to subject the asserted equality of inertial mass and gravitational mass to very accurate tests. What has to be done is to find out whether the acceleration of all bodies is the same in the same gravitational field.

Since it is impossible to measure time intervals accurately enough, we cannot measure directly the accelerations of freely falling bodies, but must employ an indirect method. There is a type of acceleration which is certainly independent of the mass of the accelerated body, "inertial acceleration." When we refer the motion of bodies to a frame of reference which is not an inertial system, we encounter accelerations which do not correspond to real forces acting on a body, but which are merely reflections of the accelerations of the frame of reference relative to some inertial system. In Chapter II, we discussed these "inertial forces" in a special case, in which the chosen frame of reference was rotating with a constant angular velocity relative to an inertial system.

The "inertial force" on a body is proportional to its "inertial mass." If we can observe bodies under the combined influence of "inertial forces" and gravitational forces, the direction of the resultant for a particular body will depend on the ratio of its "inertial mass" to its "gravitational mass." If we observe several different bodies, we have an extremely sensitive test which will tell us whether this ratio is the same for all the bodies tested.

The experimental set-up is already provided by nature: The earth is not an inertial system, but rotates around its axis with a constant angular velocity. A body which is at rest relative to the earth is, therefore, subject both to the gravitational attraction of the earth and to "centrifugal force." Its total acceleration relative to the earth will be the vector sum of the gravitational acceleration and the "centrifugal acceleration." Except for points on the equator, the two constituent accelerations are not parallel, and the direction of the resultant is a measure of the ratio between inertial and gravitational mass.

Eötvös[1] suspended two weights of different materials, but with equal gravitational masses, from the two arms of a torsion balance. If the two inertial masses had been unequal, that is, if the resultants of the two weights had not been parallel, the balance would have been subject to a torque. The absence of such a torque showed, with a relative accuracy of about 10^{-8}, that the ratio of inertial and gravitational mass is the same for various materials.

The development of the special theory of relativity showed that at least part of the inertial mass of a body had to be attributed to internal energy. In radioactive materials, the contribution to the total mass from this source was bound to be considerable. Did this part of the "inertial mass" also show up as "gravitational mass"? The question was answered by Southerns,[2] who repeated Eötvös' experiments with radioactive materials. The result was the same as before: The "gravitational mass" turned out to be equal to the "inertial mass," even though the latter was in part caused by great quantities of bound energy. The principle of equivalence was ostensibly a fundamental property of the gravitational forces.

Preparations for a relativistic theory of gravitation. Before we can hope to create a relativistic theory of gravitation, we must first attempt to reformulate Newton's theory so that action at a distance is eliminated. This can be done fairly easily.

[1] *Math. und Naturw. Ber. aus Ungarn*, **8**, 65 (1890).
[2] *Proc. Roy. Soc.*, **84A**, 325 (1910).

The gravitational attraction of one body with the mass m by several other ones can be represented by the sum of the "gravitational potentials," (10.3), of these other bodies; this sum represents the potential energy U_m of the first body divided by its mass m. The force experienced by that body is the negative gradient of its potential energy,

$$\mathbf{f} = -m \operatorname{grad} G. \tag{10.5}$$

The gravitational potential depends on the positions of the other bodies. The contribution of every mass point is given by eq. (10.3). If we introduce a "gravitational field strength,"

$$\mathbf{g} = -\operatorname{grad} G, \tag{10.6}$$

we find, just as in electrostatics, that the gravitational lines of force neither originate nor terminate outside of masses, and that, in a mass M, $4\pi\kappa M$ lines of force terminate. We conclude that the divergence of $\mathbf{g}$ is

$$\operatorname{div} \mathbf{g} = -4\pi\kappa\rho,$$

where ρ is the mass density. The potential G itself satisfies the equation

$$\operatorname{div} \operatorname{grad} G \equiv \nabla^2 G = 4\pi\kappa\rho. \tag{10.7}$$

This equation, which was first formulated by Poisson, is, then, the classical equation of the gravitational field. Eqs. (10.5) and (10.7) together are completely equivalent to the equations of Newton's theory, which is based on action at a distance.

Poisson's equation, (10.7), is not Lorentz-invariant. Wherever ρ vanishes, it seems reasonable to assume that the three dimensional Laplacian operator ∇^2 has to be replaced by its four dimensional analogue, the operator

$$\eta^{\rho\sigma} \frac{\partial^2}{\partial x^\rho \, \partial x^\sigma} = \frac{\partial^2}{\partial t^2} - c^2 \nabla^2.$$

In the presence of matter, we must remember that the mass density ρ is not a scalar, but one component of the tensor $P^{\mu\nu}$. We face the alternative of either replacing ρ by the Lorentz-invariant scalar $\eta_{\mu\nu} P^{\mu\nu}$, or replacing the nonrelativistic scalar G by a world tensor $G^{\mu\nu}$.

On inertial systems. Suppose we were confronted by the task of finding a frame of reference which is an inertial system. An inertial system, according to the definition of Chapter II, is a coördinate system, with respect to which all bodies not subjected to forces are unaccelerated. This definition by itself is not very helpful, as we have first to determine whether a given body is subjected to forces or not. According to classical mechanics, all (real) forces represent the interaction of bodies with each other. A body is, therefore, not subjected to forces if it is sufficiently far removed from all other bodies.[3]

This criterion is satisfactory from the point of view of classical mechanics. But in the theory of relativity we must try to eliminate all concepts which involve finite spatial distances. A concept such as "sufficiently far" has no Lorentz-invariant significance. The definition of an inertial system should be based on the properties of the immediate neighborhood of the observer.

We can determine an inertial system if we can predict the accelerations of test bodies, that is, if we know the gravitational and electromagnetic fields in the neighborhood. But there is only one method of measuring the field, and that is to measure the accelerations of test bodies. This is a vicious circle.

However, there is a profound difference between the electromagnetic and the gravitational field. Nothing prevents us from choosing as test bodies uncharged and unpolarized bodies and, thereby, from reducing the electromagnetic forces acting on them to zero. The effects of a gravitational field on a test body, however, cannot be eliminated, for the acceleration of a body in a gravitational field is independent of its mass.

The action of a gravitational field on a body is indistinguishable from "inertial accelerations." Both gravitational and inertial accelerations are independent of the characteristics of the test body. Therefore, we are unable to separate the gravitational from the inertial accelerations and to find an inertial system.

The equivalence of gravitational and inertial fields in this respect is a consequence of the equality of gravitational and inertial masses. In fact, the equivalence of gravitational and inertial fields gave the principle of equivalence its name.

From this point of view, inertial systems are not a particular class of coördinate systems; there is no real difference between a supposed inertial frame of reference with a gravitational field and a non-inertial frame of reference.

Einstein's "elevator." To illustrate the equivalence of inertial and non-inertial frames of reference, Einstein gives the example of a man enclosed in an elevator car. As long as the elevator is at rest, the man can determine, by one of the usual methods, the field strength of the gravitational field on the surface of the earth, which is about 981 cm sec^{-2}. He can, for instance, determine the time interval which a body

[3] Strictly speaking, "sufficiently far" means at an infinite distance. Our condition can be only approximately satisfied.

takes to drop to the ground from a point 100 cm above the ground. The gravitational field strength, in this case, is

$$g = \frac{100 \times 2}{t^2}. \tag{10.8}$$

Suppose the man had no possibility of obtaining information from outside his car. Instead of concluding that he and his car are at rest and in a gravitational field, he might also argue as follows: "All objects in my car undergo an apparent acceleration of 981 cm $\sec^{-2}$ as long as their motion is not stopped by collision with other bodies or with the floor of my car. As this acceleration does not depend on the individual characteristics of my test bodies, it is not likely that the accelerations correspond to real forces acting on the test bodies. Probably, my frame of reference (which is connected with the car) is not an inertial system, but for some reason, unknown to me, is accelerated upward relative to an inertial system at the rate of 981 cm $\sec^{-2}$. Those bodies inside my car which, at least temporarily, are not forced to participate in this accelerated motion, obey the law of inertia and remain behind until the floor of the car has caught up with them."

Imagine now that the cable of the elevator breaks and that the car, not equipped with an automatic safety device, is allowed to fall freely in the gravitational field of the earth. During this fall, the bodies inside the car undergo the same acceleration as the car itself, and, therefore, are unaccelerated relatively to the car. The observer inside the car might interpret this to indicate that the acceleration of the car has ceased and that his frame of reference is now an inertial system.

Conversely, we may consider an even more fantastic "conceptual experiment": The car is now placed in a region of space where the gravitational field vanishes. If the car is left alone and if it does not happen to rotate around an axis through its center of gravity, it will constitute an inertial system. A playful spirit decides to have some fun with the car; he begins to pull at the cable which is attached to the top of it, with a constant force. The car is no longer an inertial system. If a body inside the car is released from contact with other bodies, it will obey the law of inertia and remain behind the accelerated car, that is to say, it will "fall" to the floor. The man inside may mistake the apparent acceleration of his test bodies for the effects of a gravitational field.

The principle of general covariance. If we wish to develop a theory of gravitation which incorporates the principle of equivalence as an integral part, we must discard the concept of inertial frames of reference.

All frames of reference are equally suitable for a formulation of the laws of nature.

How shall we express this equivalence of all frames of reference in a mathematical form? Since we have always represented frames of reference by coördinate systems, and inertial systems in particular by Lorentzian coördinate systems, we conclude that we must no longer restrict ourselves to Lorentz coördinate transformations. To go over to linear coördinate transformations with arbitrary transformation coefficients is not sufficient, for the transition from one frame of reference to another which is accelerated can certainly not be represented by a coördinate transformation which is linear with respect to the time coördinate.

The history of the general theory of relativity has shown that *we have to consider the group of all continuous, differentiable coördinate transformations with non-vanishing Jacobian* in order to develop a relativistic theory of gravitation. That is why this theory of gravitation is called the *general theory of relativity.*

In the second part of Chapter V, we have considered arbitrary coördinate transformations in some detail. It is always possible to introduce in a Euclidean space a curvilinear coördinate system; all operations of vector and tensor calculus can be expressed in terms of curvilinear coördinates as well as in terms of orthogonal Cartesian coördinates. But if we introduce curvilinear coördinates and arbitrary transformations, then we must, for the formulation of many tensor relations, also introduce the metric tensor g_{mn}, the components of which are functions of the coördinates. A straight line, for instance, cannot be characterized in any simpler fashion than by the differential equations (5.99). Though we may formulate any conceivable geometrical relationship in a Euclidean space in terms of curvilinear coördinates, it is usually preferable to use Cartesian coördinates.

In a Euclidean space, Cartesian coördinate systems and orthogonal coördinate transformations permit the development of a restricted tensor calculus which contains fewer basic elements than does the general formalism. As the metric tensor degenerates into the form δ_{ik}, we can eliminate it as an independent element of the geometry. In a Riemannian space, the introduction of a Cartesian coördinate system is impossible. There we must apply the unrestricted formalism, which is covariant with respect to general coördinate transformations.

In the theory of gravitation, we encounter a similar situation. We can formulate the special theory of relativity in terms of curvilinear coördinate systems and general coördinate transformations in a four dimensional world. However, it is possible to introduce coördinate systems in which the components of the metric tensor take particular, constant values, $\eta_{\mu\nu}$, and where the components of the affine connection vanish. A formalism which is covariant only with respect to transformations leading from one coördinate system of this kind to another system of the same kind does not require the introduction of a number of geometrical concepts which are an integral part of a formalism which is covariant with respect to general coördinate transformations. The particular coördinate systems in which the components of the metric tensor take the constant values $\eta_{\mu\nu}$ are the inertial systems, and the coördinate transformations which lead from one inertial system to another inertial system are the Lorentz transformations.

The equivalence of all frames of reference must be represented by the equivalence of all coördinate systems. It must be impossible to introduce, in the presence of a gravitational field, the privileged Lorentz coördinate systems. Extending the terminology of Chapter V, we shall call a four dimensional Minkowski space *Riemannian* if it is impossible to introduce Lorentzian coördinate systems.

In a Riemannian space, the components of the metric tensor, the $g_{\mu\nu}$, are non-constant functions of the coördinates in all coördinate systems. A restriction to Lorentz transformations would not bring about a simplification of the formalism. The hypothesis that the geometry of physical space is represented best by a formalism which is covariant with respect to general coördinate transformations, and that a restriction to a less general group of transformations would not simplify that formalism, is called *the principle of general covariance*. It is the mathematical representation of the principle of equivalence. The development of a theory of gravitation which satisfies the principle of general covariance has furnished theoretical physics with the most satisfactory field theory which has so far been proposed.

The nature of the gravitational field. From the principle of equivalence, it might appear that gravitational fields are not real, that they are basically nothing more than "inertial forces." Everybody feels instinctively that that cannot be true.

If the man in the elevator car were to measure the direction of the accelerating force of the earth with great accuracy, he would find that the lines of force converge. This discovery would not enable him to separate gravitational field and inertial field, but it would tell him that the field was not wholly inertial. Because of the convergence of the lines of force, there is no frame of reference in which the gravitational field of the earth vanishes everywhere. The impossibility of introducing a frame of reference which has everywhere the properties of an inertial frame of reference is represented by the impossibility of introducing a Lorentzian coördinate system, that is, by the Riemannian character of space.

If it is impossible to introduce coördinate systems in which the components of the metric tensor assume constant, preassigned values, then the metric tensor itself becomes part of the field, and there must be field equations which restrict and determine, to some extent, the functional dependence of the $g_{\mu\nu}$ on the four world coördinates.

Then what is the physical significance of this tensor field $g_{\mu\nu}$? Let us consider a region of space in which the gravitational field vanishes. If we introduce a non-inertial coördinate system, free bodies will be accelerated with respect to the chosen coördinate system, although they move along straight world lines. If we express the law of inertia in terms of an arbitrary, curvilinear coördinate system, the equations of motion, according to eq. (5.99), are

$$\frac{dU^\alpha}{d\tau} = -\begin{Bmatrix} \alpha \\ \iota\kappa \end{Bmatrix} U^\iota U^\kappa, \tag{10.9}$$

where the $\left\{ {\alpha \atop \iota\kappa} \right\}$ are linear in the first derivatives of the $g_{\mu\nu}$,

$$\begin{Bmatrix} \alpha \\ \iota\kappa \end{Bmatrix} = \tfrac{1}{2} g^{\alpha\beta}(g_{\iota\beta,\kappa} + g_{\kappa\beta,\iota} - g_{\iota\kappa,\beta}). \tag{10.10}$$

The $g_{\mu\nu}$ appear, in a manner of speaking, as the potentials of the "inertial field." It is, therefore, reasonable to assume that, in the presence of a gravitational field, the $g_{\mu\nu}$ are again the potentials which determine the accelerations of free bodies: in other words, that the $g_{\mu\nu}$ are the potentials of the gravitational field. These gravitational potentials must satisfy differential equations which resemble the Laplacian or Poisson's equation in four dimensions. We shall find later that there is only one particular set of equations of this type which is covariant with respect to general coördinate transformations.

At any rate, we find that the theory of gravitation will have to deal with spaces which are not "quasi-Euclidean," that is, in which no inertial coördinate systems can be introduced. Before we can continue our discussion of the gravitational field, we must develop the geometry of Riemannian spaces somewhat further than we did in Chapter V. In particular, we shall have to find a mathematical criterion which tells us whether a space is Euclidean or not.

CHAPTER XI

The Riemann-Christoffel Curvature Tensor

The characterization of Riemannian spaces. According to the definition which we gave in Chapter V, a Euclidean space is one in which it is possible to introduce Cartesian coördinates; all other spaces are non-Euclidean.

Even if we knew, in a specific case, the components of the metric tensor as functions of a particular coördinate system, we could not, obviously, try out all the conceivable coördinate transformations to find whether some of them lead to Cartesian coördinates. We need a criterion which can be applied in a systematic fashion to determine whether or not a space is Euclidean.

The non-Euclidean spaces which we encounter in our daily experience are curved two dimensional surfaces imbedded in our three dimensional space. It might appear that their geometric properties cannot be characterized without taking into account their relationship to the imbedding space. Actually, at least the Euclidean or non-Euclidean character of such a two dimensional space is independent of its relationship to the three dimensional space. Let us consider, for instance, a plane, which we shall represent by a sheet of graph paper. The ruling on the paper represents a Cartesian coördinate system, so there is no doubt that the plane is Euclidean. Let us now change the relationship of the two dimensional space to the imbedding space by rolling up the paper; the ruling will still retain all the characteristics of a Cartesian coördinate system. The distance between two infinitesimally near points on the paper is given by the equation

$$ds^2 = dx^2 + dy^2,$$

both before and after we have rolled up the paper. In other words, the metric tensor has the components

$$g_{ik} = \delta_{ik} . \tag{11.1}$$

Furthermore, any line on the paper which was straight before we rolled the paper remains the shortest line which connects two points on the paper and which lies wholly in the two dimensional space.

The Euclidean character of a space depends only on the metric. And we must develop a method by which we can distinguish a Euclidean from a non-Euclidean metric.

The integrability of the affine connection. To find such a method, we shall return to the concept of the parallel displacement of vectors, which was introduced in Chapter V. An affine connection with the components Γ^l_{ik} enables us to displace a vector along a curve uniquely according to the differential laws

$$\left.\begin{aligned} \delta a^l &= -\Gamma^l_{ik} a^i \delta\xi^k, \\ \delta b_i &= +\Gamma^l_{ik} b_l \delta\xi^k. \end{aligned}\right\} \tag{11.2}$$

A metric g_{ik} determines a particular affine connection, $\{{l \atop ik}\}$, with the components

$$\{{l \atop ik}\} = \tfrac{1}{2} g^{ls}(g_{is,k} + g_{ks,i} - g_{ik,s}). \tag{11.3}$$

If the affine connection has as its components the Christoffel symbols, the result of the parallel displacement of a vector is independent of whether the law (11.2) is applied to its covariant or its contravariant representation.

Let us displace a vector parallel along a *closed* curve (Fig. 9), until it returns to the starting point. Then we shall find either that the vector obtained is identical with the vector with which we started, or that it is a different vector. If it is the same vector, regardless of the choice of the initial vector and regardless of the shape of the closed curve, the affine connection is said to be *integrable*. In such a case, we can speak of "distant parallelism," which means that when we displace a vector at a point P_1 parallel to itself along some curve to another point P_2, the components of the vector at P_2 do not depend on the choice of the path of displacement between P_1 and P_2. When the affine connection is integrable, a vector at one point generates a whole field of parallel vectors throughout the space.

Euclidicity and integrability. If the components of the affine connection are connected with the metric tensor by eqs. (11.3), we shall find that the Euclidicity of a space is directly related to the integrability of the affine connection.

When a space is Euclidean, we can introduce a Cartesian coördinate system, in which the components of the metric tensor are constants,

$$g_{ik} = \delta_{ik}\,. \tag{11.4}$$

According to eqs. (11.3), the $\left\{ {l \atop ik} \right\}$ vanish in such a coördinate system, and the δa^l, δb_i of eqs. (11.2) vanish, too. The parallel vectors have the same components at all points; this affine connection is certainly

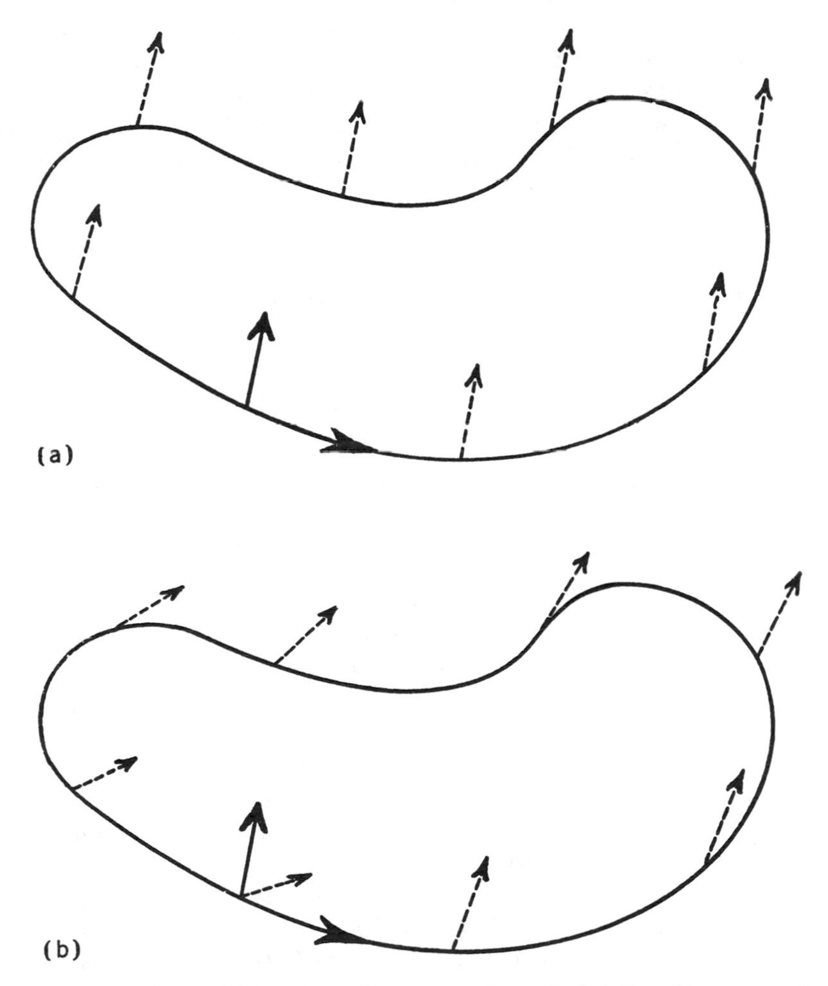

Fig. 9. The integrability of an affine connection. In (*a*), the affine connection is integrable; in (*b*), it is not.

integrable. Integrability is, by its definition, an invariant property of the affine connection, independent of the choice of the coördinate system. We conclude, therefore, that *the affine connection of a Euclidean space is always integrable.*

Conversely, we shall show, by actual construction, that we can always find a Cartesian coördinate system when the affine connection (11.3) is integrable. This statement, to be true, requires a slight generalization of our definition of a Euclidean space. So far, we have defined as Euclidean a space in which we can, by means of a real coördinate transformation, introduce a coördinate system in which the metric tensor g_{ik} takes the constant values δ_{ik}. According to this definition, the four dimensional Minkowski space is not Euclidean. The essential difference between a Euclidean space and the Minkowski space is that in a Euclidean space the quadratic form of the coördinate differentials is positive definite,

$$ds^2 = dx_i dx_i \geqslant 0, \tag{11.5}$$

for arbitrary real values of dx_i; in the Minkowski space, however, with the quadratic form

$$d\tau^2 = (dx^4)^2 - \frac{1}{c^2} dx^i dx^i, \tag{11.6}$$

we found that $d\tau^2$ can take negative as well as positive values, and the interval can be "space-like" or "time-like" (see Chapter IV, p. 41). It is, therefore, impossible to carry out a real coördinate transformation leading from eq. (11.5) to eq. (11.6).

But in their analytical properties, the forms (11.5) and (11.6) are very similar. We pointed out, at the end of Chapter V, that the $\{{}_{ik}^{\,l}\}$ belonging to the metric form (11.6) vanish; the components of a parallel displaced vector are, therefore, constants, and the parallel displacement is integrable. This will be true, in general, whenever it is possible to introduce coördinates in space so that the components of the metric tensor become constants. In such a case, we shall call the space *flat*. Flatness is the generalization of Euclidicity for spaces in which the metric is not necessarily positive definite.

With this correction in mind, we assert now that *whenever the parallel displacement defined by eqs.* (11.2), (11.3) *is integrable, the space is flat; that is, there exists a coördinate system in which the metric form is*

$$ds^2 = \sum_i \epsilon_i dx^i dx^i, \qquad \epsilon_i = \pm 1. \tag{11.7}$$

The proof will be carried out in two steps. If the components of the affine connection are symmetric in their subscripts, the integrability of the affine connection enables us to construct a coördinate system in which the components of the affine connection vanish. This fact is independent of the existence of a metric and will be proved without resorting to eq. (11.3). Then, if a metric is defined, the vanishing of the $\left\{ {l \atop ik} \right\}$ is equivalent to the constancy of the components of the metric tensor.

Let us consider at a point P a set of n covariant vectors (n being the number of dimensions), $\overset{s}{b}_i$, which are linearly independent of each other; that is, which satisfy the inequality

$$\Delta \equiv \delta^{i_1 \cdots i_n} \overset{1}{b}_{i_1} \cdots \overset{n}{b}_{i_n} \neq 0, \tag{11.8}$$

where $\delta^{i_1 \cdots i_n}$ is the contravariant Levi-Civita tensor density. Let us now displace all the n vectors $\overset{s}{b}_i$ along the same path. The change in Δ is

$$\left.\begin{aligned} \delta\Delta &= \delta^{i_1 \cdots i_n}[\Gamma^k_{i_1 s} \overset{1}{b}_k \overset{2}{b}_{i_2} \cdots \overset{n}{b}_{i_n} + \cdots + \Gamma^k_{i_n s} \overset{1}{b}_{i_1} \cdots \overset{n}{b}_k]\,\delta\xi^s \\ &= \overset{1}{b}_{i_1} \cdots \overset{n}{b}_{i_n}[\delta^{k i_2 \cdots i_n} \Gamma^{i_1}_{ks} + \cdots + \delta^{i_1 \cdots k} \Gamma^{i_n}_{ks}]\delta\xi^s. \end{aligned}\right\} \tag{11.9}$$

This expression may be considerably simplified. First of all, the bracket on the right-hand side of eq. (11.9) is skewsymmetric in all indices $i_1 \cdots i_n$. For, if we exchange, for instance, the two indices i_1 and i_2, the bracket goes over into

$$\left.\begin{aligned} &\delta^{k i_1 \cdots i_n} \Gamma^{i_2}_{ks} + \delta^{i_2 k \cdots i_n} \Gamma^{i_1}_{ks} + \cdots + \delta^{i_2 i_1 \cdots k} \Gamma^{i_n}_{ks} \\ &\qquad = -[\delta^{i_1 k \cdots i_n} \Gamma^{i_2}_{ks} + \delta^{k i_2 \cdots i_n} \Gamma^{i_1}_{ks} + \cdots + \delta^{i_1 i_2 \cdots k} \Gamma^{i_n}_{ks}]. \end{aligned}\right\} \tag{11.10}$$

Second, k can assume only the same value as the displaced i_s, because, for all other values, the Levi-Civita tensor density component vanishes. We can, therefore, replace the square bracket in eq. (11.9) by the expression

$$\delta^{k i_2 \cdots i_n} \Gamma^{i_1}_{ks} + \cdots + \delta^{i_1 \cdots k} \Gamma^{i_n}_{ks} = \delta^{i_1 \cdots i_n} \Gamma^k_{ks}, \tag{11.11}$$

and eq. (11.9) goes over into

$$\delta\Delta = \Delta \cdot \Gamma^k_{ks} \cdot \delta\xi^s. \tag{11.9a}$$

Along any path, Δ satisfies a linear, homogeneous differential equation of the first order. Therefore, it cannot vanish anywhere on that path if it does not vanish everywhere.

We conclude that, *if n vectors $\overset{s}{b}_i$ are linearly independent, they preserve that property under parallel displacements.*

We shall assume now that the affine connection is symmetric in its subscripts and integrable; then each one of the vectors $\overset{s}{b}_i$ generates a field of parallel vectors. Each of these fields satisfies differential equations of the form

$$\overset{s}{b}_{i,k} = \Gamma^l_{ik}\overset{s}{b}_l. \tag{11.12}$$

The right-hand side is symmetric in the subscripts i and k. Therefore, the antisymmetric derivatives of the $\overset{s}{b}_i$ vanish,

$$\overset{s}{b}_{i,k} - \overset{s}{b}_{k,i} = 0. \tag{11.13}$$

From this equation, we conclude that each of the n fields $\overset{s}{b}_i$ is a gradient field, and that there are n scalars $\overset{s}{b}$, so that

$$\overset{s}{b}_i = \overset{s}{b}_{,i}. \tag{11.14}$$

These n scalars $\overset{s}{b}$ may be considered as the n coördinates of a new coördinate system. Because of eq. (11.8), the Jacobian of the coördinate transformation does not vanish. Now we can show that, *in the new coördinate system, the* Γ^l_{ik} *vanish.*

Let us transform the components of the affine connection according to eq. (5.81),

$$\Gamma^{*l}_{ik} = \frac{\partial\xi^a}{\partial\xi^{*i}}\frac{\partial\xi^b}{\partial\xi^{*k}}\left(\frac{\partial\xi^{*l}}{\partial\xi^c}\Gamma^c_{ab} - \frac{\partial^2\xi^{*l}}{\partial\xi^a\,\partial\xi^b}\right). \tag{11.15}$$

According to eqs. (11.14), the derivatives $\dfrac{\partial\xi^{*l}}{\partial\xi^c}$ are the vector components $\overset{l}{b}_c$, so that the parenthesis of eq. (11.15) is

$$\frac{\partial\xi^{*l}}{\partial\xi^c}\Gamma^c_{ab} - \frac{\partial^2\xi^{*l}}{\partial\xi^a\,\partial\xi^b} = \overset{l}{b}_c\Gamma^c_{ab} - \overset{l}{b}_{a,b}. \tag{11.16}$$

On account of eq. (11.12), this expression vanishes, and, therefore, the Γ^{*l}_{ik} of eq. (11.15) vanish, too.

Returning now to a consideration of the metric tensor, we can solve the eqs. (11.3) with respect to the derivatives of g_{mn}. We add the two equations

$$\tfrac{1}{2}(g_{ik,l} + g_{il,k} - g_{kl,i}) = \begin{Bmatrix} s \\ kl \end{Bmatrix} g_{si} \qquad \text{and}$$

$$\tfrac{1}{2}(g_{ki,l} + g_{kl,i} - g_{il,k}) = \begin{Bmatrix} s \\ il \end{Bmatrix} g_{sk}, \tag{11.17}$$

and obtain:

$$g_{ik,l} = \begin{Bmatrix} s \\ kl \end{Bmatrix} g_{si} + \begin{Bmatrix} s \\ il \end{Bmatrix} g_{sk}. \tag{11.18}$$

If the $\begin{Bmatrix} s \\ kl \end{Bmatrix}$ *vanish, the* g_{ik} *are constant.*

To reduce the constant g_{ik} to the form (11.7) is a purely algebraic problem. It is solved by a standard orthogonalization and normalization process and is of no particular interest to us. Any space in which the components of the metric tensor are constants is *ipso facto* flat.

The criterion of integrability. When the affine connection of a space is symmetric and integrable, the equations of parallel displacement, (11.2), may be considered not only as ordinary differential equations along a given path, but as partial differential equations for a whole vector field. We may write them in the form

$$a^n{}_{,i} = -\Gamma^n_{si}a^s, \tag{11.19}$$

and analogous equations for covariant vector fields. These equations are overdetermined: they are n^2 equations for the n components of a vector. To have solutions, they must satisfy differential identities. The form of these identities is well known. If we differentiate eqs. (11.19) with respect to a coördinate ξ^k, we obtain

$$a^n{}_{,ik} = -\Gamma^n_{si,k}a^s - \Gamma^n_{si}a^s{}_{,k} = (-\Gamma^n_{li,k} + \Gamma^n_{si}\Gamma^s_{lk})a^l. \tag{11.20}$$

By subtracting the same expression, with the indices i and k exchanged, we obtain the conditions which must be satisfied so that the sequence of the differentiations is without effect:

$$0 = -(\Gamma^n_{li,k} - \Gamma^n_{lk,i} - \Gamma^n_{si}\Gamma^s_{lk} + \Gamma^n_{sk}\Gamma^s_{li})a^l. \tag{11.21}$$

As the values of the a^l at one point may be chosen arbitrarily, we obtain the conditions of integrability,

$$R_{ikl\cdot}{}^{n} \equiv \Gamma^n_{li,k} - \Gamma^n_{lk,i} - \Gamma^n_{si}\Gamma^s_{lk} + \Gamma^n_{sk}\Gamma^s_{li} = 0. \tag{11.22}$$

These conditions are not only necessary, but also sufficient. The proof runs along the same lines as the proof of the theorem that a covariant vector field is a gradient field when and only when its skewsymmetric derivatives vanish (see Problem 14, Chapter V).

The commutation law for covariant differentiation, the tensor character of $R_{ikl\cdot}{}^{n}$. The vanishing of the expression $R_{ikl\cdot}{}^{n}$ of eq. (11.22) is equivalent to the integrability of the affine connection, and, therefore, must be an invariant property. There are, of course, spaces which are not flat and in which the $R_{ikl\cdot}{}^{n}$ are different from zero (for example, the spherical surface). How do the quantities $R_{ikl\cdot}{}^{n}$ transform?

To answer this question, we shall derive a tensor equation in which the $R_{ikl\cdot}{}^{n}$ appear: the commutation law for covariant differentiation. We compute the expression

$$A^n{}_{;ik} - A^n{}_{;ki}\,.$$

According to the definition of covariant differentiation, we have

$$A^n{}_{;i} = A^n{}_{,i} + \Gamma^n_{li} A^l; \tag{11.23}$$

and a second covariant differentiation yields the expression

$$\left.\begin{aligned} A^n{}_{;ik} &= (A^n{}_{;i})_{,k} + \Gamma^n_{sk} A^s{}_{;i} - \Gamma^s_{ik} A^n{}_{;s} \\ &= \underline{A^n{}_{,ik}} + \Gamma^n_{li,k} A^l + \underline{\Gamma^n_{li} A^l{}_{,k} + \Gamma^n_{sk} A^s{}_{,i}} \\ &\quad + \Gamma^n_{sk} \Gamma^s_{li} A^l - \underline{\Gamma^s_{ik} A^n{}_{,s}} - \underline{\Gamma^s_{ik} \Gamma^n_{ls} A^l}\,. \end{aligned}\right\} \tag{11.24}$$

Let us assume that the components of the affine connection are symmetric in their subscripts. When we exchange the indices i and k and subtract the resulting equations from (11.24), the underscored terms cancel, because of their symmetry in i and k, and we are left with the relation

$$A^n{}_{;ik} - A^n{}_{;ki} = R_{ikl\cdot}{}^{n} A^l. \tag{11.25}$$

This equation is the commutation law for covariant differentiation. In a flat space, covariant differentiations commute like ordinary partial differentiations; we could have predicted this, for in a flat space there are coördinate systems with respect to which covariant and ordinary differentiation are the same.

When the space is not flat, the commutator depends only on the undifferentiated vector.

The commutation law for covariant differentiation of covariant vectors is

$$A_{l;ik} - A_{l;ki} = -R_{ikl\cdot}{}^{n} A_n\,. \tag{11.26}$$

The left-hand sides of eqs. (11.25) and (11.26) transform as tensors. The right-hand sides are, therefore, tensors too; and, as the factors A^l and A^n are arbitrary, it follows that *the $R_{ikl\cdot}{}^{n}$ themselves are the components of a tensor. The tensor*

$$R_{ikl\cdot}{}^{n} = \Gamma^n_{li,k} - \Gamma^n_{lk,i} - \Gamma^n_{si} \Gamma^s_{lk} + \Gamma^n_{sk} \Gamma^s_{li} \tag{11.27}$$

is called the (Riemann-Christoffel) curvature tensor.[1]

---

[1] In this book, the index notation of Levi-Civita has been adopted. Unfortunately, there is no standard for the writing of indices of the curvature tensor. Many authors write our last index first, our third index second, and our first and second indices as third and fourth indices, respectively. The notation in this book will consistently follow the definition given by eq. (11.27).

Properties of the curvature tensor. The curvature tensor is defined together with any affine connection. However, it has certain symmetry properties only on the condition that the components of the affine connection are Christoffel symbols (11.3), which are associated with a metric. We shall first consider those properties of the curvature tensor which are independent of the relationship of the Γ^l_{ik} to a metric.

(1) *$R_{ikl\cdot}{}^{n}$ is skewsymmetric in the indices i and k,*

$$R_{ikl\cdot}{}^{n} + R_{kil\cdot}{}^{n} = 0. \tag{11.28}$$

This relationship is satisfied by the expression (11.27), regardless of any symmetry properties of the Γ^l_{ik}.

When the components of the affine connection are symmetric in their subscripts, the curvature tensor satisfies another symmetry law and, furthermore, a set of differential identities.

(2) *When we rotate the first three indices cyclically, the sum of the components vanishes,*

$$R_{ikl\cdot}{}^{n} + R_{kli\cdot}{}^{n} + R_{lik\cdot}{}^{n} = 0. \tag{11.29}$$

The proof is carried out by straightforward computation of the expression (11.29).

(3) We obtain the differential identities as follows: We differentiate eqs. (11.26) covariantly with respect to a new coördinate,

$$A_{s;ikl} - A_{s;kil} = -R_{iks\cdot}{}^{n}{}_{;l}A_n - R_{iks\cdot}{}^{n}A_{n;l}\,, \tag{11.30}$$

rotate the three indices i, k, l cyclically, and add. The result is

$$\left.\begin{aligned}(A_{s;ikl} - A_{s;ilk}) + (A_{s;kli} - A_{s;kil}) + (A_{s;lik} - A_{s;lki})\\ = -A_n(R_{iks\cdot}{}^{n}{}_{;l} + R_{kls\cdot}{}^{n}{}_{;i} + R_{lis\cdot}{}^{n}{}_{;k})\\ -(R_{iks\cdot}{}^{n}A_{n;l} + R_{kls\cdot}{}^{n}A_{n;i} + R_{lis\cdot}{}^{n}A_{n;k}).\end{aligned}\right\} \tag{11.31}$$

The parentheses on the left-hand side are all commutators of the covariant differentiation. As can be readily shown, the commutation law for the covariant differentiation of a covariant tensor of rank 2 is

$$B_{lm;ik} - B_{lm;ki} = -R_{ikl\cdot}{}^{n}B_{nm} - R_{ikm\cdot}{}^{n}B_{ln}\,. \tag{11.32}$$

Applying this law to each of the parentheses on the left-hand side, we obtain, for instance, for the first parenthesis,

$$A_{s;ikl} - A_{s;ilk} = -R_{kls\cdot}{}^{n}A_{n;i} - R_{kli\cdot}{}^{n}A_{s;n}\,. \tag{11.33}$$

When we substitute these expressions in eq. (11.31), the first term on the right-hand side of eq. (11.33) cancels with a term in the last parenthesis on the right-hand side of eq. (11.31). The second term on the right-hand side of eq. (11.33) cancels together with its cyclically rotated analogues, because of eq. (11.29); and we are left with the equation

$$A_n(R_{iks\cdot}{}^{n}{}_{;l} + R_{kls\cdot}{}^{n}{}_{;i} + R_{lis\cdot}{}^{n}{}_{;k}) \equiv 0. \tag{11.34}$$

The vector A_n is arbitrary; therefore, *the curvature tensor must satisfy the identities*

$$R_{iks\cdot}{}^{n}{}_{;l} + R_{kls\cdot}{}^{n}{}_{;i} + R_{lis\cdot}{}^{n}{}_{;k} \equiv 0. \tag{11.35}$$

They are called the Bianchi identities.

The covariant form of the curvature tensor. So far, we have not made use of a metric. When a metric is defined, and when the Γ^l_{ik} are connected with the metric through eqs. (11.3), the curvature tensor satisfies additional algebraic identities. The purely covariant curvature tensor is obtained by lowering the index n of eq. (11.27),

$$R_{iklm} = R_{ikl\cdot}{}^{n}g_{mn}\,. \tag{11.36}$$

This covariant curvature tensor can be expressed in terms of the "Christoffel symbols of the first kind," which are the components of the affine connection, $\left\{ {s \atop ik} \right\}$, multiplied by g_{ls},

$$[ik, l] = g_{ls}\left\{ {s \atop ik} \right\} = \tfrac{1}{2}(g_{kl,i} + g_{il,k} - g_{ik,l}). \tag{11.37}$$

The first terms of R_{iklm} can be written in the form

$$\left.\begin{aligned} g_{mn}\left\{ {n \atop li} \right\}_{,k} &= [li, m]_{,k} - \left\{ {n \atop li} \right\} g_{mn,k} \\ &= [li, m]_{,k} - \left\{ {n \atop li} \right\} ([nk, m] + [mk, n]). \end{aligned}\right\} \tag{11.38}$$

Substituting these expressions in eq. (11.36), we obtain:

$$R_{iklm} = [li, m]_{,k} - [lk, m]_{,i} + g^{rs}([mi, r][lk, s] - [mk, s][li, r]). \tag{11.39}$$

Once we have obtained the covariant curvature tensor in this form, we can verify the following two algebraic identities, in addition to identities (1) and (2):

(4) *The covariant curvature tensor is skewsymmetric in its last index pair,*

$$R_{iklm} + R_{ikml} \equiv 0. \tag{11.40}$$

The parenthesis of eq. (11.39) is obviously skewsymmetric in m and l.

The first two terms contain only second derivatives of the components of the metric tensor, in the combination

$$[li, m]_{,k} - [lk, m]_{,i} = \tfrac{1}{2}[(g_{mi,lk} - g_{li,mk}) + (g_{lk,mi} - g_{mk,li})].$$

This expression is also skewsymmetric in m and l.

(5) *The covariant curvature tensor is symmetric in its two index pairs,*

$$R_{iklm} \equiv R_{lmik} \,. \tag{11.41}$$

This relation can be verified exactly like eq. (11.40).

In the remainder of this chapter, we shall consider only metric spaces, where the components of the affine connection are given by eq. (11.3).

Contracted forms of the curvature tensor. From the curvature tensor we can obtain tensors of lower rank by contraction. We can form the tensors $R_{ikl\cdot}^{\ \ \ i}$, $R_{ikl\cdot}^{\ \ \ k}$, $R_{ikl\cdot}^{\ \ \ n} g^{il}$, and $R_{ikl\cdot}^{\ \ \ n} g^{kl}$; all other contracted tensors vanish because of the skewsymmetry of R_{iklm} in (i, k) and in (l, m).

The four tensors of rank 2 listed above are all identical (except for the sign), because they can be obtained from each other by changing the sequence of indices in one pair, or by changing the sequence of the two pairs. *It is customary to designate the contracted tensor* $R_{ikl\cdot}^{\ \ \ i}$ *by* R_{kl}. *This tensor* R_{kl} *is symmetric in its two indices*, because of the symmetry properties of $R_{ikl\cdot}^{\ \ \ n}$. It is known as the *Ricci tensor.*

By contracting R_{kl} once more, we obtain the curvature scalar, R,

$$R = g^{kl}R_{kl}\,, \qquad R_{kl} = R_{ikl\cdot}^{\ \ \ i}\,. \tag{11.42}$$

Written in terms of the Christoffel symbols, R_{kl} takes the form

$$R_{kl} = \begin{Bmatrix} s \\ ls \end{Bmatrix}_{,k} - \begin{Bmatrix} s \\ lk \end{Bmatrix}_{,s} - \begin{Bmatrix} r \\ sr \end{Bmatrix}\begin{Bmatrix} s \\ lk \end{Bmatrix} + \begin{Bmatrix} s \\ rk \end{Bmatrix}\begin{Bmatrix} r \\ ls \end{Bmatrix}. \tag{11.43}$$

Except for the first term, the symmetry of each term with respect to the indices k and l is obvious. As for the first term, $\begin{Bmatrix} s \\ ls \end{Bmatrix}$ can be expressed in the form

$$\begin{Bmatrix} s \\ ls \end{Bmatrix} = \tfrac{1}{2}g^{rs}(g_{rl,s} + g_{rs,l} - g_{ls,r}).$$

The first and the third term in the parenthesis, taken together, are skewsymmetric in r and s and vanish when multiplied by g^{rs}. There remains only the second term,

$$\begin{Bmatrix} s \\ ls \end{Bmatrix} = \tfrac{1}{2}\, g^{rs} g_{rs,l} = \tfrac{1}{2}\,\frac{g_{,l}}{g} = (\log \sqrt{g})_{,l}\,, \qquad g = |\, g_{ab} \,|. \tag{11.44}$$

The first term of eq. (11.43) is, therefore,

$$\left\{ {s \atop ls} \right\}_{,k} = (\log \sqrt{g})_{,lk}, \tag{11.45}$$

and is symmetric in l and k.

The contracted Bianchi identities. By contracting the Bianchi identities (11.35) twice, we obtain identities which contain only the contracted curvature tensor. Contracting eq. (11.35) first with respect to i and n, we get

$$R_{ks;l} + R_{kls\cdot;r}^{\ \ \ \ r} + R_{lrs\cdot;k}^{\ \ \ \ r} \equiv 0.$$

By changing in the last term the sequence of l and r, we obtain, because of eq. (11.28),

$$R_{ks;l} + R_{kls\cdot;r}^{\ \ \ \ r} - R_{ls;k} \equiv 0$$

or

$$R_{k\cdot;l}^{\ s} + R_{kl\cdot\cdot;r}^{\ \ sr} - R_{l\cdot;k}^{\ s} \equiv 0.$$

As the next step, we change the sequence of the contravariant indices s and r in the second term,

$$R_{kl\cdot\cdot}^{\ \ sr} = -R_{kl\cdot\cdot}^{\ \ rs},$$

and contract with respect to the indices k and s. We obtain

$$R_{;l} - 2R_{l\cdot;r}^{\ r} \equiv 0$$

or

$$(R^{ls} - \tfrac{1}{2}g^{ls}R)_{;s} \equiv 0. \tag{11.46}$$

The expression in the parenthesis is often denoted by G^{ls},

$$G^{ls} = R^{ls} - \tfrac{1}{2}g^{ls}R. \tag{11.47}$$

The number of algebraically independent components of the curvature tensor. The components of the covariant curvature tensor, R_{iklm}, satisfy the algebraic relations (11.28), (11.29) (both with the index n lowered), (11.40), and (11.41). The number of the algebraically independent components is thereby reduced, and we shall show in this section that their number in an n dimensional space is

$$N = \tfrac{1}{12} n^2(n^2 - 1). \tag{11.48}$$

In a two dimensional space, the curvature tensor has only one significant component; the scalar R is already sufficient to characterize the curvature completely.[2] In a three dimensional space, there are six algebraically independent components. This is also the number of independent components of the contracted tensor, R_{kl}; the contracted tensor characterizes the uncontracted tensor completely.[3] In a four dimensional space, N equals 20, while the contracted curvature tensor has only 10 independent components. *Unless a space has at least four components, its curvature is completely characterized by the contracted forms R_{kl}.*

We shall now derive eq. (11.48). We shall divide the components of R_{iklm} into three groups: those components where each index of the first pair has the same value as an index of the second pair, such as R_{1212}; those where only one index value is represented twice, such as R_{1213}; and those where all four indices are different, R_{1234}, and so forth. Obviously, not all four indices can be equal.

In the first type, with only two indices different, the first and the second index pairs must be identical, as the two indices of a pair must be different (because of eqs. (11.28) and (11.40)). These components are of the type R_{ikik} (do not sum!). R_{ikki} differs from R_{ikik} only with respect to the sign. There are as many components R_{ikik} as there are different index pairs (i, k), with $i \neq k$.

The index i can take n different values. k is different from i, therefore, for any given i, k can take only $(n - 1)$ different values. Since the sequence of i and k is of no consequence, we must divide the product $n(n - 1)$ by 2. The number of different index pairs (i, k), $i \neq k$ is, therefore,

$$N_P = \tfrac{1}{2}n(n - 1), \tag{11.49}$$

and the number of algebraically independent components with two different indices is also

$$N_I = \tfrac{1}{2}n(n - 1). \tag{11.50}$$

The cyclic identities (11.29) do not further decrease this number, because they are independent of the other algebraic identities only when

---

[2] It can be shown that in a two dimensional space the curvature tensor depends on R, as follows:

$$R_{ikl\cdot}^{\cdot\cdot\cdot n} = \tfrac{1}{2}(\delta_i^n g_{kl} - \delta_k^n g_{il})R.$$

[3] In a three dimensional space, the $R_{ikl\cdot}^{\cdot\cdot\cdot n}$ depend on the R_{kl}, as follows:

$$R_{ikl\cdot}^{\cdot\cdot\cdot n} = \delta_i^n R_{kl} - \delta_k^n R_{il} + g_{kl} R_{i\cdot}^{\cdot n} - g_{il} R_{k\cdot}^{\cdot n} - \tfrac{1}{2}(\delta_i^n g_{kl} - \delta_k^n g_{il})R.$$

all four indices are different. If two of the four indices i, k, l, m are equal, eqs. (11.29) are either of the form

$$R_{ikli} + \langle R_{klii} \rangle + R_{liki} = 0,$$

or

$$R_{ikim} + R_{kiim} + \langle R_{iikm} \rangle = 0.$$

Either equation is satisfied because of eqs. (11.28), (11.40), and (11.41).

Let us now consider the second group of the components, those with three different indices. All these components can be brought into the form R_{ikim} by applying eqs. (11.28) and (11.40). There are n different choices for the value of i. Of the remaining $(n - 1)$ numbers, we must pick two different ones for k and m. According to eq. (11.49), there are $\frac{1}{2}(n - 1)(n - 2)$ different choices for (k, m), and the number of algebraically independent components of the second type is

$$N_{II} = \tfrac{1}{2}n(n - 1)(n - 2). \tag{11.51}$$

Again the cyclic identities do not further decrease this number.

In the third group, all four indices are different. We may first pick the first index pair in $\frac{1}{2}n(n - 1)$ different ways. Out of the remaining $(n - 2)$ values, we must choose the second index pair, which can be done in $\frac{1}{2}(n - 2)(n - 3)$ different ways. Because of eq. (11.41), the sequence of the two pairs does not matter; we must, therefore, divide once more by 2. There are, then,

$$\tfrac{1}{2} \cdot \tfrac{1}{2}n(n - 1) \cdot \tfrac{1}{2}(n - 2)(n - 3)$$

different ways of picking two completely different index pairs.

In this case, the number of algebraically independent components is further decreased by the existence of the identities (11.29). Of the three components R_{1234}, R_{2314}, and R_{3124}, for instance, each has a different combination of index pairs, but any one can be expressed in terms of the other two. The number of algebraically independent components of R_{iklm} with four different indices is, therefore,

$$N_{III} = \tfrac{2}{3} \cdot \tfrac{1}{2} \cdot \tfrac{1}{2}n(n - 1) \cdot \tfrac{1}{2}(n - 2)(n - 3) = \tfrac{1}{12}n(n - 1)(n - 2)(n - 3). \tag{11.52}$$

The total number of algebraically independent components of R_{iklm} is the sum of the three numbers N_I, N_{II}, and N_{III}. This is the expression (11.48).

CHAPTER XII

The Field Equations of the General Theory of Relativity

The ponderomotive equations of the gravitational field. In this chapter, we shall formulate the field equations and the ponderomotive equations of the gravitational field.

Unfortunately, we cannot treat the ponderomotive equations fully at this point. We must, for the present, restrict ourselves to the motions of small particles which contribute only negligible amounts to the field.

The principle of equivalence determines the law of motion of such particles. Their motion under the influence of the gravitational field must be indistinguishable from inertial motion, that is, their paths are geodesic world lines,

$$\frac{d^2\xi^\mu}{d\tau^2} = -\begin{Bmatrix} \mu \\ \rho\sigma \end{Bmatrix} \frac{d\xi^\rho}{d\tau}\frac{d\xi^\sigma}{d\tau}, \qquad d\tau^2 = g_{\rho\sigma}\,d\xi^\rho\,d\xi^\sigma. \tag{12.1}$$

This law of motion is more involved than, for instance, the law of motion of electrically charged particles in the special theory of relativity. While eq. (7.49) is linear in the field intensities, eq. (12.1) is not linear in the $g_{\mu\nu}$ and their derivatives. This nonlinearity is characteristic of equations which are covariant with respect to general coördinate transformations; it is, thus, a consequence of the equivalence principle.

The representation of matter in the field equations. Before we set up the differential equations for the gravitational field, we shall briefly consider the representation of gravitating matter in the equations and their solutions.

Just as the gravitational field is generated by gravitating matter, so is the electromagnetic field generated by electric charges. These charges can be represented in two entirely different ways. When Maxwell set up his field equations, the atomic character of electric charges was not yet known. Maxwell assumed that the charge was distributed continuously throughout a charged insulating body, or on the surface of a conductor, and so forth. Correspondingly, he introduced the concepts of charge density and current density. These four densities are represented by our world vector I^ρ, which enters into the system of Maxwell's field equations.

In a similar fashion, we can set up field equations in which gravitating matter is represented by the world tensor $P^{\mu\nu}$, the stress-energy tensor. Ten differential expressions of the second order, which are formed from the components of the metric tensor, must equal the ten quantities $P^{\mu\nu}$. These ten expressions must, of course, transform like the $P^{\mu\nu}$, that is, as the components of a symmetric tensor of rank 2. Only then will the field equations be covariant.

When physicists discovered that electric charges were necessarily connected with small, individual particles, electrons and ions (and today we can add, mesons), Lorentz described the electromagnetic properties of matter by means of a new model. According to his point of view, the greatest part of the space is free of electric charges. The electric charges are point-like and constitute singularities of the field. Outside the point charges, there is the electromagnetic field, which satisfies Maxwell's equations for charge-free space. At the location of each point charge, the equations are not satisfied—each point charge constitutes a singularity of the field. Although the field equations are not satisfied at certain points, the charge contained in each of these singular regions is conserved, because the field equations are satisfied everywhere around the singularity. If we enclose a singularity by a closed surface, then the charge in the interior is given by the integral

$$\epsilon = \frac{1}{4\pi} \oint_S (\mathbf{E} \cdot \mathbf{dS}),$$

and the change of ϵ is determined by the integral

$$\frac{d\epsilon}{dt} = \frac{1}{4\pi} \oint_S \left(\frac{\partial \mathbf{E}}{\partial t} \cdot \mathbf{dS} \right).$$

As long as the field equations are satisfied everywhere on S (that is, as long as no electric current flows through S), we may substitute for $\frac{\partial \mathbf{E}}{\partial t}$ the expression c curl $\mathbf{H}$, according to eq. (7.4), the right-hand side of which is assumed to vanish. But the integral of a curl over a closed surface vanishes, according to Stokes' law; and we find that ϵ does not change, even though no assumptions have been made regarding the behavior of the field in the interior.

Despite the assumption of singular regions, the field outside these regions remains determined to a high degree. That is why Lorentz was able to explain the older theory of Maxwell, which assumed a continuous distribution of charge and current, as an approximation of his own theory, in which point charges were singularities of the field.

We can apply the point of view of Lorentz' electron theory to the representation of matter in the theory of gravitation. Instead of representing matter by means of the stress-energy tensor $P^{\mu\nu}$, we can assume that the gravitating matter is concentrated in small regions of space, and that elsewhere space is free of gravitating matter. The differential equations of the gravitational field will hold only outside the mass concentrations: they will be field equations of empty space. The mass concentrations themselves, the "mass points," will be singularities of the field.

We may consider the representation of matter by the tensor field $P^{\mu\nu}$ as a method of averaging over a great number of mass points and their states of motion, just as the concept of charge density is to be considered as the average number of elementary charges per unit volume. But, on the other hand, the description of matter by means of mass points may also be used as a convenient approximation when the components of the tensor $P^{\mu\nu}$ are different from zero only in small, isolated regions of space. This condition is realized in the solar system, where the bulk of the matter is concentrated in the interior of celestial bodies, while outside of these regions all components of $P^{\mu\nu}$ vanish. Each of these regions can be replaced by one mass point, and the treatment of the system is thereby greatly simplified.

Both representations of matter—by mass points and by a continuous medium—break down in the face of a sufficiently detailed treatment, for neither does justice to the quantum effects of atomic physics. But the usual fields of application of a theory of gravitation—astronomical problems—furnish us with both kinds of examples. If we wish to determine the balance of stresses in the interior of a star, or if we wish to get an overall picture of the behavior of a nebula which consists of millions of individual stars, we may treat matter as a continuous medium. If, on the other hand, the problem is one of computing the motions of a small number of celestial bodies, for instance, the bodies composing the solar system, matter must be represented by mass points.

Regardless of whether we describe matter as a continuous medium or by means of mass points, we shall assume that the number of field equations equals the number of field variables, ten. Furthermore, the equations must be of the second differential order in the $g_{\mu\nu}$, for they must involve the inhomogeneities of the gravitational field strength;

and they must be covariant with respect to general coördinate transformations.

If we treat matter as a continuous medium, the tensor field $P^{\mu\nu}$ must equal everywhere a certain other tensor field (which we have yet to find) which consists of differential expressions of the second order in the $g_{\mu\nu}$. On the other hand, if we choose the mass point representation of matter, then the same differential expressions must vanish everywhere, except in certain isolated regions, the locations of the mass points. In these regions, the solutions of the field equations become singular.

The differential identities. A physical law, such as the equations of the gravitational field, cannot be derived by purely logical processes. However, the range of possible field equations has already been limited by our assumptions that the field equations be ten differential equations of the second order in the $g_{\mu\nu}$ and that they be covariant with respect to general coördinate transformations. In this section, we shall formulate a further condition for the field equations, which will exclude all possibilities but one.

The ten differential equations for the $g_{\mu\nu}$ cannot be fully independent of each other, but must satisfy four identities. This condition is intimately connected with the condition of general covariance. Let us assume that we have obtained a set of ten covariant equations for the $g_{\mu\nu}$, and that we know one solution of these equations. Then we can obtain apparently new solutions of the same equations by merely carrying out arbitrary coördinate transformations. The transformed components of the metric tensor, $\overset{*}{g}_{\mu\nu}$, will be other functions of $\xi^{*\rho}$ than the original $g_{\mu\nu}$ are of ξ^{ρ}. These formally different solutions are actually equivalent representations of the same physical case, for their diversity reflects merely the variety of possible frames of reference with respect to which the same gravitational field can be described. The actual diversity of gravitational fields is much smaller than the number of formally different solutions of the field equations.

To restrict the variety of *formal* solutions, one may subject the coördinate system to auxiliary conditions. As the coördinate transformations contain four arbitrary functions (in a four dimensional continuum), it is possible to set up four equations for the $g_{\mu\nu}$, which must not be covariant and which must be chosen so that, if we start with any set of $g_{\mu\nu}$, we can satisfy them by merely carrying out a coördinate transformation. Such equations are called *coördinate conditions.*

By adjoining to the ten covariant field equations four coördinate conditions, we obtain a set of fourteen equations, which have the same variety of inequivalent solutions as the ten field equations alone, though the number of *formally* different solutions is smaller.

Fourteen fully independent equations for ten variables would have very few solutions, which represent either only a flat metric or at least a lesser variety of actually different cases than is required by the variety of conceivable distributions of matter in space. The fourteen equations must, therefore, satisfy four identities.

The four coördinate conditions are, to a high degree, arbitrary. They can be any equations, involving the $g_{\mu\nu}$, which are not covariant and which can be satisfied by any metric if only a suitable coördinate system is chosen. Since the choice of particular coördinate conditions has no effect on the nature of the solutions, it is necessary that the identities involve only the covariant field equations and that they be independent of the coördinate conditions.

The preceding argument shows that the ten field equations, because of their covariance, must satisfy four identities. But we have not as yet any clue to the form of the equations and the nature of their identities. We can obtain such a clue from the properties of the tensor $P^{\mu\nu}$. If matter is treated as a continuous medium, the $P^{\mu\nu}$ form the right-hand sides of the field equations of the gravitational field, just as the components of the current world vector form the right-hand sides of Maxwell's equations. Just as the conservation law of electric charges is expressed in the equation

$$I^{\rho}{}_{;\rho} = 0,$$

so the conservation laws of energy and momentum are expressed in the equations

$$P^{\mu\nu}{}_{;\nu} = 0. \tag{12.2}$$

We shall, therefore, expect the ten left-hand sides of the field equations to be the components of a symmetric tensor of rank two, and the four identities to have the form of divergences.

The field equations. In Chapter XI, we have encountered a tensor expression with just these properties. It is the tensor $G^{\mu\nu}$, defined by eq. (11.47). It is possible to show that there is no other tensor with ten components which depends only on the $g_{\mu\nu}$ and the divergence of which vanishes identically. We shall, therefore, choose as the field equations of the gravitational field the equations

$$G^{\mu\nu} + \alpha P^{\mu\nu} = 0, \qquad G^{\mu\nu} = R^{\mu\nu} - \tfrac{1}{2}g^{\mu\nu}R, \tag{12.3}$$

if matter is to be represented by the tensor $P^{\mu\nu}$; in the mass point representation of matter, the field equations of the gravitational field will be

$$G^{\mu\nu} = 0 \tag{12.4}$$

outside the mass points, but will not be satisfied at the locations of the mass points themselves. The constant α of eq. (12.3) will be determined later.

The field equations (12.4) satisfy the identities

$$G^{\mu\rho}{}_{;\rho} \equiv 0, \tag{12.5}$$

while the equations (12.3) yield eqs. (12.2),

$$0 = (G^{\mu\rho} + \alpha P^{\mu\rho})_{;\rho} \equiv \alpha P^{\mu\rho}{}_{;\rho} \,. \tag{12.6}$$

The linear approximation and the standard coördinate conditions. The proposed field equations and the ponderomotive law of gravitation are nonlinear with respect to the field variables $g_{\mu\nu}$. But we know that a linear theory—Newton's theory—accounts, with a considerable degree of accuracy, for the motions of bodies under the influence of forces. We must, therefore, assume that the gravitational fields (that is, the deviations of the actual metric from a flat metric) encountered in celestial mechanics and elsewhere are so weak that the nonlinear character of the field equations leads only to secondary effects.

The metric units, on which we usually base our measurements, are chosen so that the gravitational accelerations encountered in nature are of the order of magnitude of unity, while the speed of light, c, is a large quantity. For the purpose of the theory of gravitation, it is preferable to employ different units, in which the speed of light in flat space equals unity rather than 3×10^{10}. We shall keep the centimeter as the unit of length, but shall measure both time and proper time in units which are one 3×10^{10}th part of a second. In these units, the flat metric has the components

$$\epsilon_{\rho\sigma} = \begin{Bmatrix} -1, & 0\,, & 0\,, & 0 \\ 0\,, & -1, & 0\,, & 0 \\ 0\,, & 0\,, & -1, & 0 \\ 0\,, & 0\,, & 0\,, & +1 \end{Bmatrix}.^{1} \tag{12.7}$$

[1] For the remainder of this book, the notation $\epsilon_{\rho\sigma}$ will always be used for the flat metric when the new units of time are employed, while $\eta_{\rho\sigma}$ will denote, as before, the flat metric in terms of metric units.

The fact that the velocities of most material bodies are small compared with the speed of light is expressed in the new units by the condition that the U^i, the spatial components of U^ρ, are small compared with unity.

Using our new units of time, we shall assume that it is possible to introduce coördinate systems so that the components of the metric tensor can be expanded into a series,

$$g_{\rho\sigma} = \epsilon_{\rho\sigma} + \lambda \underset{1}{h}_{\rho\sigma} + \lambda^2 \underset{2}{h}_{\rho\sigma} + \cdots , \tag{12.8}$$

where λ is the parameter of expansion and a small constant.

The contravariant metric tensor will have the components

$$\left.\begin{aligned} g^{\rho\sigma} &= \epsilon^{\rho\sigma} + \lambda \underset{1}{h}^{\rho\sigma} + \lambda^2 \underset{2}{h}^{\rho\sigma} + \cdots, \\ \epsilon^{\rho\sigma} &= \epsilon_{\rho\sigma}, \\ \underset{1}{h}^{\rho\sigma} &= -\epsilon^{\rho\alpha}\epsilon^{\sigma\beta}\underset{1}{h}_{\alpha\beta}. \end{aligned}\right\} \tag{12.9}$$

The determinant of the metric tensor has the value

$$g \equiv | g_{\rho\sigma} | = -(1 + \lambda\epsilon^{\rho\sigma}\underset{1}{h}_{\rho\sigma} + \cdots). \tag{12.10}$$

Let us now consider the ponderomotive law. It is

$$\frac{dU^\rho}{d\tau} = - g^{\rho\sigma}[\iota\kappa, \sigma]\, U^\iota U^\kappa . \tag{12.11}$$

The Christoffel symbols $[\iota\kappa, \sigma]$ are small quantities of the first order in λ. If we neglect quantities of higher order, we may replace $g^{\rho\sigma}$ by $\epsilon^{\rho\sigma}$. Furthermore, as long as the velocities are small compared with c, we may neglect terms which contain components U^i as factors, while U^4 is approximately equal to unity. We shall, therefore, replace eq. (12.11) by the approximate equation

$$\frac{dU^\rho}{d\tau} \sim - \epsilon^{\rho\sigma}[44, \sigma] \sim - \tfrac{1}{2}\epsilon^{\rho\sigma}\lambda(2\underset{1}{h}_{4\sigma,4} - \underset{1}{h}_{44,\sigma}). \tag{12.12}$$

Finally, if the field does not change quickly with time—if it is created by mass points which themselves move only at moderate velocities—the derivatives with respect to ξ^4 are small, compared with derivatives with respect to the spatial coördinates ξ^i, and may be neglected. We find, as the first three equations (12.12),

$$\frac{dU^i}{d\tau} \sim - \tfrac{1}{2}\lambda \underset{1}{h}_{44,i} . \tag{12.13}$$

Upon comparing this equation with the ponderomotive law of classical mechanics, eq. (10.5), we find that $+\frac{1}{2}\lambda \underset{1}{h}_{44}$ takes the place of the Newtonian gravitational potential. This remark will help us later to interpret solutions of the field equations.

Let us now proceed to the linear approximation of the field equations. As we shall limit ourselves to linear expressions, we shall be able to simplify the form of these equations considerably. In the tensor

$$\left.\begin{aligned} G_{\mu\nu} &\equiv R_{\mu\nu} - \tfrac{1}{2}g_{\mu\nu}R \\ &= \begin{Bmatrix} \rho \\ \nu\rho \end{Bmatrix}_{,\mu} - \begin{Bmatrix} \rho \\ \mu\nu \end{Bmatrix}_{,\rho} - \begin{Bmatrix} \sigma \\ \rho\sigma \end{Bmatrix}\begin{Bmatrix} \rho \\ \mu\nu \end{Bmatrix} + \begin{Bmatrix} \sigma \\ \mu\rho \end{Bmatrix}\begin{Bmatrix} \rho \\ \nu\sigma \end{Bmatrix} - \tfrac{1}{2}g_{\mu\nu}g^{\alpha\beta}\left(\begin{Bmatrix} \rho \\ \alpha\rho \end{Bmatrix}_{,\beta}\right. \\ &\qquad \left. - \begin{Bmatrix} \rho \\ \alpha\beta \end{Bmatrix}_{,\rho} - \begin{Bmatrix} \sigma \\ \rho\sigma \end{Bmatrix}\begin{Bmatrix} \rho \\ \alpha\beta \end{Bmatrix} + \begin{Bmatrix} \sigma \\ \alpha\rho \end{Bmatrix}\begin{Bmatrix} \rho \\ \beta\sigma \end{Bmatrix}\right), \end{aligned}\right\} \quad (12.14)$$

we may neglect all the terms which are not linear in the $\underset{1}{h}_{\mu\nu}$. This refers to all terms which are not linear in the Christoffel symbols; in the remaining terms, we may replace all undifferentiated $g_{\mu\nu}$ and $g^{\mu\nu}$ by $\epsilon_{\mu\nu}$ and $\epsilon^{\mu\nu}$. We obtain the *"linearized"* expressions

$$\left.\begin{aligned} G_{\mu\nu} &\sim \frac{\lambda}{2}\,[\underset{1}{h}_{,\nu\mu} + \epsilon^{\rho\sigma}(\underset{1}{h}_{\mu\nu,\rho\sigma} - \underset{1}{h}_{\mu\rho,\nu\sigma} - \underset{1}{h}_{\nu\rho,\mu\sigma}) \\ &\qquad - \epsilon_{\mu\nu}\epsilon^{\rho\sigma}(\underset{1}{h}_{,\rho\sigma} - \epsilon^{\iota\kappa}\underset{1}{h}_{\rho\iota,\sigma\kappa})], \\ \underset{1}{h} &= \epsilon^{\rho\sigma}\underset{1}{h}_{\rho\sigma}. \end{aligned}\right\} \quad (12.15)$$

Eq. (12.15) can be somewhat simplified by the introduction of the quantities

$$\left.\begin{aligned} \underset{1}{\gamma}_{\mu\nu} &= \underset{1}{h}_{\mu\nu} - \tfrac{1}{2}\epsilon_{\mu\nu}\underset{1}{h}, \\ \underset{1}{h}_{\mu\nu} &= \underset{1}{\gamma}_{\mu\nu} - \tfrac{1}{2}\epsilon_{\mu\nu}\underset{1}{\gamma}. \end{aligned}\right\} \quad (12.16)$$

If we express the linearized $G_{\mu\nu}$ in terms of the $\underset{1}{\gamma}_{\mu\nu}$, we obtain

$$\left.\begin{aligned} G_{\mu\nu} &\sim \frac{\lambda}{2}\,(\epsilon^{\rho\sigma}\underset{1}{\gamma}_{\mu\nu,\rho\sigma} - \sigma_{\mu,\nu} - \sigma_{\nu,\mu} + \epsilon_{\mu\nu}\epsilon^{\rho\sigma}\sigma_{\rho,\sigma}), \\ \sigma_\mu &= \epsilon^{\rho\sigma}\underset{1}{\gamma}_{\mu\rho,\sigma}. \end{aligned}\right\} \quad (12.17)$$

Still it is difficult to obtain solutions of the field equations (12.3) or (12.4), for each component of the linearized $G_{\mu\nu}$, eq. (12.17), contains several components $\underset{1}{\gamma}_{\mu\nu}$, and all ten field equations must be solved simultaneously. This situation, however, can be greatly improved if we make use of the possibility of introducing *coördinate conditions. We shall show that we can always carry out a coördinate transformation so that the expressions σ_μ vanish.*

Let us consider coördinate transformations of the type

$$\xi^{*\alpha} = \xi^\alpha + \lambda v^\alpha(\xi^\rho), \tag{12.18}$$

which change the coördinate values only by amounts which are proportional to the parameter λ. The inverse transformations are

$$\xi^\alpha = \xi^{*\alpha} - \lambda v^\alpha(\xi^\rho) \sim \xi^{*\alpha} - \lambda v^\alpha(\xi^{*\rho}), \tag{12.19}$$

up to quantities of the first order in λ.

The components of the metric tensor (12.8) transform according to the law

$$\left.\begin{aligned} \overset{*}{g}_{\mu\nu} &= \frac{\partial\xi^\alpha}{\partial\xi^{*\mu}}\frac{\partial\xi^\beta}{\partial\xi^{*\nu}}\, g_{\alpha\beta} \sim (\delta^\alpha_\mu - \lambda v^\alpha{}_{,\mu^*})(\delta^\beta_\nu - \lambda v^\beta{}_{,\nu^*})(\epsilon_{\alpha\beta} + \lambda \underset{1}{h}_{\alpha\beta}) \\ &\sim \epsilon_{\mu\nu} + \lambda(\underset{1}{h}_{\mu\nu} - \epsilon_{\alpha\nu} v^\alpha{}_{,\mu^*} - \epsilon_{\alpha\mu} v^\alpha{}_{,\nu^*}), \end{aligned}\right\} \tag{12.20}$$

up to first order quantities. The transformation law for $\underset{1}{h}_{\mu\nu}$ is, therefore,

$$\underset{1}{\overset{*}{h}}_{\mu\nu} = \underset{1}{h}_{\mu\nu} - \epsilon_{\alpha\nu} v^\alpha{}_{,\mu^*} - \epsilon_{\alpha\mu} v^\alpha{}_{,\nu^*} . \tag{12.20a}$$

The transformation law of the quantities $\underset{1}{\gamma}_{\mu\nu}$ is

$$\underset{1}{\overset{*}{\gamma}}_{\mu\nu} = \underset{1}{\gamma}_{\mu\nu} - \epsilon_{\alpha\nu} v^\alpha{}_{,\mu^*} - \epsilon_{\alpha\mu} v^\alpha{}_{,\nu^*} + \epsilon_{\mu\nu} v^\rho{}_{,\rho^*} , \tag{12.21}$$

and the expressions σ_μ transform according to the law

$$\left.\begin{aligned} \overset{*}{\sigma}_\mu &= \sigma_\mu - \epsilon_{\mu\alpha}\epsilon^{\rho\sigma} v^\alpha{}_{,\rho^*\sigma^*} = \sigma_\mu - \epsilon^{\rho\sigma} v_{\mu,\rho\sigma} , \\ v_\mu &= \epsilon_{\mu\alpha} v^\alpha, \end{aligned}\right\} \tag{12.22}$$

again up to quantities of the first order.

We find that we obtain a coördinate system in which the σ_μ vanish if we carry out a coördinate transformation (12.18) in which the v^α satisfy the differential equations

$$\epsilon^{\rho\sigma} v_{\alpha,\rho\sigma} = \sigma_\alpha . \tag{12.23}$$

These differential equations, Poisson's equations in four dimensions, always have solutions.

In the linear approximation, the field equations may be replaced by the equations

$$\left.\begin{aligned} \lambda \epsilon^{\rho\sigma} \underset{1}{\gamma}_{\mu\nu,\rho\sigma} + 2\alpha P_{\mu\nu} &= 0, \\ \epsilon^{\rho\sigma} \underset{1}{\gamma}_{\mu\rho,\sigma} &= 0, \end{aligned}\right\} \tag{12.3a}$$

and

$$\left.\begin{aligned} \epsilon^{\rho\sigma} \underset{1}{\gamma}_{\mu\nu,\rho\sigma} &= 0, \\ \epsilon^{\rho\sigma} \underset{1}{\gamma}_{\mu\rho,\sigma} &= 0. \end{aligned}\right\} \tag{12.4a}$$

In the equations of the second differential order, the variables are now completely separated; the discussion of their solutions is thereby greatly facilitated.

Solutions of the linearized field equations. Let us first consider static solutions of the field equations, that is, solutions which are independent of ξ^4. If we assume that the field variables depend only on the three coördinates ξ^s, the linearized equations reduce to the equations

$$\left.\begin{aligned} -\lambda \nabla^2 \underset{1}{\gamma}_{\mu\nu} + 2\alpha P_{\mu\nu} &= 0, \\ \underset{1}{\gamma}_{\mu s,s} &= 0, \end{aligned}\right\} \tag{12.24}$$

and

$$\left.\begin{aligned} \nabla^2 \underset{1}{\gamma}_{\mu\nu} &= 0, \\ \underset{1}{\gamma}_{\mu s,s} &= 0. \end{aligned}\right\} \tag{12.25}$$

Ordinarily, the component $P_{44} = P^{44} = \rho$ is large, compared with the other components of $P_{\mu\nu}$. We shall, therefore, treat the case

$$\left.\begin{aligned} -\lambda \nabla^2 \underset{1}{\gamma}_{44} + 2\alpha\rho &= 0, \\ \nabla^2 \underset{1}{\gamma}_{\mu i} &= 0, \\ \underset{1}{\gamma}_{\mu s,s} &= 0. \end{aligned}\right\} \tag{12.26}$$

These equations can be solved by the assumption that of all the quantities $\underset{1}{\gamma}_{\mu\nu}$ only $\underset{1}{\gamma}_{44}$ does not vanish. $\underset{1}{\gamma}_{44}$ itself satisfies Poisson's equation in three dimensions; the solution is given by the integral

$$\lambda \underset{1}{\gamma_{44}}(\mathbf{r}) = -\frac{\alpha}{2\pi} \int_{V'} \frac{\rho(\mathbf{r}')\, dV'}{|\mathbf{r} - \mathbf{r}'|}. \tag{12.27}$$

We found that $\frac{1}{2}\lambda \underset{1}{h_{44}}$ must be considered as the quantity which assumes the role of the classical gravitational potential G of eqs. (10.5) and (10.7). Because only $\underset{1}{\gamma_{44}}$ of the $\underset{1}{\gamma_{\mu\nu}}$ does not vanish, $\underset{1}{h_{44}}$ has the value

$$\underset{1}{h_{44}} = \underset{1}{\gamma_{44}} - \tfrac{1}{2}\epsilon_{44}\epsilon^{44}\underset{1}{\gamma_{44}} = \tfrac{1}{2}\underset{1}{\gamma_{44}}. \tag{12.28}$$

We find, therefore, for $\underset{1}{h_{44}}$ the differential equation

$$-\lambda\nabla^2 \underset{1}{h_{44}} + \alpha\rho = 0. \tag{12.29}$$

By comparing this equation with eq. (10.7), we find that the constant α has the value

$$\alpha = 8\pi\kappa. \tag{12.30}$$

A theory of gravitation in which matter is represented by continuous media remains incomplete unless we know the equations of state of the media. If matter is rarefied to such an extent that there is no interaction between neighboring volume elements, then we may assume that $P^{\mu\nu}$ may be replaced by

$$P^{\mu\nu} = \rho U^\mu U^\nu,$$

where U^μ is subject to the ponderomotive law (12.1), and the change of ρ is determined by the conservation laws:

$$(\rho U^\mu U^\nu)_{;\nu} = 0,$$

$$(\rho U^\mu U^\nu)_{;\nu} U_\mu \equiv (\rho U^\nu)_{;\nu} = 0.$$

In all other cases, we must make assumptions regarding the internal forces of matter. Whether these assumptions are compatible with the theory of relativity may not be easy to decide. It is impossible, for instance, to conceive of rigid solid bodies or of an incompressible liquid. Either type of material would transmit elastic waves with an infinite speed of propagation, contrary to the fundamental assumption of the theory of relativity—that signals cannot be transmitted with a velocity greater than c. If we have a relativistic theory describing the interaction of the individual particles which make up the material, we can compute the equation of state, which will then not contradict the principles of relativity. Actually, such a program has so far been carried out for very few types of molecular interaction.

The field of a mass point. Let us now consider the representation of matter by mass points; that is, let us consider the linearized field equations (12.4a), p. 184.

First, we shall set up the field which is produced by a mass point at rest. This field will be static and spherically symmetric. We shall choose the mass point as the point of origin. Solutions of the Laplacian equation which vanish at infinity and which have no singularities outside the point of origin are all derivatives of the function $\frac{1}{r}$, or linear combinations of such derivatives. To solve the first set of equations (12.25), we shall make the assumption

$$\left.\begin{aligned} \lambda \underset{1}{\gamma}_{44} &= \frac{a}{r}, \\ \lambda \underset{1}{\gamma}_{4s} &= b\cdot\left(\frac{1}{r}\right)_{,s}, \\ \lambda \underset{1}{\gamma}_{st} &= c\cdot\left(\frac{1}{r}\right)_{,st} + f\cdot\frac{1}{r}\cdot\delta_{st}, \end{aligned}\right\} \tag{12.31}$$

where a, b, c, and f are constants to be determined. Let us now satisfy the other set of equations (12.25), the coördinate conditions. We find

$$\left.\begin{aligned} \lambda\sigma_4 &= -b\cdot\left(\frac{1}{r}\right)_{,ss} \equiv 0, \\ \lambda\sigma_s &= -c\cdot\left(\frac{1}{r}\right)_{,stt} - f\cdot\left(\frac{1}{r}\right)_{,s} \equiv -f\cdot\left(\frac{1}{r}\right)_{,s}. \end{aligned}\right\} \tag{12.32}$$

The constant f must vanish, while the other constants remain arbitrary. However, upon closer examination, we find that the terms which contain the constants b and c are dependent on the choice of the coördinate system. By carrying out a coördinate transformation (12.18), we can remove these terms if we choose the functions v^α as follows:

$$\left.\begin{aligned} \lambda v^4 &= \frac{b}{r}, \\ 2\lambda v^s &= -\left(\frac{c}{r}\right)_{,s}. \end{aligned}\right\} \tag{12.33}$$

In accordance with the transformation law (12.21), we find that we are left with the solution

$$\left.\begin{aligned} \lambda \underset{1}{\gamma}_{44} &= \frac{a}{r}, \\ \underset{1}{\gamma}_{4s} &= 0, \qquad \underset{1}{\gamma}_{st} = 0. \end{aligned}\right\} \tag{12.31a}$$

The remaining arbitrary constant, a, must be related to the mass which produces the field. The Newtonian potential which is produced by a mass M is

$$G = -\frac{\kappa M}{r}.$$

Because of eq. (12.28), and because $\frac{1}{2}\lambda \underset{1}{h}_{44}$ corresponds to G, we find that a determines the mass M by the equation

$$M = -\frac{a}{4\kappa}. \tag{12.34}$$

Gravitational waves. So far, we have treated only those solutions of the field equations which have counterparts in the classical theory of gravitation. However, there are solutions which are typical for a field theory. The most important of these are the "gravitational waves," rapidly variable fields, which must originate whenever mass points undergo accelerations.

Let us consider plane wave fields which depend only on ξ^4 and ξ^1. There are waves progressing in the positive Ξ^1-direction and waves which propagate in the opposite direction. The most general wave which propagates in the positive Ξ^1-direction has the components

$$\left.\begin{aligned} \underset{1}{\gamma}_{44} &= \underset{1}{\gamma}_{44}(\xi^1 - \xi^4), \\ \underset{1}{\gamma}_{4s} &= \underset{1}{\gamma}_{4s}(\xi^1 - \xi^4), \\ \underset{1}{\gamma}_{rs} &= \underset{1}{\gamma}_{rs}(\xi^1 - \xi^4). \end{aligned}\right\} \tag{12.35}$$

The field equations are automatically satisfied. The coördinate conditions are

$$\underset{1}{\gamma}_{\mu 4,4} - \underset{1}{\gamma}_{\mu 1,1} = -(\underset{1}{\gamma}'_{\mu 4} + \underset{1}{\gamma}'_{\mu 1}) = 0, \tag{12.36}$$

where the prime denotes differentiation with respect to the argument $(\xi^1 - \xi^4)$. We obtain the conditions

$$\left.\begin{aligned} \underset{1}{\gamma}'_{44} &= \underset{1}{\gamma}'_{11} = -\underset{1}{\gamma}'_{14}, \\ \underset{1}{\gamma}'_{12} &= -\underset{1}{\gamma}'_{42}, \\ \underset{1}{\gamma}'_{13} &= -\underset{1}{\gamma}'_{43}, \end{aligned}\right\} \tag{12.36a}$$

while the remaining components, $\underset{1}{\gamma}_{22}$, $\underset{1}{\gamma}_{23}$, and $\underset{1}{\gamma}_{33}$ remain arbitrary functions of the argument $(\xi^1 - \xi^4)$.

Again, it turns out that several of these components do not correspond to a physical wave field, but can be eliminated by a coördinate transformation. If we carry out a coördinate transformation (12.18), and let the v^α depend only on the argument $(\xi^1 - \xi^4)$, the transformation law (12.21) takes the form

$$\left.\begin{aligned} \underset{1}{\overset{*}{\gamma}}_{11} &= \underset{1}{\gamma}_{11} + v^{1\prime} + v^{4\prime}, \\ \underset{1}{\overset{*}{\gamma}}_{12} &= \underset{1}{\gamma}_{12} + v^{2\prime}, \qquad \underset{1}{\overset{*}{\gamma}}_{13} = \underset{1}{\gamma}_{13} + v^{3\prime}, \\ \underset{1}{\overset{*}{\gamma}}_{14} &= \underset{1}{\gamma}_{14} - v^{4\prime} - v^{1\prime}, \\ \underset{1}{\overset{*}{\gamma}}_{22} &= \underset{1}{\gamma}_{22} - v^{1\prime} + v^{4\prime}, \qquad \underset{1}{\overset{*}{\gamma}}_{33} = \underset{1}{\gamma}_{33} - v^{1\prime} + v^{4\prime}, \\ \underset{1}{\overset{*}{\gamma}}_{23} &= \underset{1}{\gamma}_{23}, \\ \underset{1}{\overset{*}{\gamma}}_{24} &= \underset{1}{\gamma}_{24} - v^{2\prime}, \qquad \underset{1}{\overset{*}{\gamma}}_{34} = \underset{1}{\gamma}_{34} - v^{3\prime}, \\ \underset{1}{\overset{*}{\gamma}}_{44} &= \underset{1}{\gamma}_{44} + v^{4\prime} + v^{1\prime}. \end{aligned}\right\} \tag{12.37}$$

By a suitable choice of the four functions v^α, we can obtain a coördinate system in which all components with at least one index 1 or 4 vanish, and in which the expression $(\underset{1}{\gamma}_{22} + \underset{1}{\gamma}_{33})$ is also equal to zero. The only waves which cannot be eliminated by coördinate transformations are those in which

$$\underset{1}{\gamma}_{22} = -\underset{1}{\gamma}_{33} \neq 0, \tag{12.38}$$

and those in which

$$\underset{1}{\gamma}_{23} \neq 0. \tag{12.39}$$

These two types of wave can be transformed into each other if the spatial coördinates are rotated around the Ξ^1-axis by an angle of $\frac{\pi}{4}$ radians (45°).

The gravitational waves have no counterpart in classical theory. Unfortunately, the intensity of those waves which are presumably produced by oscillating systems, double stars, planets, and so forth, is not strong enough to be observed by any method known to date.

Einstein and Rosen[2] investigated the wave solutions of the rigorous, nonlinear field equations. They found that there are no plane waves, but that there are cylindrical waves. Though they obtained this result by strictly formal methods, a physical explanation can be given. The gravitational waves, just like electrodynamic waves, carry energy.[3] This energy density in turn creates a *stationary* gravitational field which deforms the metric, and the gravitational waves must be superimposed on this deformed metric. A plane wave would be connected with a constant finite energy density everywhere, and the deviation of the metric from flatness, therefore, would increase toward infinity in all directions. Cylindrical waves, on the other hand, have a singularity in the axis of symmetry, and there are solutions in which the amplitude of the waves approaches zero and the amplitude of the stationary field becomes infinite for infinite values of the coördinate ρ (which, in a Euclidean space, denotes the distance from the axis). †

Our discussion of the "linearized" field equations of the relativistic theory of gravitation indicates that these equations possess solutions which correspond to Newtonian fields; in addition, there are solutions which have no counterpart in the classical theory, gravitational waves which propagate with the velocity of light. Now that we have found that the relativistic equations have solutions which are approximated by the classical theory of gravitation, we shall consider some of the formal properties of the relativistic field equations.

The variational principle. The classical field equations, (10.7), for empty space can be represented as the Euler-Lagrange equation of a variational problem (or *Hamiltonian principle*),

$$\delta \int_V (\text{grad } G)^2 \, dV = 0, \tag{12.40}$$

where the integral is to be extended over a three dimensional volume, V; the variation of G will vanish on the boundary of the domain V of integration, but is arbitrary in its interior. The variation of the integral (12.40) can be represented as follows:

---

[2] "On Gravitational Waves," *Journal Franklin Inst.*, **223**, 43 (1937).

[3] The concept of energy in the general theory of relativity will be treated later in this chapter.

†See Appendix B.

$$\delta \int_V (\text{grad } G)^2 \, dV = 2 \int_V (\text{grad } G \cdot \delta \text{ grad } G) \, dV$$

$$= 2 \int_V (\text{grad } G \text{ grad } (\delta G)) \, dV$$

$$= 2 \int_V \text{div } (\text{grad } G \cdot \delta G) \, dV - 2 \int_V \nabla^2 G \, \delta G \, dV$$

$$= 2 \oint_S \delta G (\text{grad } G \cdot \mathbf{dS}) - 2 \int_V \nabla^2 G \, \delta G \, dV.$$

The first integral of the last expression vanishes, because δG vanishes on the boundary. We have, therefore,

$$\delta \int_V (\text{grad } G)^2 \, dV = -2 \int_V \nabla^2 G \cdot \delta G \cdot dV; \tag{12.41}$$

and, as δG is arbitrary in the interior of V, it follows that the integral $\int_V (\text{grad } G)^2 dV$ is stationary only if G satisfies the equation

$$\nabla^2 G = 0. \tag{12.42}$$

Likewise, the relativistic field equations (12.4) can be represented as the Euler-Lagrange equations of a Hamiltonian principle. The integral in this case is a four dimensional integral

$$\left. \begin{aligned} I &= \int_D R \sqrt{-g} \, d\xi^1 \, d\xi^2 \, d\xi^3 \, d\xi^4, \\ \delta I &= 0. \end{aligned} \right\} \tag{12.43}$$

The variations of the $g_{\mu\nu}$ (and their first derivatives) must again vanish on the boundary of the four dimensional domain D, but are arbitrary in its interior.

The integral I is an invariant. The integrand,

$$\mathfrak{R} = \sqrt{-g} \, R, \tag{12.44}$$

is a density of the weight 1, and transforms according to the law

$$\mathfrak{R}^* = \left| \frac{\partial \xi^\alpha}{\partial \xi^{*\beta}} \right| \mathfrak{R};$$

and I transforms, therefore, as follows:

$$I^* = \int \mathfrak{R}^* \, d\xi^{*1} \, d\xi^{*2} \, d\xi^{*3} \, d\xi^{*4} = \int \mathfrak{R} \det \left| \frac{\partial \xi^\alpha}{\partial \xi^{*\beta}} \right| d\xi^{*1} \, d\xi^{*2} \, d\xi^{*3} \, d\xi^{*4}.$$

It is shown, in the theory of multiple integrals, that the integrand of a multiple integral is multiplied by the Jacobian of the transformation whenever new parameters of integration are introduced; in other words, I^* is the same integral as I,

$$I^* = I.$$

The Euler-Lagrange equations express the conditions which must be satisfied if a certain integral is to be stationary with respect to variations of the variables which make up the integrand. If the integral itself is invariant with respect to coördinate transformations, its Euler-Lagrange equations express conditions which cannot depend on the choice of coördinates; in other words, the Euler-Lagrange equations of an invariant Hamiltonian principle are themselves covariant differential equations.

Let us now express the variation of the integral

$$\left.\begin{aligned} I = \int \Bigg(\left\{ {\rho \atop \mu\rho} \right\}_{,\nu} &- \left\{ {\rho \atop \mu\nu} \right\}_{,\rho} \\ &- \left\{ {\sigma \atop \sigma\rho} \right\} \left\{ {\rho \atop \mu\nu} \right\} + \left\{ {\sigma \atop \mu\rho} \right\} \left\{ {\rho \atop \nu\sigma} \right\} \Bigg) g^{\mu\nu} \sqrt{-g}\, \mathbf{d}\xi, \\ \mathbf{d}\xi = d\xi^1\, d\xi^2\, d\xi^3\, d\xi^4, \end{aligned}\right\} \tag{12.45}$$

in terms of the variations of the $g^{\mu\nu}$.[4] We shall divide the variation of the integral into two parts, in this manner:

$$\left.\begin{aligned} \delta \int R_{\mu\nu} g^{\mu\nu} \sqrt{-g}\, \mathbf{d}\xi = \int \delta R_{\mu\nu} \sqrt{-g}\, g^{\mu\nu}\, \mathbf{d}\xi \\ + \int R_{\mu\nu}\, \delta(\sqrt{-g}\, g^{\mu\nu})\, \mathbf{d}\xi. \end{aligned}\right\} \tag{12.46}$$

First we shall express the variation of $R_{\mu\nu}$ in terms of the variations of the Christoffel symbols,

$$\left.\begin{aligned} \delta R_{\mu\nu} = \left(\delta \left\{ {\rho \atop \mu\rho} \right\} \right)_{,\nu} - \left(\delta \left\{ {\rho \atop \mu\nu} \right\} \right)_{,\rho} - \left\{ {\rho \atop \mu\nu} \right\} \delta \left\{ {\sigma \atop \rho\sigma} \right\} - \left\{ {\sigma \atop \rho\sigma} \right\} \delta \left\{ {\rho \atop \mu\nu} \right\} \\ + \left\{ {\rho \atop \nu\sigma} \right\} \delta \left\{ {\sigma \atop \mu\rho} \right\} + \left\{ {\sigma \atop \mu\rho} \right\} \delta \left\{ {\rho \atop \nu\sigma} \right\}. \end{aligned}\right\} \tag{12.47}$$

We know that the Christoffel symbols are not tensors, as they transform according to the transformation law (5.81). However, if two

---

[4] It is advantageous to introduce the contravariant components, $g^{\mu\nu}$, as the independent variables. As the $g^{\mu\nu}$ and the $g_{\mu\nu}$ determine each other uniquely, the final result of the computation is not affected by this choice.

different affine connections are defined on the same space, their difference, $\Gamma^{\lambda}_{\iota\kappa} - \bar{\Gamma}^{\lambda}_{\iota\kappa}$, transforms as a mixed tensor of rank 3, since the last term in eq. (5.81) cancels. The variation of the Christoffel symbols, $\delta\left\{{\rho \atop \mu\nu}\right\}$, is the difference between two affine connections, the varied and the unvaried Christoffel symbols, and is, therefore, a tensor.

As the left-hand side of eq. (12.47) is a tensor, the right-hand side can contain only *covariant* derivatives of the tensor $\delta\left\{{\rho \atop \mu\nu}\right\}$, and, in fact, straightforward computation shows that the right-hand side of eq. (12.47) is

$$\delta R_{\mu\nu} = \left(\delta\left\{{\rho \atop \mu\rho}\right\}\right)_{;\nu} - \left(\delta\left\{{\rho \atop \mu\nu}\right\}\right)_{;\rho}. \tag{12.48}$$

This simplification of eq. (12.47) was first pointed out by Palatini. As the covariant derivatives of the metric tensor vanish, we may multiply by $g^{\mu\nu}$ under the differentiation,

$$\left.\begin{aligned} g^{\mu\nu}\,\delta R_{\mu\nu} &= \left(g^{\mu\nu}\,\delta\left\{{\rho \atop \mu\rho}\right\}\right)_{;\nu} - \left(g^{\mu\nu}\,\delta\left\{{\rho \atop \mu\nu}\right\}\right)_{;\rho} \\ &= \left(g^{\mu\sigma}\,\delta\left\{{\rho \atop \mu\rho}\right\} - g^{\mu\nu}\,\delta\left\{{\sigma \atop \mu\nu}\right\}\right)_{;\sigma}. \end{aligned}\right\} \tag{12.49}$$

This expression is the covariant divergence of a vector. In Chapter V, Problem 10(*b*), we stated that a covariant divergence, $V^{\sigma}{}_{;\sigma}$, can be written in the form

$$V^{\sigma}{}_{;\sigma} = \frac{1}{\sqrt{g}}\,(\sqrt{g}\,V^{\sigma})_{,\sigma}. \tag{12.50}$$

Applying this formula, we obtain for the integrand of the first integral on the right-hand side of eq. (12.46)

$$\sqrt{-g}\,g^{\mu\nu}\,\delta R_{\mu\nu} = \left[\sqrt{-g}\left(g^{\mu\sigma}\,\delta\left\{{\rho \atop \mu\rho}\right\} - g^{\mu\nu}\,\delta\left\{{\sigma \atop \mu\nu}\right\}\right)\right]_{,\sigma}. \tag{12.51}$$

The integrand is an ordinary divergence. According to Gauss' theorem (which holds in n dimensional space just as well as in three dimensional space), the integral

$$\int \sqrt{-g}\,g^{\mu\nu}\,\delta R_{\mu\nu}\,\mathbf{d}\boldsymbol{\xi}$$

can be transformed into a surface integral over the boundary; this integral vanishes because the variations $\delta\left\{{\sigma \atop \mu\nu}\right\}$ vanish everywhere on the boundary.

There remains the second integral of eq. (12.46). $\delta(\sqrt{-g}\, g^{\mu\nu})$ is simply

$$\delta(\sqrt{-g}\, g^{\mu\nu}) = \sqrt{-g}(\delta g^{\mu\nu} - \tfrac{1}{2} g_{\rho\sigma} g^{\mu\nu}\, \delta g^{\rho\sigma}); \tag{12.52}$$

and this, multiplied by $R_{\mu\nu}$, gives

$$R_{\mu\nu}\delta(\sqrt{-g}\, g^{\mu\nu}) = \sqrt{-g}(R_{\mu\nu} - \tfrac{1}{2} g_{\mu\nu} R)\delta g^{\mu\nu} = \sqrt{-g}\, G_{\mu\nu}\, \delta g^{\mu\nu}. \tag{12.53}$$

We find, therefore, that the variation of I is given by the expression

$$\delta \int R\sqrt{-g}\, \mathrm{d}\xi = \int G_{\mu\nu}\, \delta g^{\mu\nu} \sqrt{-g}\, \mathrm{d}\xi. \tag{12.54}$$

The equations (12.4) are the Euler-Lagrange equations of the Hamiltonian principle (12.43).

The combination of the gravitational and electromagnetic fields. So far, we have treated only the gravitational field in the absence of the electromagnetic field. When an electromagnetic field is present, we can obtain the field equations of the combined fields by replacing $P^{\mu\nu}$ in the equations (12.3) by the expressions (8.31).

We have, then,

$$\left.\begin{aligned} G_{\mu\nu} - \frac{\kappa}{c^6}(2\varphi_{\mu\rho}\varphi_{\nu}{}^{\rho} - \tfrac{1}{2} g_{\mu\nu}\varphi_{\rho\sigma}\varphi^{\rho\sigma}) &= 0, \\ \varphi^{\mu\rho}{}_{;\rho} &= 0. \end{aligned}\right\} \tag{12.55}$$

These field equations are the Euler-Lagrange equations which are obtained when the integral

$$I = \int_D \left(R - \frac{\kappa}{c^6}\varphi_{\rho\sigma}\varphi^{\rho\sigma}\right)\sqrt{-g}\, \mathrm{d}\xi \tag{12.56}$$

is varied with respect to the 14 variables $g^{\mu\nu}$ and φ_μ.

The ponderomotive law of a charged mass point is

$$\frac{dU^\mu}{d\tau} + \left\{ {\mu \atop \rho\sigma} \right\} U^\rho U^\sigma - \frac{e}{mc^3}\varphi^{\mu\rho} U_\rho = 0. \tag{12.57}$$

The conservation laws in the general theory of relativity.[5] The energy-momentum tensor $P^{\mu\nu}$ in the field equations (12.3) represents the

---

[5] In this section, the transfer of the energy concept to the field of general relativity is discussed. As the concepts of energy and momentum are not of great importance in the general theory, the student may omit this section without losing the connection with the following chapters.

energy and momentum densities and stresses of a continuum, apart from the energy, momentum, and stresses which may be associated with the gravitational field. It satisfies the divergence relations

$$P^{\mu\rho}{}_{;\rho} = 0. \tag{12.58}$$

These equations are covariant; they transform as the components of a vector. But it is precisely for this reason that they are not what can properly be called conservation laws. In a proper conservation law, the change of a certain three dimensional volume integral with time is determined by the surface integral of certain other expressions, representing a flux, taken over the spatial boundary of the volume. In other words, a proper conservation law has the form

$$\frac{d}{dt}\left(\int_V P\, d\xi^1\, d\xi^2\, d\xi^3\right) = -\oint_S (\mathbf{F}\cdot\mathbf{dS}),$$

or, if we apply Gauss' law to the surface integral,

$$\int_V \left[\frac{\partial P}{\partial t} + \operatorname{div} \mathbf{F}\right] d\xi^1\, d\xi^2\, d\xi^3 = 0, \qquad P_{,4} + F^s{}_{,s} = 0. \tag{12.59}$$

The conservation law of electrical charges in the general theory of relativity has this form,

$$(\sqrt{-g}\,\varphi^{\mu\rho})_{,\rho\mu} = 0, \tag{12.60}$$

and the expressions $\int_V (\sqrt{-g}\,\varphi^{4\rho})_{,\rho}\, \mathbf{d}\xi$, $\int_S (\sqrt{-g}\,\varphi^{1\rho})_{,\rho}\, d\xi^2\, d\xi^3$, and so forth, represent the charge contained in the volume V, the current through a face parallel to the X^2, X^3-surface, and so on.

While the covariant divergence of a vector is equivalent to the ordinary divergence of a vector density, the covariant divergence of a symmetric tensor of rank 2 is not equivalent to an ordinary divergence of some density. However, it has been possible to find a set of expressions which satisfy four ordinary conservation laws without being the components of a tensor.

The equations

$$\left.\begin{aligned} 0 = \sqrt{-g}\, P_{\mu\cdot;\nu}^{\cdot\nu} &= \sqrt{-g}\left(P_{\mu\cdot,\nu}^{\cdot\nu} + \begin{Bmatrix}\nu\\ \rho\nu\end{Bmatrix} P_{\mu\cdot}^{\cdot\rho} - \begin{Bmatrix}\rho\\ \mu\nu\end{Bmatrix} P_{\rho\cdot}^{\cdot\nu}\right) \\ &= (\sqrt{-g}\, P_{\mu\cdot}^{\cdot\nu})_{,\nu} - \sqrt{-g}\begin{Bmatrix}\rho\\ \mu\nu\end{Bmatrix} P_{\rho\cdot}^{\cdot\nu}, \end{aligned}\right\} \tag{12.61}$$

have the form of a set of four conservation laws, except for the last term. We shall now show how this last term, $-\sqrt{-g}\left\{{\rho \atop \mu\nu}\right\} P_{\rho\cdot}^{\cdot\nu}$, can be brought into the form of an ordinary divergence. First of all, because of the field equations (12.3), the $P_{\rho\cdot}^{\ \nu}$ can be replaced by $-\frac{1}{\alpha}G_{\rho\cdot}^{\ \nu}$. We must now consider the expression $\frac{1}{\alpha}\sqrt{-g}\{{}_{\mu\nu}^{\rho}\}G_{\rho\cdot}^{\ \nu}$, which contains only the $g_{\mu\nu}$ and their derivatives. Replacing the Christoffel symbol by the derivatives of the metric tensor, we have

$$\sqrt{-g}\left\{{\rho \atop \mu\nu}\right\}G_{\rho\cdot}^{\ \nu} = \sqrt{-g}\,[\mu\nu, \rho]G^{\rho\nu} = \tfrac{1}{2}\sqrt{-g}\,g_{\nu\rho,\mu}G^{\rho\nu}. \tag{12.62}$$

In the proof that this last expression is the ordinary divergence of a set of 16 quantities, we shall make use of the fact that the expressions $\sqrt{-g}\,G_{\mu\nu}$ are the Euler-Lagrange equations of a variational principle.

First, we shall show that there is a variational principle which contains only *first* derivatives of the metric tensor and which has the same Euler-Lagrange equations as the integral (12.43).

An (ordinary) divergence, added to the integrand $R\sqrt{-g}$, contributes to the integral (12.43) an expression which can be written in the form of a surface integral because of Gauss' theorem. When we carry out a variation of the integrand so that the variations of the variables and their derivatives vanish at the boundary, the contribution of the added divergence will remain unchanged. The Euler-Lagrange equations will, therefore, remain the same if we add a divergence to $(\sqrt{-g}R)$ in eq. (12.43).

Let us now consider the integral in the form (12.45). The first two terms can be changed as follows:

$$\left.\begin{aligned}\left(\left\{{\rho \atop \mu\rho}\right\}_{,\nu} - \left\{{\rho \atop \mu\nu}\right\}_{,\rho}\right)\sqrt{-g}\,g^{\mu\nu} \\ = \left(\sqrt{-g}\,g^{\mu\nu}\left\{{\rho \atop \mu\rho}\right\}\right)_{,\nu} - \left(\sqrt{-g}\,g^{\mu\nu}\left\{{\rho \atop \mu\nu}\right\}\right)_{,\rho} \\ - (\sqrt{-g}\,g^{\mu\nu})_{,\nu}\left\{{\rho \atop \mu\rho}\right\} + (\sqrt{-g}\,g^{\mu\nu})_{,\rho}\left\{{\rho \atop \mu\nu}\right\}.\end{aligned}\right\} \tag{12.63}$$

The first two terms on the right-hand side are divergences. The Euler-Lagrange equations will, therefore, remain unchanged if we subtract them from $(\sqrt{-g}\,R)$. In the remaining terms, we replace the derivatives of the metric tensor everywhere by Christoffel symbols, according to the equation

$$g_{\mu\nu,\rho} \equiv [\mu\rho, \nu] + [\nu\rho, \mu]. \tag{12.64}$$

When they are combined with the other terms of the integrand of (12.45), we obtain the equation

$$\left.\begin{aligned} \delta I &= \delta \mathbf{W}, \qquad \mathbf{W} = \int \mathfrak{H} \cdot d\xi, \\ \mathfrak{H} &= \sqrt{-g}\, g^{\mu\nu}\left(\begin{Bmatrix} \sigma \\ \rho\sigma \end{Bmatrix}\begin{Bmatrix} \rho \\ \mu\nu \end{Bmatrix} - \begin{Bmatrix} \sigma \\ \mu\rho \end{Bmatrix}\begin{Bmatrix} \rho \\ \nu\sigma \end{Bmatrix}\right). \end{aligned}\right\} \tag{12.65}$$

The function $\mathfrak{H}$ is, of course, not a scalar density, and $\mathbf{W}$ is not invariant with respect to coördinate transformations. But the Euler-Lagrange equations belonging to the integral $\mathbf{W}$ are covariant equations. If we consider $\mathfrak{H}$ as a function of the variables $g^{\mu\nu}$ and $g^{\mu\nu}{}_{,\rho}$, then $G_{\mu\nu}$ must have the form

$$\sqrt{-g}\, G_{\mu\nu} \equiv \frac{\partial \mathfrak{H}}{\partial g^{\mu\nu}} - \left(\frac{\partial \mathfrak{H}}{\partial g^{\mu\nu}{}_{,\rho}}\right)_{,\rho}. \tag{12.66}$$

Let us multiply this equation by $g^{\mu\nu}{}_{,\alpha}$. We obtain the equation

$$\left.\begin{aligned} \sqrt{-g}\, G_{\mu\nu} g^{\mu\nu}{}_{,\alpha} &= \frac{\partial \mathfrak{H}}{\partial g^{\mu\nu}} g^{\mu\nu}{}_{,\alpha} - \left(\frac{\partial \mathfrak{H}}{\partial g^{\mu\nu}{}_{,\rho}}\right)_{,\rho} g^{\mu\nu}{}_{,\alpha} \\ &= \frac{\partial \mathfrak{H}}{\partial g^{\mu\nu}} g^{\mu\nu}{}_{,\alpha} - \left(\frac{\partial \mathfrak{H}}{\partial g^{\mu\nu}{}_{,\rho}} g^{\mu\nu}{}_{,\alpha}\right)_{,\rho} + \frac{\partial \mathfrak{H}}{\partial g^{\mu\nu}{}_{,\rho}} g^{\mu\nu}{}_{,\rho\alpha}. \end{aligned}\right\} \tag{12.67}$$

The first and the last terms on the right-hand side are together the derivative of $\mathfrak{H}$ with respect to ξ^α ($\mathfrak{H}$ depends on the coördinates only indirectly, by way of the $g^{\mu\nu}$ and $g^{\mu\nu}{}_{,\alpha}$). We have, therefore,

$$\sqrt{-g}\, G_{\mu\nu} g^{\mu\nu}{}_{,\alpha} = \left(\delta^\rho_\alpha \mathfrak{H} - \frac{\partial \mathfrak{H}}{\partial g^{\mu\nu}{}_{,\rho}} g^{\mu\nu}{}_{,\alpha}\right)_{,\rho} \equiv -(\sqrt{-g}\, \mathbf{t}^\rho_\alpha)_{,\rho}. \tag{12.68}$$

Herewith, the proof is essentially completed. For the expression on the left-hand side is

$$\sqrt{-g}\, G_{\mu\nu} g^{\mu\nu}{}_{,\alpha} = -\sqrt{-g}\, G^{\mu\nu} g_{\mu\nu,\alpha}, \tag{12.69}$$

and we find that eqs. (12.61) can be brought into the form

$$0 = \left[\sqrt{-g}\left(P_{\mu\cdot}^{\cdot\nu} + \frac{1}{2\alpha} \mathbf{t}^\nu_\mu\right)\right]_{,\nu} = \left[\sqrt{-g}\left(P_{\mu\cdot}^{\cdot\nu} + \frac{1}{16\pi\kappa} \mathbf{t}^\nu_\mu\right)\right]_{,\nu}. \tag{12.70}$$

The expressions

$$\mathbf{T}^\nu_\mu = P_{\mu\cdot}^{\cdot\nu} + \frac{1}{16\pi\kappa} \mathbf{t}^\nu_\mu \tag{12.71}$$

are not the components of a tensor. But because they satisfy the four conservation laws (12.70), they are called the components of the *stress-energy pseudo-tensor* of general relativity.

The expressions $\mathbf{t}_\mu^\nu$, the stress-energy components of the gravitational field, contain only *first* derivatives of the $g_{\mu\nu}$; one might say, they are *algebraic* functions of the gravitational "field intensities." It is characteristic of the general theory of relativity that expressions of this type cannot have tensor character. It is always possible to find frames of reference relative to which the "gravitational field strength" vanishes locally, and then the stress-energy components $\mathbf{t}_\mu^\nu$ vanish locally, too.

Conversely, in a perfectly flat space it is possible to choose a frame of reference relative to which we observe "inertial forces." According to the principle of equivalence, we cannot distinguish these "inertial forces" locally from a gravitational field, and the components $\mathbf{t}_\mu^\nu$ will not vanish in a noninertial coördinate system.

CHAPTER XIII

Rigorous Solutions of the Field Equations of the General Theory of Relativity

The field equations of the general theory of relativity are nonlinear equations. So far, we have solved only their linear approximations. In this chapter, we shall consider cases in which it has been possible to solve the rigorous equations in a closed form.

There is no general method of finding rigorous solutions of the field equations. However, the equations have been solved in a few cases in which the number of variables is reduced by symmetry conditions.

The solution of Schwarzschild. Let us first consider the solution which represents a mass point at rest. We shall assume that this solution has spherical symmetry and that none of the variables depend on ξ^4. If we introduce the variable

$$r = \sqrt{\xi^{1^2} + \xi^{2^2} + \xi^{3^2}}, \tag{13.1}$$

the most general line element with these properties takes the form

$$\left.\begin{aligned} d\tau^2 &= A(r)\,d\xi^{4^2} + 2B(r)\chi_s\,d\xi^4\,d\xi^s - C(r)\delta_{rs}\,d\xi^r\,d\xi^s \\ &\qquad\qquad + D(r)\chi_r\chi_s\,d\xi^r\,d\xi^s, \\ \chi_r &= \frac{\xi^r}{r}, \end{aligned}\right\} \tag{13.2}$$

where A, B, C, and D are functions of r. This line element does not change its form if we carry out a spatial rotation of the coördinates ξ^s around any axis which goes through the point of origin.

Without destroying either the static character or the spherical symmetry of the line element (13.2), we can eliminate two of the four unknown functions A, B, C, and D by suitable coördinate transformations. First we can eliminate the terms which contain products of spatial coördinate differentials, $d\xi^s$ and $d\xi^4$, by carrying out a coördinate transformation

$$\xi^{*4} = \xi^4 + f(r), \qquad \xi^{*s} = \xi^s. \tag{13.3}$$

The components g_{4s} transform according to the equation

$$\overset{*}{g}_{4s} = g_{44} \frac{\partial \xi^4}{\partial \xi^{*s}} + g_{4s},$$

or

$$B^* = B - \frac{df}{dr} A. \tag{13.4}$$

By choosing f so that it satisfies the equation

$$\frac{df}{dr} = \frac{B}{A}, \tag{13.5}$$

we can eliminate the term which contains B.

Let us now consider a metric with the components

$$\left.\begin{aligned} g_{44} &= A, \qquad g_{4s} = 0, \\ g_{rs} &= -C\delta_{rs} + D\chi_r \chi_s. \end{aligned}\right\} \tag{13.6}$$

By carrying out a transformation of the spatial coördinates,

$$\left.\begin{aligned} \xi^{*s} &= g(r)\xi^s, \\ \xi^s &= \psi(r^*)\xi^{*s}, \qquad r = \psi \cdot r^*, \qquad \overset{*}{\chi}_s = \chi_s, \end{aligned}\right\} \tag{13.7}$$

we shall be able to obtain a new coördinate system in which the metric has also the form (13.6), but in which the function C is constant and equal to unity. The components g_{rs} transform according to the law

$$\left.\begin{aligned} \overset{*}{g}_{rs} &= \frac{\partial \xi^i}{\partial \xi^{*r}} \frac{\partial \xi^k}{\partial \xi^{*s}} g_{ik} \\ &= \left(\psi\delta^i_r + r^* \frac{d\psi}{dr^*} \chi_i \chi_r\right)\left(\psi\delta^k_s + r^* \frac{d\psi}{dr^*} \chi_k \chi_s\right)(-C\delta_{ik} + D\chi_i \chi_k) \\ &= -\psi^2 C\delta_{rs} + \left[\left(\psi + r^* \frac{d\psi}{dr^*}\right)^2 D \right. \\ &\qquad \left. - r^* \frac{d\psi}{dr^*}\left(2\psi + r^* \frac{d\psi}{dr^*}\right) C\right] \chi_r \chi_s. \end{aligned}\right\} \tag{13.8}$$

We find that we obtain the coördinate system in which C equals unity, if we choose as the function ψ in eq. (13.7)

$$\psi = C^{-1/2}. \tag{13.9}$$

There remain only two unknown functions, A and D. Instead of these two functions, we shall introduce two other functions of r, because it has been found that the field equations are easier to handle if we set them up for these functions, μ and ν. The new functions are defined by the equations

$$\left.\begin{aligned} g_{44} &= A = e^{\mu}, \\ g_{4s} &= 0, \\ g_{rs} &= -\delta_{rs} + D\chi_r\chi_s \\ &\qquad = -\delta_{rs} + (1 - e^{\nu})\chi_r\chi_s, \qquad \nu = \log(1 - D). \end{aligned}\right\} \quad (13.10)$$

The contravariant metric tensor has the components

$$\left.\begin{aligned} g^{44} &= e^{-\mu}, \\ g^{4s} &= 0, \\ g^{rs} &= -\delta_{rs} + (1 - e^{-\nu})\chi_r\chi_s. \end{aligned}\right\} \quad (13.11)$$

We shall now compute the Christoffel symbols and the components of $G_{\mu\nu}$. The Christoffel symbols of the first kind are

$$\left.\begin{aligned} [44, s] &= -\tfrac{1}{2}\mu' e^{\mu}\chi_s, \\ [4s, 4] &= +\tfrac{1}{2}\mu' e^{\mu}\chi_s, \\ [rs, t] &= \chi_t\left[\frac{1 - e^{\nu}}{r}(\delta_{rs} - \chi_r\chi_s) - \tfrac{1}{2}\nu' e^{\nu}\chi_r\chi_s\right], \end{aligned}\right\} \quad (13.12)$$

where the primes denote differentiation with respect to r. The components with an odd number of indices 4 vanish. The Christoffel symbols of the second kind have the values

$$\left.\begin{aligned} \begin{Bmatrix} s \\ 44 \end{Bmatrix} &= \tfrac{1}{2}\mu' e^{\mu-\nu}\chi_s, \\ \begin{Bmatrix} 4 \\ 4s \end{Bmatrix} &= \tfrac{1}{2}\mu'\chi_s \\ \begin{Bmatrix} t \\ rs \end{Bmatrix} &= \chi_t\left[\frac{1 - e^{-\nu}}{r}(\delta_{rs} - \chi_r\chi_s) + \tfrac{1}{2}\nu'\chi_r\chi_s\right]. \end{aligned}\right\} \quad (13.13)$$

Again the components with an odd number of indices 4 vanish.

The components of the contracted curvature tensor, $R_{\mu\nu}$, are

$$\left.\begin{aligned}
R_{44} &= -e^{\mu-\nu}\left\{\frac{1}{2}\mu'' + \frac{1}{r}\mu' + \frac{1}{4}\mu'(\mu' - \nu')\right\},\\
R_{4s} &= 0,\\
R_{rs} &= \left\{\frac{1}{2}\mu'' - \frac{1}{r}\nu' + \frac{1}{4}\mu'(\mu' - \nu')\right\}\chi_r\chi_s\\
&\qquad + \left\{\frac{1}{r^2} - e^{-\nu}\left[\frac{1}{r^2} + \frac{1}{2r}(\mu' - \nu')\right]\right\}(\chi_r\chi_s - \delta_{rs}),
\end{aligned}\right\} \quad (13.14)$$

and the components of $G_{\mu\nu}$ are

$$\left.\begin{aligned}
G_{44} &= e^{\mu}\left\{e^{-\nu}\left(\frac{1}{r^2} - \frac{\nu'}{r}\right) - \frac{1}{r^2}\right\},\\
G_{4s} &= 0,\\
G_{rs} &= -\left\{\frac{\mu'}{r} + \frac{1 - e^{\nu}}{r^2}\right\}\chi_r\chi_s + \left\{\frac{1}{2}\mu'' + \frac{1}{4}\mu'(\mu' - \nu')\right.\\
&\qquad \left. + \frac{1}{2r}(\mu' - \nu')\right\}e^{-\nu}(\chi_r\chi_s - \delta_{rs}).
\end{aligned}\right\} \quad (13.15)$$

Let us first consider the case of a purely gravitational field, where eqs. (12.4) are satisfied. We must solve the following three equations for two variables:

$$\left.\begin{aligned}
\nu' - \frac{1}{r}(1 - e^{\nu}) &= 0,\\
\mu' + \frac{1}{r}(1 - e^{\nu}) &= 0,\\
\mu'' + \frac{1}{2}\mu'(\mu' - \nu') + \frac{1}{r}(\mu' - \nu') &= 0.
\end{aligned}\right\} \quad (13.16)$$

These three equations are not completely independent of each other, for they fulfill the contracted Bianchi identities, eqs. (11.46).

The first and the second of these equations show that the sum of μ' and ν' vanishes,

$$\mu' + \nu' = 0. \quad (13.17)$$

The first equation, for ν, can be solved. Let us introduce the quantity x,

$$x = e^{-\nu}, \qquad \nu = -\log x. \quad (13.18)$$

We obtain, instead of the first equation (13.16), the equation for x:

$$\frac{dx}{dr} + \frac{x-1}{r} = 0,$$

and as the solution

$$\left.\begin{aligned} x &= 1 - \frac{\alpha}{r}, \qquad \nu = -\log\left(1 - \frac{\alpha}{r}\right), \\ D &= -\frac{\alpha}{r-\alpha}, \end{aligned}\right\} \tag{13.19}$$

where α is a constant of integration. Because of eq. (13.17), the function μ must have the form

$$\mu = \log\left(1 - \frac{\alpha}{r}\right) + \beta, \tag{13.20}$$

where β is another constant. The solution (13.19), (13.20) also satisfies the third equation (13.16). The metric tensor then has the components

$$\left.\begin{aligned} g_{44} &= e^{\beta}\left(1 - \frac{\alpha}{r}\right), \\ g_{4s} &= 0, \\ g_{rs} &= -\delta_{rs} + \left(1 - \frac{1}{1 - \frac{\alpha}{r}}\right)\chi_r\chi_s = -\delta_{rs} - \frac{\alpha}{r-\alpha}\chi_r\chi_s . \end{aligned}\right\} \tag{13.21}$$

For large values of r, the metric tensor must approach the values $\epsilon_{\mu\nu}$, eq. (12.7). This is the case only if we choose for β the value zero. The remaining constant, α, must characterize the mass of the particle which creates the field (13.21).

In accordance with eq. (12.13), the Newtonian potential of the mass point which creates the field (13.21) is

$$G = -\frac{1}{2}\frac{\alpha}{r}. \tag{13.22}$$

On the other hand, G depends on the mass according to the equation

$$G = -\frac{\kappa m}{r}; \tag{13.23}$$

we find that the constant α is determined by the mass m, according to the equation

$$\alpha = 2\kappa m. \tag{13.24}$$

The gravitational field of a mass point is, therefore, represented by the expressions

$$\left.\begin{aligned} g_{44} &= 1 - \frac{2\kappa m}{r}, \\ g_{4s} &= 0, \\ g_{rs} &= -\delta_{rs} - \frac{2\kappa m}{r - 2\kappa m}\, x_r x_s . \end{aligned}\right\} \tag{13.25}$$

This solution was found by Schwarzschild.[1]

Schwarzschild's solution is significant because it is the only solution of the field equations in empty space which is static, which has spherical symmetry, and which goes over into the flat metric at infinity. Other solutions of the field equations for empty space with these properties can be carried over into Schwarzschild's solution merely by a coördinate transformation. Birkhoff has even shown that all spherically symmetric solutions of the field equations for empty space which satisfy the boundary conditions at infinity are equivalent to Schwarzschild's field, that is, their time dependence can be eliminated by a suitable coördinate transformation.[2]

Therefore, if we consider a concentration of matter of finite dimensions which is spherically symmetric, we know that the gravitational field outside the region filled with matter must be Schwarzschild's field. Inside this region, the matter might even be pulsating (in a spherically symmetric manner) without modifying the gravitational field outside. It is, of course, assumed that there is no flux of matter or electromagnetic radiation in the outside space.

The "Schwarzschild singularity." The expression (13.23), the solution of the classical field equation, (10.7), has a singularity at the point $r = 0$. The Schwarzschild field has a similar singularity at the same point. In addition, it has a singular spherical surface at $r = 2\kappa m$. On this surface, the component g_{44} vanishes, while some of the spatial components become infinite.

Robertson has shown that, if a Schwarzschild field could be realized, a test body which falls freely toward the center would take only a finite proper time to cross the "Schwarzschild singularity," even though the coördinate time is infinite; and he has concluded that at least part of the singular character of the surface $r = 2\kappa m$ must be attributed to the choice of the coördinate system.†

[1] *Berl. Ber.*, 1916, p. 189.

[2] Birkhoff, *Relativity and Modern Physics*, Harvard University Press, 1923, p. 253.

†See Appendix B.

In nature, mass is never sufficiently concentrated to permit a Schwarzschild singularity to occur in empty space. Einstein investigated the field of a system of many mass points, each of which is moving along a circular path, r = const., under the influence of the field created by the ensemble.[3] If the axes of the circular paths are assumed to be oriented at random, the whole system or cluster is spherically symmetric. The purpose of the investigation was to find out whether the constituent particles can be concentrated toward the center so strongly that the total field exhibits a Schwarzschild singularity. The investigation showed that even before the critical concentration of particles is reached, some of the particles (those on the outside) begin to move with the velocity of light, that is, along zero world lines. It is, therefore, impossible to concentrate the particles of the cluster to such a degree that the field has a singularity. (The singularities connected with each individual mass point are, of course, not considered.)

Einstein chose this example so that he would not have to consider thermodynamical questions, or to introduce a pressure, for the particles of his cluster do not undergo collisions, and their individual paths are explicitly known. In this respect, Einstein's cluster has properties which are nowhere encountered in nature. Nevertheless, it appears reasonable to believe that Einstein's result can be extended to conglomerations of particles where the motions of the individual particles are not artificially restricted as in Einstein's example.

The field of an electrically charged mass point. We shall now treat a mass point which carries an electric charge. The electrostatic field will be characterized by a scalar potential φ_4, which is a function of r. The covariant components of the electromagnetic field are

$$\varphi_{4s} = \varphi_{4,s} = \varphi_4'\chi_s\,, \qquad \varphi_{rs} = 0. \tag{13.26}$$

The components of the electromagnetic stress-energy tensor are

$$\left.\begin{aligned}
M_{\mu\nu} &= \frac{1}{4\pi}\,[\tfrac{1}{4}\,g_{\mu\nu}\,\varphi_{\rho\sigma}\varphi^{\rho\sigma} - \varphi_{\mu\rho}\,\varphi_{\nu\cdot}^{\;\rho}],\\
M_{44} &= -\frac{1}{8\pi}\,\varphi_{4s}\,\varphi_{4r}\,g^{sr} = \frac{1}{8\pi}\,(\varphi_4')^2 e^{-\nu},\\
M_{4s} &= 0\\
M_{rs} &= \frac{1}{4\pi}\,[\tfrac{1}{2}g_{rs}\,\varphi_{4t}\,\varphi^{4t} - \varphi_{r4}\,\varphi_{s\cdot}^{\;4}]\\
&= \frac{1}{8\pi}\,(\varphi_4')^2 e^{-(\mu+\nu)}[-e^{\nu}\chi_r\chi_s + (\delta_{rs} - \chi_r\chi_s)].
\end{aligned}\right\} \tag{13.27}$$

---

[3] *Annals of Mathematics*, **40**, 922 (1939).

By combining eqs. (12.3), (12.30), and (13.15), we obtain the equations

$$\left.\begin{aligned} e^{\mu}\left\{e^{-\nu}\left(\frac{1}{r^2}-\frac{\nu'}{r}\right)-\frac{1}{r^2}\right\}+\kappa e^{-\nu}(\varphi_4')^2 &= 0,\\ -\left(\frac{\mu'}{r}+\frac{1-e^{\nu}}{r^2}\right)-\kappa(\varphi_4')^2\cdot e^{-\mu} &= 0,\\ -\left\{\frac{1}{2}\mu''+\frac{1}{2r}(\mu'-\nu')+\tfrac{1}{4}\mu'(\mu'-\nu')\right\}e^{-\nu}+\kappa(\varphi_4')^2e^{-(\mu+\nu)} &= 0. \end{aligned}\right\} \quad (13.28)$$

In addition, we have the electromagnetic equation,

$$\varphi^{4\rho}{}_{;\rho} = 0.$$

We shall compute this expression in the form

$$\varphi^{4s}{}_{,s} + \varphi^{4m}\left\{ {\rho \atop m\,\rho} \right\} = 0.$$

For φ^{4s}, we obtain

$$\varphi^{4s} = g^{44}g^{rs}\varphi_{4r} = -e^{-(\mu+\nu)}\varphi_4'\chi_s,$$

and we have

$$[e^{-(\mu+\nu)}\varphi_4'\chi_s]_{,s} + \tfrac{1}{2}e^{-(\mu+\nu)}\varphi_4'(\mu'+\nu') = 0,$$

or

$$\varphi_4'' + \frac{2}{r}\varphi_4' - \tfrac{1}{2}(\mu'+\nu')\varphi_4' = 0. \quad (13.29)$$

This last equation has a first integral,

$$r^2e^{-\frac{1}{2}(\mu+\nu)}\varphi_4' = -\epsilon, \quad (13.30)$$

where ϵ is a constant of integration, the charge.

With the help of this integral, we can eliminate φ_4' from the field equations (13.28). We have

$$\left.\begin{aligned} e^{-\nu}\left(\frac{1}{r^2}-\frac{\nu'}{r}\right)-\frac{1}{r^2}+\frac{\kappa\epsilon^2}{r^4} &= 0,\\ e^{-\nu}\left(\frac{1}{r^2}+\frac{\mu'}{r}\right)-\frac{1}{r^2}+\frac{\kappa\epsilon^2}{r^4} &= 0,\\ -e^{-\nu}\left[\frac{1}{2}\mu''+\frac{1}{2r}(\mu'-\nu')+\frac{1}{4}\mu'(\mu'-\nu')\right]+\frac{\kappa\epsilon^2}{r^4} &= 0. \end{aligned}\right\} \quad (13.31)$$

Again, the combination of the first two equations yields eq. (13.17). The introduction of the variable x, eq. (13.18), into the first equation of the set (13.31) leads to the differential equation

$$\frac{dx}{dr} + \frac{x - 1}{r} + \frac{\kappa\epsilon^2}{r^3} = 0, \tag{13.32}$$

with the solution

$$x = 1 - \frac{2\kappa m}{r} + \frac{\kappa\epsilon^2}{r^2}. \tag{13.33}$$

The metric tensor has the components

$$\left.\begin{aligned} g_{44} &= 1 - \frac{2\kappa m}{r} + \frac{\kappa\epsilon^2}{r^2}, \\ g_{4s} &= 0, \\ g_{rs} &= -\delta_{rs} + \left(1 - \frac{1}{1 - \frac{2\kappa m}{r} + \frac{\kappa\epsilon^2}{r^2}}\right)\chi_r\chi_s\,. \end{aligned}\right\} \tag{13.34}$$

Eq. (13.30) takes the form

$$\varphi_4' = -\frac{\epsilon}{r^2}, \tag{13.30a}$$

with the solution

$$\varphi_4 = \frac{\epsilon}{r}. \tag{13.35}$$

The solutions with rotational symmetry. Weyl and Levi-Civita succeeded in finding those static solutions which have only rotational, but not spherical, symmetry.[4] If it is understood at the outset that "static" indicates both the independence of the $g_{\mu\nu}$ of ξ^4 and the vanishing of the components g_{4s}, it can be shown that any metric tensor with rotational symmetry can be brought into the form

$$\left.\begin{aligned} &g_{44} = e^{\mu}, \quad g_{4s} = 0, \quad g_{43} = 0, \\ &g_{33} = -e^{\nu-\mu}, \quad g_{3s} = 0, \\ &g_{rs} = e^{-\mu}[-\delta_{rs} + (1 - e^{-\nu})\chi_r\chi_s], \end{aligned}\right\} \quad \left.\begin{aligned} &r, s = 1, 2 \text{ (but not 3)}, \\ &\chi_r = \frac{\xi^r}{\rho}, \\ &\rho^2 = \xi^{1^2} + \xi^{2^2}. \end{aligned}\right\} \tag{13.36}$$

[4] Weyl, *Annalen d. Physik*, **54,** 117 (1917); **59,** 185 (1919). Bach and Weyl, *Mathematische Zeitschrift*, **13,** 142 (1921). Levi-Civita, *Rend. Acc. dei Lincei*, *Several Notes* (1918–1919).

μ and ν are functions of the two variables

$$\rho = \sqrt{\xi^{1^2} + \xi^{2^2}} \text{ and}$$
$$z = \xi^3.$$

The components of $G_{\mu\nu}$ take the form

$$\left.\begin{aligned}
G_{44} &= e^{2\mu-\nu}\left\{-\left(\frac{\partial^2\mu}{\partial\rho^2} + \frac{1}{\rho}\frac{\partial\mu}{\partial\rho} + \frac{\partial^2\mu}{\partial z^2}\right)\right. \\
&\qquad \left. + \frac{1}{2}\left(\frac{\partial^2\nu}{\partial\rho^2} + \frac{\partial^2\nu}{\partial z^2}\right) + \frac{1}{4}\left[\left(\frac{\partial\mu}{\partial\rho}\right)^2 + \left(\frac{\partial\mu}{\partial z}\right)^2\right]\right\}, \\
G_{33} &= \frac{1}{2\rho}\frac{\partial\nu}{\partial\rho} - \frac{1}{4}\left[\left(\frac{\partial\mu}{\partial\rho}\right)^2 - \left(\frac{\partial\mu}{\partial z}\right)^2\right], \\
G_{3s} &= \left\{-\frac{1}{2\rho}\frac{\partial\nu}{\partial z} + \frac{1}{2}\frac{\partial\mu}{\partial\rho}\frac{\partial\mu}{\partial z}\right\}\chi_s, \\
G_{rs} &= \left\{-\frac{1}{2\rho}\frac{\partial\nu}{\partial\rho} + \frac{1}{4}\left[\left(\frac{\partial\mu}{\partial\rho}\right)^2 - \left(\frac{\partial\mu}{\partial z}\right)^2\right]\right\}\chi_r\chi_s \\
&\qquad + \left\{-\frac{1}{2}\left(\frac{\partial^2\nu}{\partial\rho^2} + \frac{\partial^2\nu}{\partial z^2}\right) - \frac{1}{4}\left[\left(\frac{\partial\mu}{\partial\rho}\right)^2\right.\right. \\
&\qquad\qquad \left.\left. + \left(\frac{\partial\mu}{\partial z}\right)^2\right]\right\} e^{-\nu}\cdot(\delta_{rs} - \chi_r\chi_s), \\
G_{4s} &= 0, \qquad G_{43} = 0.
\end{aligned}\right\} \quad (13.37)$$

In a purely gravitational field, we have the equations

$$\left.\begin{aligned}
\kappa_1 &\equiv \frac{\partial^2\mu}{\partial\rho^2} + \frac{1}{\rho}\frac{\partial\mu}{\partial\rho} + \frac{\partial^2\mu}{\partial z^2} = 0, \\
\kappa_2 &\equiv \frac{\partial\nu}{\partial\rho} - \frac{\rho}{2}\left[\left(\frac{\partial\mu}{\partial\rho}\right)^2 - \left(\frac{\partial\mu}{\partial z}\right)^2\right] = 0, \\
\kappa_3 &= \frac{\partial\nu}{\partial z} - \rho\frac{\partial\mu}{\partial\rho}\frac{\partial\mu}{\partial z} = 0, \\
\kappa_4 &\equiv \frac{\partial^2\nu}{\partial\rho^2} + \frac{\partial^2\nu}{\partial z^2} + \frac{1}{2}\left[\left(\frac{\partial\mu}{\partial\rho}\right)^2 + \left(\frac{\partial\mu}{\partial z}\right)^2\right] = 0.
\end{aligned}\right\} \quad (13.38)$$

These four equations have two identities, the contracted Bianchi identities with the indices 3 and s ($s = 1, 2$). It turns out that the last equation (13.38) is identical in the remaining three equations,

$$\kappa_4 \equiv \frac{\partial\kappa_2}{\partial\rho} + \frac{\partial\kappa_3}{\partial z} + \rho\frac{\partial\mu}{\partial\rho}\kappa_1, \quad (13.39)$$

while the second and the third equations have this identity with κ_1:

$$\frac{\partial \kappa_2}{\partial z} - \frac{\partial \kappa_3}{\partial \rho} \equiv \rho \frac{\partial \mu}{\partial z} \kappa_1 . \tag{13.40}$$

The two functions μ and ν must vanish at infinity. Furthermore, the form of the components g_{rs}, eq. (13.36), indicates that the g_{rs} become singular (that is, indeterminate) along the Ξ^3-axis unless $(1 - e^{-\nu})$ vanishes there, that is, unless ν vanishes at $\rho = 0$.

Of the equations (13.38), the first one, κ_1, is a linear, homogeneous equation for μ only; and, moreover, it is the Laplacian equation in cylindrical coördinates for functions with rotational symmetry. We know that the solutions of the Laplacian equation, apart from the solution $\mu = 0$, satisfy the boundary conditions at infinity only if they have singularities somewhere for finite coördinate values. Singularities off the Ξ^3-axis are necessarily circular, while singularities on the Ξ^3-axis may be pointlike and of the form $\sum_i [\rho^2 + (z - a_i)^2]^{-1/2}$, or nth derivatives of such "poles" with respect to z.

However, not all of these solutions are compatible with the differential equations for ν, κ_2 and κ_3. Because of eq. (13.40), we know that if the equation κ_1 is satisfied in any simply connected domain of the ρ, z-space ($\rho \geqslant 0$), the equations κ_2, κ_3 have solutions. But in the presence of singularities, the ρ, z-space is no longer simply connected.

Let us first consider singularities off the Ξ^3-axis. If we take a solution μ of the equation κ_1 with an arbitrary circular singularity, the closed line integral around this singularity in the ρ, z-plane,

$$\left.\begin{aligned} &\oint \left(\frac{\partial \nu}{\partial \rho} d\rho + \frac{\partial \nu}{\partial z} dz\right) \\ &\qquad = \oint \left\{\frac{\rho}{2}\left[\left(\frac{\partial \mu}{\partial \rho}\right)^2 - \left(\frac{\partial \mu}{\partial z}\right)^2\right] d\rho + \rho \frac{\partial \mu}{\partial \rho}\frac{\partial \mu}{\partial z} dz\right\}, \end{aligned}\right\} \tag{13.41}$$

will, in general, not vanish. But the function ν will not be single-valued outside the singularity, unless the integral (13.41) vanishes; in other words, the vanishing of the integral (13.41) is a necessary condition for the existence of a solution.

Let us now turn to singularities on the Ξ^3-axis. Outside of this singularity, $\dfrac{\partial \nu}{\partial z}$ vanishes on the Ξ^3-axis, and if the function ν is assumed to vanish at one point on the Ξ^3-axis, it will vanish everywhere on the

Ξ^3-axis, up to the singularity. The singularity itself must satisfy the condition that the line integral of the differential

$$d\nu \equiv \frac{\rho}{2}\left[\left(\frac{\partial\mu}{\partial\rho}\right)^2 - \left(\frac{\partial\mu}{\partial z}\right)^2\right] d\rho + \rho\,\frac{\partial\mu}{\partial\rho}\frac{\partial\mu}{\partial z}\,dz \qquad (13.42)$$

over a small half-circle, around the singularity, from the Ξ^3-axis and back to it, must vanish.

Let us consider a typical singularity on the Ξ^3-axis,

$$\overset{(1)}{\mu} = (\rho^2 + z^2)^{-1/2}. \qquad (13.43)$$

The derivatives of $\overset{(1)}{\mu}$ are

$$\frac{\partial\overset{(1)}{\mu}}{\partial\rho} = -\frac{\rho}{r^3},$$

$$\frac{\partial\overset{(1)}{\mu}}{\partial z} = -\frac{z}{r^3},$$

$$r = \sqrt{\rho^2 + z^2}.$$

The differential (13.42) has the form

$$d\nu = \left(\frac{\rho}{2}\frac{\rho^2 - z^2}{r^6}\,d\rho + \frac{\rho^2 z}{r^6}\,dz\right). \qquad (13.44)$$

Let us carry the integration out along a small half-circle. For that purpose, we shall introduce the angle φ,

$$\left.\begin{aligned} \rho &= r\cos\varphi, & d\rho &= -z\cdot d\varphi,\\ z &= r\sin\varphi, & dz &= \rho\cdot d\varphi, \end{aligned}\right\} r = \text{const.}$$

Substituting these expressions in eq. (13.44), we obtain

$$d\nu = \frac{1}{2r^2}\cos\varphi\sin\varphi\,d\varphi, \qquad r = \text{const.} \qquad (13.45)$$

Let us integrate this expression over the half-circle, that is, from $-\frac{\pi}{2}$ to $+\frac{\pi}{2}$. We have

$$[\nu]_{\varphi=-\pi/2}^{+\pi/2} = \frac{1}{2r^2}\int_{-\pi/2}^{+\pi/2}\cos\varphi\sin\varphi\,d\varphi = \frac{1}{4r^2}[\sin^2\varphi]_{-\pi/2}^{+\pi/2} = 0. \qquad (13.46)$$

The solution $\overset{(1)}{\mu}$, eq. (13.43), is compatible with the regularity conditions for ν.

Let us now consider the case of *two* singularities. At the point of one, the other can be expanded into a power series in ρ^2 and z, and we shall assume that around the point of origin μ has the form

$$\overset{(2)}{\mu} = \frac{a}{r} + \sum_{m,n} a_{m,n}\,\rho^{2m} z^n. \tag{13.47}$$

Before we compute the integral again, let us remark that there are only certain of the expansion coefficients which can enter into the integral along the half-circle. The value of the integral is, of course, independent of the size of the half-circle, that is, of the value of r, as long as the circle does not enclose any other singularities but the one at the origin ($r = 0$). Therefore, all the expansion coefficients $a_{m,n}$ which would make the value of the integral depend on r need not be considered. Furthermore, the regular part of $\overset{(2)}{\mu}$ by itself cannot give rise to a nonvanishing integral, and we need to consider only the cross products of the singular and the regular part of $\overset{(2)}{\mu}$. The derivatives of the singular part of $\overset{(2)}{\mu}$ decrease as r^{-2} (for a given value of φ). They are multiplied by ρ (increasing as r^{+1}), by the coördinate differentials (increasing as r^{+1}), and by derivatives of the regular part of $\overset{(2)}{\mu}$. We are, therefore, interested only in those powers of the expansion the derivatives of which depend only on φ, but not on r. The only power with this property is $\rho^0 z^{+1}$. We shall, therefore, replace eq. (13.47) by

$$\overset{(2)}{\mu} = \frac{a}{r} + bz + \cdots. \tag{13.47a}$$

We shall compute the expression

$$d\nu = -\rho\,\frac{\partial}{\partial z}\left(\frac{a}{r}\right)\frac{\partial}{\partial z}(bz)\,d\rho + \rho\,\frac{\partial}{\partial \rho}\left(\frac{a}{r}\right)\frac{\partial}{\partial z}(bz)\,dz + \cdots. \tag{13.48}$$

The terms written out are the only ones which contribute to the integral. They are

$$d\nu = \frac{ab}{r^3}\,\rho(z\,d\rho - \rho\,dz) = -ab\cos\varphi\,d\varphi = -ab\,d(\sin\varphi). \tag{13.49}$$

This expression is to be integrated from $-\frac{\pi}{2}$ to $+\frac{\pi}{2}$. This integral does not vanish.

We find that at the location of one singularity, the derivative with respect to z of the regular part of μ must vanish. *This excludes the simultaneous existence of several pointlike singularities on the* Ξ^3*-axis.* It looks as if the field equations themselves exclude motions (or the lack of motion) of mass points which are incompatible with the equations of motion. In Chapter XV, we shall find that this is really the case.

CHAPTER XIV

The Experimental Tests of the General Theory of Relativity

As we found in the first part of this book, there are many experimental confirmations of the Lorentz covariant physical laws. The most convincing arguments in favor of the general theory of relativity, however, remain, so far, theoretical. Before we delve into the experimental evidence in favor of the general theory of relativity, it might be well to summarize these theoretical arguments.

Only a theory of gravitation which is covariant with respect to general coördinate transformations can explain the principle of equivalence and make it an integral part of its structure. A theory of gravitation which accounts for this principle must be considered more satisfactory than other theories—which, though compatible with the principle of equivalence, do not require it and could be maintained with slight modifications if the "gravitating mass" and "inertial mass" were to be considered different and independent quantities.

Moreover, the general theory of relativity presents us with the most nearly perfect example of a field theory which is yet known. In the next chapter we shall find that the laws of motion in the general theory of relativity are not independent of the field equations, but completely determined by them.

Let us now turn to the experimental tests of the general theory of relativity. There are three instances in which the general theory of relativity leads to observable effects. Each of these effects has been observed; however, two of them are just outside the limits of experimental error, so that the quantitative agreement between observations and theoretical predictions is still doubtful.†

The general theory of relativity accounts for the advance in the perihelion of Mercury, which was known before the new theory was formulated. Furthermore, the theory predicted correctly the deflection of light rays which pass near the surface of the sun, and the red shift of spectral lines of light originating in dense stars.

†See Appendix B.

The advance of the perihelion of Mercury. Let us consider the motion of a small body in a Schwarzschild field which is produced by a much larger body. It is advantageous to introduce polar coördinates by the coördinate transformation

$$\left.\begin{aligned} \xi^{*1} &= r = \sqrt{\xi^{1^2} + \xi^{2^2} + \xi^{3^2}}, \\ \xi^{*2} &= \theta = \text{arc tan}\left(\frac{\xi^3}{\sqrt{\xi^{1^2} + \xi^{2^2}}}\right), \\ \xi^{*3} &= \varphi = \text{arc tan}\,(\xi^2/\xi^1), \\ \xi^{*4} &= \xi^4. \end{aligned}\right\} \tag{14.1}$$

In terms of these coördinates, the metric tensor (13.25) goes over into the form

$$\left.\begin{aligned} g_{44} &= 1 - \frac{2\kappa m}{r}, \\ g_{11} &= -\frac{1}{1 - \dfrac{2\kappa m}{r}}, \\ g_{22} &= -r^2 \\ g_{33} &= -r^2 \cos^2\theta, \end{aligned}\right\} \tag{14.2}$$

all other components being zero.

The equations of motion of a small particle in this Schwarzschild field are

$$\frac{d^2\xi^\mu}{d\tau^2} + \begin{Bmatrix} \mu \\ \rho\sigma \end{Bmatrix} \frac{d\xi^\rho}{d\tau}\frac{d\xi^\sigma}{d\tau} = 0. \tag{14.3}$$

Let us now compute the Christoffel symbols, $\{{}^{\mu}_{\rho\sigma}\}$. If we introduce, for the sake of brevity, the expression

$$e^\mu = 1 - 2\kappa m/r, \tag{14.4}$$

the Christoffel symbols which do not vanish are

$$\left.\begin{aligned}
&\begin{Bmatrix}4\\14\end{Bmatrix} = \tfrac{1}{2}\mu',\\
&\begin{Bmatrix}1\\44\end{Bmatrix} = \tfrac{1}{2}e^{2\mu}\mu', \quad \begin{Bmatrix}1\\11\end{Bmatrix} = -\tfrac{1}{2}\mu', \quad \begin{Bmatrix}1\\22\end{Bmatrix} = -e^{\mu}r,\\
&\qquad\qquad\qquad\qquad\qquad\qquad \begin{Bmatrix}1\\33\end{Bmatrix} = -e^{\mu}\, r\cos^2\theta,\\
&\begin{Bmatrix}2\\12\end{Bmatrix} = \frac{1}{r}, \quad \begin{Bmatrix}2\\33\end{Bmatrix} = \cos\theta\sin\theta,\\
&\begin{Bmatrix}3\\13\end{Bmatrix} = \frac{1}{r}, \quad \begin{Bmatrix}3\\23\end{Bmatrix} = -\tan\theta.
\end{aligned}\right\} \tag{14.5}$$

If we simplify our mechanical problem by assuming that the whole motion takes place in the plane $\theta = 0$, we obtain as the equations of motion the differential equations

$$\left.\begin{aligned}
\frac{d^2t}{d\tau^2} + \mu'\frac{dt}{d\tau}\frac{dr}{d\tau} &= 0,\\
\frac{d^2r}{d\tau^2} + \tfrac{1}{2}e^{2\mu}\mu'\left(\frac{dt}{d\tau}\right)^2 - \tfrac{1}{2}\mu'\left(\frac{dr}{d\tau}\right)^2 - e^{\mu}r\left(\frac{d\varphi}{d\tau}\right)^2 &= 0,\\
\frac{d^2\varphi}{d\tau^2} + \frac{2}{r}\frac{dr}{d\tau}\frac{d\varphi}{d\tau} &= 0.
\end{aligned}\right\} \tag{14.6}$$

One integral of these equations is furnished by the definition of the proper time differential,

$$e^{\mu}\left(\frac{dt}{d\tau}\right)^2 - e^{-\mu}\left(\frac{dr}{d\tau}\right)^2 - r^2\left(\frac{d\varphi}{d\tau}\right)^2 = 1. \tag{14.7}$$

The first equation (14.6) has an integral,

$$e^{\mu}\frac{dt}{d\tau} = k. \tag{14.8}$$

The last equation (14.6) has also an integral,

$$r^2\frac{d\varphi}{d\tau} = h, \tag{14.9}$$

the integral of angular momentum. The integral (14.8) corresponds to the energy integral. The three equations (14.7), (14.8), and (14.9) replace the equations of the second order, (14.6). Finally, we may eliminate the coördinate time t with the help of eq. (14.8), and obtain the two equations

$$\left.\begin{aligned}\left(\frac{dr}{d\tau}\right)^2 + r^2\left(\frac{d\varphi}{d\tau}\right)^2 - \frac{2\kappa m}{r} &= k^2 - 1 + 2\kappa m r\left(\frac{d\varphi}{d\tau}\right)^2,\\ r^2\frac{d\varphi}{d\tau} &= h,\end{aligned}\right\} \quad (14.10)$$

where e^μ has been replaced by its value, $1 - \frac{2\kappa m}{r}$. These equations differ from the classical equations of motion of a body in a Newtonian field only in that the last term in the first equation (14.10) does not occur in the nonrelativistic equations, and that all derivatives are taken with respect to proper time rather than with respect to coördinate time. The classical integrals of energy and angular momentum are

$$\left.\begin{aligned}\dot{r}^2 + r^2\dot{\varphi}^2 - \frac{2\kappa m}{r} &= \frac{2E}{m'},\\ r^2\dot{\varphi} &= \frac{I}{m'},\end{aligned}\right\} \quad (14.11)$$

where m' is the mass of the moving body.

The equations (14.10) cannot be solved in a closed form. But we may solve them approximately, so that the first approximation corresponds to the classical path of a body; the second approximation will then reveal the deviation of the solutions of the relativistic equations (14.10) from the classical equations (14.11).

We multiply the first equation (14.10) by $\left(\frac{d\tau}{d\varphi}\right)^2$ and substitute this factor itself from the second equation. We obtain the one differential equation

$$\left(\frac{dr}{d\varphi}\right)^2 = -(1 - k^2)\frac{r^4}{h^2} + \frac{2\kappa m}{h^2}r^3 - r^2 + 2\kappa m r. \quad (14.12)$$

Into this equation we introduce the function $u = \frac{1}{r}$, and obtain the equation

$$\left(\frac{du}{d\varphi}\right)^2 = -\frac{1 - k^2}{h^2} + \frac{2\kappa m}{h^2}u - u^2 + 2\kappa m u^3. \quad (14.13)$$

By differentiating this equation with respect to φ, we obtain an equation of the second order,

$$\frac{d^2u}{d\varphi^2} + u = \frac{\kappa m}{h^2}(1 + 3h^2u^2). \quad (14.14)$$

The second term in the round bracket, $3h^2u^2$, is the one which distinguishes the relativistic equation from the corresponding classical equation. According to eq. (14.9), this term has the significance

$$3h^2 u^2 = 3\left(r\frac{d\varphi}{d\tau}\right)^2; \tag{14.15}$$

in other words, it is approximately proportional to the square of the velocity component which is perpendicular to the radius vector. As we are using the "relativistic units" for time, in which the velocity of light equals unity, the velocity of a star, for instance, is small, compared with unity. The relativistic term in eq. (14.14) has, therefore, the character of a higher order correction.

The solution of the equation

$$\frac{d^2 u_0}{d\varphi^2} + u_0 = \frac{\kappa m}{h^2} \tag{14.16}$$

is

$$u_0 = \frac{\kappa m}{h^2}[1 + \epsilon \cos(\varphi - \omega)], \tag{14.17}$$

where ϵ and ω are the constants of integration. ϵ is the eccentricity of the ellipse, while the value of ω determines the position of the perihelion.

Those solutions of eq. (14.14) which are approximated by ellipses are also periodic solutions. Eq. (14.13) associates with every value of u two values of $\frac{du}{d\varphi}$, which differ only with respect to the sign. The solutions will be periodic if the right-hand side of eq. (14.13) has two zeros for positive values of u and is positive between these two zeros. The solution will then oscillate between these two zeros. The period of the approximate solution (14.17) is equal to 2π; that is, the paths are closed. The period of the rigorous solutions of eq. (14.14), however, will differ from 2π by a small amount.

Let us expand the periodic solutions of the equation

$$u'' + u = a(1 + \lambda u^2) \tag{14.18}$$

into a Fourier series,

$$u = \alpha_0 + \alpha_1 \cos \rho\varphi + \alpha_2 \cos 2\rho\varphi + \cdots . \tag{14.19}$$

If λ is a small constant, the solution will be approximated by

$$u_0 = a(1 + \epsilon \cos \varphi). \tag{14.20}$$

We shall, therefore, assume that α_0 is approximately equal to a, and that α_2 and the following coefficients are at least of the order of λ. In other words, we shall replace eq. (14.19) by the series

$$u = a + \lambda\beta_0 + a\epsilon \cos \rho\varphi + \lambda \sum_2^\infty \beta_\nu \cos \nu\rho\varphi. \tag{14.21}$$

Let us substitute this assumption into eq. (14.18) and neglect all terms which are multiplied by the second and higher powers of λ. For u'', we obtain

$$u'' = -\rho^2\left[a\epsilon \cos \rho\varphi + \lambda \sum_2^\infty \nu^2\beta_\nu \cos \nu\rho\varphi\right];$$

and for λu^2, we have

$$\lambda u^2 \sim \lambda a^2[1 + 2\epsilon \cos \rho\varphi + \epsilon^2 \cos^2 \rho\varphi]$$

$$= \lambda a^2\left[1 + \frac{\epsilon^2}{2} + 2\epsilon \cos \rho\varphi + \frac{\epsilon^2}{2} \cos 2\rho\varphi\right]$$

Eq. (14.18) becomes

$$\left.\begin{aligned} a + \lambda\beta_0 + a\epsilon(1 - \rho^2) \cos \rho\varphi - \lambda \sum_2^\infty (\nu^2 \rho^2 - 1)\beta_\nu \cos \nu\rho\varphi \\ \sim a\left[1 + \lambda a^2\left(1 + \frac{\epsilon^2}{2} + 2\epsilon \cos \rho\varphi + \frac{\epsilon^2}{2} \cos 2\rho\varphi\right)\right]. \end{aligned}\right\} \tag{14.18a}$$

By comparing the terms which are constant, those which are multiplied by $\cos \rho\varphi$, and those multiplied by $\cos 2\rho\varphi$, we obtain the equations

$$\left.\begin{aligned} \beta_0 &= a^3\left(1 + \frac{\epsilon^2}{2}\right), \\ 1 - \rho^2 &= 2\lambda a^2, \\ -(4\rho^2 - 1)\beta_2 &= +a^3\frac{\epsilon^2}{2}. \end{aligned}\right\} \tag{14.22}$$

The only equation of interest to us is the second one, which determines ρ. We find that ρ is nearly equal to unity,

$$\rho = \sqrt{1 - 2\lambda a^2} \sim 1 - \lambda a^2. \tag{14.23}$$

Substituting for λ and a the values given in eq. (14.14), we find for ρ

$$\rho \sim 1 - 3\frac{\kappa^2 m^2}{h^2}. \tag{14.24}$$

The angle between two succeeding perihelions is, therefore,

$$\Phi \sim 2\pi\left(1 + 3\,\frac{\kappa^2 m^2}{h^2}\right) = 2\pi + 6\pi\,\frac{\kappa^2 m^2}{h^2}. \tag{14.25}$$

The precession of the perihelion of planets by $6\pi\,\frac{\kappa^2 m^2}{h^2}$ radians per revolution can be observed in the case of Mercury, where it amounts to about 43″ per century. The observed and the predicted values of the precession agree well within the experimental error of the astronomical observations.

The special theory of relativity also leads to a precession effect when a body moves in a field with a potential $\frac{a}{r}$. But this precession is numerically different from the one predicted by the general theory of relativity.

Let us return, for a moment, to Sommerfeld's treatment of the hydrogen atom, Chapter IX. The equations (9.27) correspond to the eqs. (14.7), (14.8), and (14.9) of this chapter. We can write them in the form:

$$\left.\begin{aligned} mc^2\frac{dt}{d\tau} &= E + \frac{e^2}{r}, \\ r^2\frac{d\theta}{d\tau} &= h', \\ \left(\frac{dt}{d\tau}\right)^2 - \frac{1}{c^2}\left[\left(\frac{dr}{d\tau}\right)^2 + r^2\left(\frac{d\theta}{d\tau}\right)^2\right] &= 1. \end{aligned}\right\} \tag{14.26}$$

In the last equation, we shall replace $\frac{dt}{d\tau}$ by the expression which is furnished by the first equation. Thus, we obtain an equation free of t,

$$\left(\frac{dr}{d\tau}\right)^2 + r^2\left(\frac{d\theta}{d\tau}\right)^2 = c^2\left[\left(\frac{E + \frac{e^2}{r}}{mc^2}\right)^2 - 1\right]. \tag{14.27}$$

We multiply this equation by

$$\left(\frac{d\tau}{d\theta}\right)^2 = \frac{r^4}{h'^2},$$

and obtain the equation

$$\left(\frac{dr}{d\theta}\right)^2 = c^2\,\frac{r^4}{h'^2}\left[\left(\frac{E + \frac{e^2}{r}}{mc^2}\right)^2 - 1\right] - r^2. \tag{14.28}$$

If we again introduce $u = 1/r$ as a new variable, the differential equation for u becomes

$$\left(\frac{du}{d\theta}\right)^2 = \frac{c^2}{h'^2}\left[\left(\frac{E + e^2 u}{mc^2}\right)^2 - 1\right] - u^2. \tag{14.29}$$

Differentiation with respect to θ yields the equation of the second order:

$$\frac{d^2 u}{d\theta^2} + \left[1 - \frac{e^4}{m^2 c^2 h'^2}\right] u = \frac{e^2 E}{m^2 h'^2 c^2}. \tag{14.30}$$

This equation has the following solutions:

$$\left.\begin{aligned} u &= \frac{e^2 E}{m^2 h'^2 c^2}\left\{1 + \epsilon \cos\left[\sqrt{1 - \frac{e^4}{m^2 c^2 h'^2}}\,(\theta - \omega)\right]\right\} \\ &\sim \frac{e^2 E}{m^2 h'^2 c^2}\left\{1 + \epsilon \cos\left[\left(1 - \frac{e^4}{2m^2 c^2 h'^2}\right)(\theta - \omega)\right]\right\}. \end{aligned}\right\} \tag{14.31}$$

The precession of the perihelion, therefore, amounts to

$$\delta\omega \sim \frac{\pi e^4}{m^2 c^2 h'^2} \tag{14.32}$$

per revolution. To compare this precession with the one obtained in eq. (14.24), we must replace e^2, the coefficient of Coulomb's law, by $\kappa m m'$, the coefficient of Newton's law of gravitation. Furthermore, the constant h' equals $\frac{h}{c}$ [h is the constant appearing in eq. (14.24)], because the τ of eq. (14.25) is measured in metric units. We obtain, instead of eq. (14.32),

$$\delta\omega \sim \pi \frac{\kappa^2 m^2}{h^2}, \tag{14.32a}$$

one-sixth of the precession predicted by the general theory of relativity.

The deflection of light in a Schwarzschild field. Light rays travel along geodesic zero lines. These lines are no longer the solutions of a variational principle, for in the case of zero lines, the variation of the integrand $\sqrt{g_{\iota\kappa}\xi^{\iota\prime}\xi^{\kappa\prime}}$ of eq. (5.93) is not a linear function of the variations $\delta\xi^\kappa$ and $\delta\xi^{\kappa\prime}$. However, there are zero lines with a tangential vector the covariant derivative of which in the direction of the tangential vector vanishes. This property characterizes non-zero geodesics and can be used as the defining property of both zero and non-zero geodesic lines. In a flat metric and in a Lorentzian coördinate system, these zero geodesics are "straight" zero lines, that is, ξ^1, ξ^2, and ξ^3 are linear functions of ξ^4.

In the case of zero lines, the tangential vector is a zero vector, and its magnitude cannot be normalized. We must, therefore, replace the parameter τ, which we have used so far, by a parameter s, which remains to a certain degree undetermined. The differential equations of the geodesic zero lines then take the form

$$\frac{d^2\xi^\mu}{ds^2} + \begin{Bmatrix} \mu \\ \rho\sigma \end{Bmatrix} \frac{d\xi^\rho}{ds}\frac{d\xi^\sigma}{ds} = 0, \qquad g_{\rho\sigma}\frac{d\xi^\rho}{ds}\frac{d\xi^\sigma}{ds} = 0. \tag{14.3a}$$

If the metric is that of a Schwarzschild field, these equations assume the form (14.6), except that τ must be replaced everywhere by s. Of the first integrals, (14.7), (14.8), and (14.9), eq. (14.7) has to be modified insofar as the right-hand side is now 0, not 1. The three integrals are

$$\left.\begin{aligned} e^\mu\left(\frac{dt}{ds}\right)^2 - e^{-\mu}\left(\frac{dr}{ds}\right)^2 - r^2\left(\frac{d\varphi}{ds}\right)^2 &= 0, \\ e^\mu\frac{dt}{ds} &= k, \\ r^2\frac{d\varphi}{ds} &= h. \end{aligned}\right\} \tag{14.33}$$

If we combine these three equations into one, by means of the same method employed before, we obtain again a relationship between r and φ,

$$\left(\frac{dr}{d\varphi}\right)^2 = \frac{k^2}{h^2}r^4 - r^2\left(1 - \frac{2\kappa m}{r}\right); \tag{14.34}$$

and if we introduce again the variable u,

$$\left(\frac{du}{d\varphi}\right)^2 = \frac{k^2}{h^2} - u^2 + 2\kappa m u^3. \tag{14.35}$$

The last term on the right-hand side represents the influence of the gravitational field on the path of the light rays. The solutions of the equation

$$\left(\frac{du_0}{d\varphi}\right)^2 = \frac{k^2}{h^2} - u_0^2$$

are

$$u_0 = \frac{1}{R}\cos(\varphi - \varphi_0), \qquad R = \frac{h}{k}, \tag{14.36}$$

where R is the distance of the light path from the point of origin of the coördinate system, and φ_0 is the constant of integration.

The angular distance between two zeros of u_0 (that is, between the two directions in which the light path goes toward infinity) is π. If $u(\varphi)$ is a solution of eq. (14.35), we are interested in the deviation of the angular distance between two zeros of $u(\varphi)$ from π. This deviation is twice the deviation of the angular distance between the maximum of u, $\bar{u}$ and the nearest zero from $\frac{\pi}{2}$.

The determining equation for $\bar{u}$ is

$$0 = \frac{k^2}{h^2} - \bar{u}^2 + 2\kappa m\bar{u}^3. \tag{14.37}$$

If we subtract this equation from eq. (14.35), we obtain

$$\left(\frac{du}{d\varphi}\right)^2 = (\bar{u}^2 - u^2) - 2\kappa m(\bar{u}^3 - u^3) \tag{14.38}$$

and

$$\frac{d\varphi}{du} = [(\bar{u}^2 - u^2) - 2\kappa m(\bar{u}^3 - u^3)]^{-1/2}. \tag{14.39}$$

The angular distance between the maximum of u and the nearest zero equals the integral of the right-hand side of eq. (14.39), taken from $u = 0$ to $u = \bar{u}$. The integral cannot be solved in a closed form. However, we know that the same integral, extended over the right-hand side of the equation

$$\frac{d\varphi}{du_0} = (\bar{u}_0^2 - u_0^2)^{-1/2}, \tag{14.40}$$

equals $\frac{\pi}{2}$. The deviation of the first integral from $\frac{\pi}{2}$ is, therefore, given by the integral

$$\tfrac{1}{2}\delta\varphi = \int_{u=0}^{\bar{u}} \{[(\bar{u}^2 - u^2) - 2\kappa m(\bar{u}^3 - u^3)]^{-1/2} - (\bar{u}^2 - u^2)^{-1/2}\}\, du. \tag{14.41}$$

Since we assume that the relativistic term (the one depending on m) is small in comparison with the classical term, we shall replace the expression $[f(x + \epsilon) - f(x)]$ by $\epsilon \cdot f'(x)$. By doing this, we obtain the integral

$$\tfrac{1}{2}\delta\varphi \sim \int_{u=0}^{\bar{u}} \frac{\kappa m(\bar{u}^3 - u^3)}{(\bar{u}^2 - u^2)^{3/2}}\, du, \tag{14.42}$$

which can be solved in a closed form,

$$\left.\begin{aligned}\underline{\int_{u=0}^{\bar{u}} \frac{\kappa m(\bar{u}^3 - u^3)}{(\bar{u}^2 - u^2)^{3/2}}\,du} &= \kappa m\bar{u} \int_{x=0}^{1} \frac{1 - x^3}{(1 - x^2)^{3/2}}\,dx \\ &= \kappa m\bar{u} \int_{\theta=0}^{\pi/2} \frac{1 - \sin^3\theta}{\cos^3\theta}\,d(\sin\theta) = \kappa m\bar{u} \int_{\theta=0}^{\pi/2} \frac{1 - \sin^3\theta}{\cos^2\theta}\,d\theta \\ &= \kappa m\bar{u}\,\Big[\tan\theta - \cos\theta - \sec\theta\Big]_{\theta=0}^{\pi/2} = \underline{2\kappa m\bar{u}}.\end{aligned}\right\} \quad (14.43)$$

The total deviation of the angular distance between two successive zeros of u from π equals, therefore,

$$\delta\varphi \sim 4\kappa m\bar{u} \sim \frac{4\kappa m}{R}\,. \quad (14.44)$$

This deflection of light rays which pass near great masses can be observed during eclipses of the sun, when fixed stars in the apparent neighborhood become visible. The predicted deflection amounts to not more than about 1.75″ and is just outside the limits of experimental error. A quantitative agreement between the predicted and the observed effects cannot be regarded as significant.

The gravitational shift of spectral lines. The interior forces of a free atom are not sensibly affected by the inhomogeneities of the surrounding gravitational field. If such a free atom goes over from one quantum state to another, the frequency of the emitted photon, *measured in proper time units of the atom*, will also be independent of the surrounding gravitational field.

Let us now consider the atoms which form the (gaseous) outer layer of a hot fixed star. Those atoms which emit any of the normal spectral lines must be falling freely at the time of emission, and their velocities relative to the star will be distributed at random. The mean frequency emitted will correspond to the emission by an atom which is momentarily at rest relative to the star.

The gravitational field of the star is described by a coördinate system in which the units of coördinate time and proper time are not identical. The proper frequency of an oscillatory process of an atom is the number of beats per unit proper time,

$$\nu_0 = \frac{dN}{d\tau}\,; \quad (14.45)$$

and the coördinate frequency is the number of beats per unit coördinate time,

$$\nu = \frac{dN}{d\xi^4}\,. \tag{14.46}$$

The two are related by the equation

$$\frac{dN}{d\xi^4} = \frac{dN}{d\tau}\cdot\frac{d\tau}{d\xi^4} = \frac{dN}{d\tau}\sqrt{g_{\iota\kappa}\frac{d\xi^\iota}{d\xi^4}\frac{d\xi^\kappa}{d\xi^4}}\,. \tag{14.47}$$

If we introduce a coördinate system in which the star is at rest, and in which the mean velocity of the atoms of the outer layer is zero, eq. (14.47) reduces, for this mean frequency, to the relationship

$$\nu = \sqrt{g_{44}}\,\nu_0\,; \tag{14.48}$$

for the three differential quotients, $\dfrac{d\xi^s}{d\xi^4}$ vanish.

The coördinate frequency ν is the frequency which will be observed by an observer who is at rest relative to a star and stationed at a great distance, so that g_{44} at his location equals unity. For the coördinate time required to transmit light signals from the surface of the star to such an observer is constant (because of the static character of the Schwarzschild field), and he will receive periodic signals at the same coördinate frequency at which they are being emitted at the surface of the star.

If the radius of the star is R and its mass m, the value of g_{44} on the surface of the star is $\left(1 - 2\dfrac{\kappa m}{R}\right)$, and eq. (14.48) becomes

$$\nu = \sqrt{1 - 2\frac{\kappa m}{R}}\,\nu_0 \sim \left(1 - \frac{\kappa m}{R}\right)\nu_0\,. \tag{14.49}$$

The "gravitational shift" of the spectral lines is, therefore,

$$\delta\nu \sim -\frac{\kappa m}{R}\,\nu_0\,. \tag{14.50}$$

In the case of the sun, the shift is barely observable, but appears to be in agreement with eq. (14.47). However, in the case of the companion to Sirius, which is an extremely dense star, the red shift is about 30 times as great as in the case of the sun. In this case, the agreement between theory and observation is satisfactory.

CHAPTER XV

The Equations of Motion in the General Theory of Relativity

Force laws in classical physics and in electrodynamics. Newtonian physics is based on the motions of mass points. The force acting on a mass point is the resultant of the actions of all other mass points in the world on the one considered. This force is uniquely determined by the positions of the other mass points, and it is finite as long as none of the other mass points coincide with the one considered.

The development of electrodynamics shows that the force acting on a body is not so simply determined as Newton thought. The action of one charge on another depends not only on the distance between them, but on their relative states of motion as well. A change in the state of motion of one charge brings about a change in its action on the other charge. But this secondary change does not take place momentarily; rather, the disturbance of the electromagnetic field spreads with a finite velocity, which is equal to the speed of light, c. Therefore, the force on a charged mass point is not determined by the position of all other point charges, or even by their positions and velocities, but by the electromagnetic field in the immediate neighborhood of the particle considered.

We cannot split up this electromagnetic field into partial fields, each representing the action of one particle. For the field itself is not uniquely determined by the motions of the charges. It is true that the field is determined by the distribution of the charges and their velocities if we impose on the field those boundary and initial conditions which exclude waves which travel toward a singular point rather than emanate from it (these are formal solutions of Maxwell's field equations). But it is doubtful whether these conditions are really satisfied in nature, or whether we impose them on the electromagnetic field just because they are suggested by our mechanical superstition that disturbances must always originate in mass points.

At any rate, the field can be treated adequately only as a unit, not as the sum total of the contributions of individual point charges.

Nevertheless, the ponderomotive law requires the splitting up of the total field into two parts: the field which is created by the particle considered, and the remainder. The field of the particle becomes infinite at the location of the point charge. We must discard the singular part of the field for the formulation of the force law. The remainder, the field which would exist if the particle were absent, determines the force acting on the particle.

The splitting process is by no means a uniquely determined mathematical operation. A particle with a given charge and in a given state of motion may give rise to a diversity of fields, depending on whether the field is formed by "retarded potentials" (outgoing waves) or "advanced potentials" (incoming waves) or a mixture of both types of wave. As far as practical applications are concerned, this ambiguity is not very serious if the particle considered is only moderately accelerated, that is, if the stationary field (Coulomb's field and Ampère's field) outweighs in magnitude whatever radiative field may be connected with the particle.

Also, the splitting operation, though not uniquely determined, has a reasonable meaning. The field equations are linear homogeneous equations; the sum of two solutions, therefore, is again a solution. Thus, if we subtract from the total field a field which might be connected with the particle, the difference is again a solution of Maxwell's equations, which remains regular at the location of the particle. But even though we must split up the field in order to formulate the equations of motion, this operation runs counter to the field concept itself.

The law of motion in the general theory of relativity. In the case of the theory of gravitation, the situation appears to be even worse. Mass points are represented by singularities of the gravitational field; and to formulate the law of force, we must know the field which would exist at the location of the mass point if the mass point itself were absent. Unlike Maxwell's field equations of electromagnetism, the field equations of gravitation are nonlinear, and the difference between two solutions is not itself a solution of the field equations. We can, therefore, treat the motion of a mass point only in a field which is practically unaffected by the presence of that mass point, such as the motion of a planet in the field of the sun. A genuine two-body problem, that of a double star, for instance, can be treated only by somewhat precarious methods of approximation which are not based on a rigorous theory. That is why several authors obtained contradictory results even in the lowest approximation in which the relativistic effects first become apparent.[1]

[1] Robertson, *Annals of Mathematics*, **39**, 101 (1938), footnotes on pp. 103 and 104.

In electrodynamics, the law of motion of charges does not follow from the field equations; in other words, it may happen that the charges do not obey Lorentz's force law, although Maxwell's field equations are satisfied. In fact, this is the case—according to classical electrodynamics—whenever the charges are subject to nonelectric forces in addition to the Lorentz forces.

In the case of gravitation, the situation is different. The mass points may be subject to forces other than gravitational forces. But then these nongravitational forces cause stresses, field energy densities, and field energy flux; that is, they give rise to a stress-energy tensor $P^{\mu\nu}$, and thereby modify the field equations to which the gravitational field is subject. It is, therefore, conceivable that the relationship between the field equations and the law of motion is closer in the theory of gravitation than it is in electrodynamics.

There are actually indications that the field equations for the $g_{\mu\nu}$ cannot be satisfied in the space surrounding the point singularities unless these singularities move along those world lines which are determined by the law of motion. At the end of Chapter XIII, we showed that there are no static solutions with rotational symmetry which correspond to two isolated, uncharged mass points at rest. In 1937 A. Einstein, L. Infeld, and B. Hoffmann proved that the motions of mass points are really determined by the field equations.[2]

The approximation method. In the relativistic theory of gravitation, as well as in electrodynamics, particles, the carriers of electric charge and mass, are represented by singularities of the field variables. In the presence of singularities, the change of the field in the course of time is not completely determined. Let us consider, for the moment, a Schwarzschild field. Near the point of origin, there is a singular region, the spherical surface $r = 2\kappa m$. Let us enclose this singular region by a small, two dimensional surface, S, so that outside this surface the field is regular everywhere and satisfies the field equations.

The Schwarzschild field is static. But if we know nothing about the structure of the field inside S, there is a possibility that radiation—electromagnetic or gravitational waves—may suddenly enter the outside world from the interior of S. If this radiation happens to be directed in one direction mainly, we must expect a "recoil," an acceleration of the whole singular region in the opposite direction.

In other words, we cannot expect a law of motion to hold for just

---

[2] *Annals of Mathematics*, **39**, 65 (1938). A later paper by A. Einstein and L. Infeld, *Annals of Mathematics*, **41**, 455 (1940), contains important improvements over the first paper.

any singularity. The motion of a singularity can be determined only if the singularity does not emit radiation. The law of motion, therefore, can be derived only if we are willing to make the assumptions that each singularity will remain a simple pole (the field equations also have solutions which correspond to mass dipoles, and so forth), and that there will be no spontaneous radiation.

These assumptions cannot be formulated conveniently in an invariant manner. There are, for instance, gravitational waves connected with the accelerated motion of a mass point. From a physical point of view, it is easy to explain what distinguishes "spontaneous" waves from "gravitational *bremsstrahlung*." But a mathematical formulation of the difference is almost impossible.

To avoid these difficulties, Einstein and his collaborators were forced to derive the laws of motion with the help of *an approximation method* by which the mathematical formulation of the necessary assumptions is comparatively easy.

Their method of approximating the equations is similar to the usual method, which was sketched in Chapter XII, and which leads, in its first step, to the "linearized" field equations. However, it differs from this approximation method in one important respect. The assumption that no spontaneous radiation occurs is equivalent to the assumption that the field variables do not change more rapidly with time than is necessitated by the motion of the singularities. This motion is slow, in comparison with the speed of light. It is, therefore, assumed that differentiation with respect to ξ^4 (measured in relativistic units) affects the order of magnitude of a quantity.

We can formulate this assumption in a slightly different way. If we introduce metric units again, that is, if the metric tensor of the flat space is $\eta_{\rho\sigma}$ instead of $\epsilon_{\rho\sigma}$, the velocities of material bodies are to be of the order of unity, and the differentiation of field variables with respect to ξ^4 is not to affect their order of magnitude. On the other hand, c^{-2} is now to be considered as a small quantity which becomes the parameter with respect to which the field variables are expanded into power series (the λ of Chapter XII). The $g_{\mu\nu}$, for instance, take the form

$$\left.\begin{aligned} g_{44} &= 1 + c^{-2}\underset{1}{h}_{44} + c^{-4}\underset{2}{h}_{44} + c^{-6}\underset{3}{h}_{44} + \cdots, \\ g_{4s} &= \qquad\qquad c^{-4}\underset{2}{h}_{4s} + c^{-6}\underset{3}{h}_{4s} + \cdots, \\ g_{rs} &= \quad - c^{-2}\delta_{rs} + c^{-4}\underset{2}{h}_{rs} + c^{-6}\underset{3}{h}_{rs} + \cdots. \end{aligned}\right\} \tag{15.1}$$

The term $c^{-2}\underset{1}{h}_{4s}$ is omitted, since it would lead to a value for g^{4s} of the order of magnitude of unity and destroy the linear character of the first approximation. As in Chapter XII, we shall introduce the quantities $\gamma_{\mu\nu}$ by the equations

$$\gamma_{\mu\nu} = h_{\mu\nu} - \tfrac{1}{2}\eta_{\mu\nu}\eta^{\rho\sigma}h_{\rho\sigma}\,, \qquad h_{\mu\nu} = g_{\mu\nu} - \eta_{\mu\nu}\,. \tag{15.2}$$

We shall also expand the $\gamma_{\mu\nu}$ into power series with respect to c^{-2},

$$\left.\begin{aligned} \gamma_{44} &= c^{-2}\underset{1}{\gamma_{44}} + c^{-4}\underset{2}{\gamma_{44}} + \cdots, \\ \gamma_{4s} &= \qquad\quad c^{-4}\underset{2}{\gamma_{4s}} + c^{-6}\underset{3}{\gamma_{4s}} + \cdots, \\ \gamma_{rs} &= \qquad\quad c^{-4}\underset{2}{\gamma_{rs}} + c^{-6}\underset{3}{\gamma_{rs}} + \cdots. \end{aligned}\right\} \tag{15.3}$$

It will be assumed that the quantities $\underset{2}{\gamma_{rs}}$ vanish. The reason is that a mass point at rest is represented by a solution of the field equations of which the first approximation is given by eqs. (12.31a), (12.34), or, in metric units, by the expressions

$$\left.\begin{aligned} \underset{1}{\gamma_{44}} &= -\frac{4\kappa M}{r}, \\ \underset{2}{\gamma_{4s}} &= 0, \\ \underset{2}{\gamma_{rs}} &= 0. \end{aligned}\right\} \tag{15.4}$$

If we wish to obtain the field of a mass point in motion in the same approximation, we must carry out a Lorentz transformation; the $g_{\mu\nu}$ transform according to the law

$$g^*_{\mu\nu} = \gamma_\mu{}^\alpha \gamma_\nu{}^\beta\, g_{\alpha\beta}, \tag{15.5}$$

where the $\gamma_\mu{}^\alpha$ have the significance which we gave them in Chapter V, eq. (5.111). The $h_{\mu\nu}$,

$$h_{\mu\nu} = g_{\mu\nu} - \eta_{\mu\nu}\,,$$

transform according to the same law as the $g_{\mu\nu}$, for the transformation law is linear, and the $\eta_{\mu\nu}$ remain unchanged. The $\gamma_{\mu\nu}$ transform according to the same law, as may be verified by straightforward computation. Applying this transformation law to eq. (15.4), we obtain the following expressions:

$$\left.\begin{aligned} \overset{*}{\gamma}_{44} &= (\gamma_4{}^4)^2\,\gamma_{44}\,, \\ \overset{*}{\gamma}_{4s} &= \gamma_4{}^4\,\gamma_s{}^4\,\gamma_{44}\,, \\ \overset{*}{\gamma}_{rs} &= \gamma_r{}^4\,\gamma_s{}^4\,\gamma_{44}\,. \end{aligned}\right\} \tag{15.6}$$

The coefficient $\gamma_4{}^4$ differs from unity only by quantities which are small and of the order c^{-2}. The coefficients $\gamma_s{}^4$ are given by the expressions

$$\gamma_s{}^4 = \eta_{sr}\eta^{44}\gamma^r{}_4 = -\frac{1}{c^2}\gamma^s{}_4 = -\frac{1}{c^2}\frac{u_s}{\sqrt{1 - u^2/c^2}}; \qquad (15.7)$$

they are, therefore, small quantities of the order of c^{-2}. We find that the transformation produces $\overset{*}{\gamma}_{4s}$ of the order c^{-4}, but not $\overset{*}{\gamma}_{rs}$ of the same order. The $\underset{2}{\gamma_{rs}}$ of the field of a mass point vanish even when the mass point is in motion.

It is reasonable to believe that, in the lowest approximation, the whole field is simply the sum of the fields of the various mass points. We are, therefore, justified in starting the power expansion of the γ_{rs} with c^{-6} as its first term.

If we substitute the power expansion (15.1), (15.3) in the field equations, the field equations themselves are decomposed into series of equations which belong to successive powers of c^{-2}. The method of Einstein, Infeld, and Hoffmann consists of setting up and solving these successive sets of equations. Each set of equations contains a set of quantities $\underset{n}{\gamma_{\rho\sigma}}$, which did not occur in any preceding set of equations, and, in addition, other $\underset{n}{\gamma_{\rho\sigma}}$ which have been determined previously. The "new" quantities always appear *linearly*; so that, at each stage of the approximation method, linear nonhomogeneous differential equations must be solved.

The determination of the motion of singularities takes place in the following manner. At each stage of the approximation method, there are ten linear, nonhomogeneous equations to be solved for ten $\underset{n}{\gamma_{\rho\sigma}}$. The left-hand sides of these equations, which contain the yet unknown variables, are not completely independent of each other, but satisfy four differential identities. Unless the right-hand sides (which contain only previously determined quantities) satisfy the same relations, the differential equations are incompatible with each other. Thus, each stage of the approximation method imposes conditions on the preceding stages. In the absence of singularities, these conditions turn out to be empty; but in the presence of singularities, these conditions are exactly the equations of motion.

Einstein and his collaborators carried out the approximation method to the stage at which the relativistic effects become apparent. We shall go no further than the approximation which yields the classical equations of motion, since this is the stage at which the connection between the field equations and the equations of motion first becomes apparent.

We shall expand the curvature tensor components, Christoffel symbols, and the covariant and contravariant components of the metric tensor into power series with respect to c^{-2}, and we shall identify the coefficients of the expansions by small numbers below each symbol; the number n will characterize the coefficient of c^{-2n}. For instance, $\underset{0}{G_{44}}$ will denote that part of G_{44} which is multiplied by c^0, $\underset{3}{h_{rs}}$ that part of h_{rs} which is multiplied by c^{-6}, and so forth.

By substituting the metric tensor (15.1), we shall find that the power expansions of the $G_{\mu\nu}$ start with these powers of c^{-2}: G_{44} with c^0, G_{4s} and G_{rs} with c^{-2}. We shall require of each component $G_{\rho\sigma}$ the first significant term, that is, $\underset{0}{G_{44}}$, $\underset{1}{G_{4s}}$, $\underset{1}{G_{rs}}$, and, in addition, $\underset{2}{G_{rs}}$. To obtain these expressions, we shall compute the $g_{\mu\nu}$, the $g^{\mu\nu}$, the Christoffel symbols of the first and of the second kind, and the $R_{\rho\sigma}$, each up to the power of c^{-2} required.

The first approximation and the mass conservation law. In the first approximation, the following components of the $h_{\mu\nu}$ are needed:

$$\left.\begin{aligned} \underset{1}{h_{44}} &= \tfrac{1}{2}\underset{1}{\gamma_{44}} \\ \underset{2}{h_{4s}} &= \underset{2}{\gamma_{4s}} \\ \underset{2}{h_{rs}} &= \tfrac{1}{2}\delta_{rs}\underset{1}{\gamma_{44}} \end{aligned}\right\}. \tag{15.8}$$

With their help, we can form these Christoffel symbols of the first kind,

$$\left.\begin{aligned} \underset{1}{[44, 4]} &= \tfrac{1}{4}\underset{1}{\gamma_{44,4}}, \\ \underset{1}{[44, t]} &= -\tfrac{1}{4}\underset{1}{\gamma_{44,t}}, \\ \underset{1}{[4r, 4]} &= \tfrac{1}{4}\underset{1}{\gamma_{44,r}}, \\ \underset{2}{[4r, t]} &= \tfrac{1}{2}(\underset{2}{\gamma_{4t,r}} - \underset{2}{\gamma_{4r,t}}) + \tfrac{1}{4}\delta_{rt}\underset{1}{\gamma_{44,4}}, \\ \underset{2}{[rs, 4]} &= \tfrac{1}{2}(\underset{2}{\gamma_{4r,s}} + \underset{2}{\gamma_{4s,r}}) - \tfrac{1}{4}\delta_{rs}\underset{1}{\gamma_{44,4}}, \\ \underset{2}{[rs, t]} &= \tfrac{1}{4}(\delta_{rt}\underset{1}{\gamma_{44,s}} + \delta_{st}\underset{1}{\gamma_{44,r}} - \delta_{rs}\underset{1}{\gamma_{44,t}}), \end{aligned}\right\} \tag{15.9}$$

and these Christoffel symbols of the second kind,

$$\left.\begin{aligned}
\underset{1}{\left\{ {4 \atop 44} \right\}} &= \tfrac{1}{4}\underset{1}{\gamma}_{44,4}\,, \\
\underset{0}{\left\{ {t \atop 44} \right\}} &= \tfrac{1}{4}\underset{1}{\gamma}_{44,t}\,, \\
\underset{1}{\left\{ {4 \atop 4r} \right\}} &= \tfrac{1}{4}\underset{1}{\gamma}_{44,r}\,, \\
\underset{1}{\left\{ {t \atop 4r} \right\}} &= \tfrac{1}{2}(\underset{2}{\gamma}_{4r,t} - \underset{2}{\gamma}_{4t,r}) - \tfrac{1}{4}\delta_{rt}\underset{1}{\gamma}_{44,4}\,, \\
\underset{2}{\left\{ {4 \atop rs} \right\}} &= \tfrac{1}{2}(\underset{2}{\gamma}_{4r,s} + \underset{2}{\gamma}_{4s,r}) - \tfrac{1}{4}\delta_{rs}\underset{1}{\gamma}_{44,4}\,, \\
\underset{1}{\left\{ {t \atop rs} \right\}} &= \tfrac{1}{4}(\delta_{rs}\underset{1}{\gamma}_{44,t} - \delta_{rt}\underset{1}{\gamma}_{44,s} - \delta_{st}\underset{1}{\gamma}_{44,r}).
\end{aligned}\right\} \tag{15.10}$$

For the components of the Ricci tensor, we obtain the expressions

$$\left.\begin{aligned}
\underset{0}{R}_{44} &= -\underset{0}{\left\{ {t \atop 44} \right\}}_{,t} = -\tfrac{1}{4}\underset{1}{\gamma}_{44,tt}\,, \\
\underset{1}{R}_{4s} &= \underset{1}{\left\{ {4 \atop 44} \right\}}_{,s} + \underset{1}{\left\{ {t \atop 4t} \right\}}_{,s} - \underset{1}{\left\{ {4 \atop 4s} \right\}}_{,4} - \underset{1}{\left\{ {t \atop 4s} \right\}}_{,t} \\
&= -\tfrac{1}{2}\underset{1}{\gamma}_{44,4s} + \tfrac{1}{2}(\underset{2}{\gamma}_{4t,s} - \underset{2}{\gamma}_{4s,t})_{,t}\,, \\
\underset{1}{R}_{rs} &= \underset{1}{\left\{ {4 \atop r4} \right\}}_{,s} + \underset{1}{\left\{ {t \atop rt} \right\}}_{,s} - \underset{1}{\left\{ {t \atop rs} \right\}}_{,t} \\
&= -\tfrac{1}{4}\delta_{rs}\underset{1}{\gamma}_{44,tt}\,, \\
\underset{0}{R} &= \underset{0}{R}_{44} - \underset{1}{R}_{tt} = +\tfrac{1}{2}\underset{1}{\gamma}_{44,tt}\,;
\end{aligned}\right\} \tag{15.11}$$

and the field equations of the lowest approximation become

$$\left.\begin{aligned}
\underset{0}{G}_{44} &\equiv -\tfrac{1}{2}\underset{1}{\gamma}_{44,tt} = 0, \\
\underset{1}{G}_{4s} &\equiv -\tfrac{1}{2}\underset{1}{\gamma}_{44,4s} + \tfrac{1}{2}(\underset{2}{\gamma}_{4t,s} - \underset{2}{\gamma}_{4s,t})_{,t} = 0, \\
\underset{1}{G}_{rs} &\equiv 0.
\end{aligned}\right\} \tag{15.12}$$

As the solution of the first equation, we shall choose the expression

$$\left.\begin{aligned} \underset{1}{\gamma_{44}} &= -\sum_{k=1}^{N} \frac{\overset{k}{\mu}(\xi^4)}{\overset{k}{r}} \equiv \psi, \\ \text{with} \\ \overset{k}{r} &= \{[\xi^1 - \overset{k}{y}{}^1(\xi^4)]^2 + [\xi^2 - \overset{k}{y}{}^2(\xi^4)]^2 + [\xi^3 - \overset{k}{y}{}^3(\xi^4)]^2\}^{1/2}. \end{aligned}\right\} \quad (15.13)$$

This solution represents N mass points. We have excluded all higher poles. The mass of the kth mass point is, according to eq. (12.34),

$$\overset{k}{M} = \frac{1}{4\kappa}\overset{k}{\mu}. \quad (15.14)$$

This mass is assumed to depend on the time coördinate, ξ^4, but we shall find very soon that the masses are actually constant. The $3N$ functions $\overset{k}{y}{}^s$ of ξ^4 determine the locations of the N mass points at every time ξ^4.

We shall now prove, with the help of the three equations G_{4s}, that the parameters $\overset{k}{\mu}$ (and, therefore, the masses $\overset{k}{M}$) are constants. This proof will be carried out with the help of a method which, at the next stage of the approximation method, will enable us to obtain the (classical) equations of motion. First, we shall prove this lemma: *When an expression depends on several indices and is skewsymmetric with respect to two of them, let us say to s and t, then its ordinary divergence with respect to one of the two skewsymmetric indices, let us say t, is equivalent to a curl, and the other skewsymmetric index which is not a dummy (in our case, s), characterizes the component of the curl.* It is assumed, as usual, that s and t run from 1 to 3. The proof of this statement is straightforward. Let us assume that our expression is $A_{ik\ldots st}$ and skewsymmetric in s and t. Then we denote

$$\begin{aligned} A_{ik\ldots 12} &= -A_{ik\ldots 21} && \text{by } B_3, \\ A_{ik\ldots 23} &= -A_{ik\ldots 32} && \text{by } B_1, \quad \text{and} \\ A_{ik\ldots 31} &= -A_{ik\ldots 13} && \text{by } B_2. \end{aligned}$$

The expression $A_{ik\ldots 1t,t}$ is then equal to

$$A_{ik\ldots 1t,t} = B_{3,2} - B_{2,3} = (\text{curl } \mathbf{B})_1, \quad (15.15)$$

and analogous equations hold for the other two divergence expressions.

Let us now return to our equations $\underset{1}{G_{4s}}$. They contain an expression which is the divergence of a skewsymmetric expression, namely, $\frac{1}{2}(\underset{2}{\gamma_{4s,t}} - \underset{2}{\gamma_{4t,s}})_{,t}$. This expression is equivalent to a curl, and the index s is the component index of that curl. *According to Stokes's law, the integral of a curl, taken over a closed surface, vanishes.* In other words, the integral

$$\frac{1}{2}\oint (\underset{2}{\gamma}_{4s,t} - \underset{2}{\gamma}_{4t,s})_{,t}\cdot\cos(s, n)\, dS \equiv 0. \qquad (15.16)$$

$\cos(s, n)$ shall denote the cosine of the angle between the ξ^s-coördinate and the normal of the surface element $\mathbf{dS}$. Let us now apply this result to a closed surface which encloses the pth singularity, but which is chosen so that on the surface itself there are no singularities. On the surface, but not in its interior, the field equations must be satisfied, and we have, therefore,

$$\left.\begin{aligned} 0 &= \oint_S G_{4s} \cos(s, n)\, dS = -\frac{1}{2}\oint_S \underset{1}{\gamma}_{44,4s} \cos(s, n)\, dS \\ &= -\frac{1}{2}\frac{\partial}{\partial\xi^4}\left\{\oint (\text{grad}\ \underset{1}{\gamma}_{44}\cdot d\mathbf{S})\right\}. \end{aligned}\right\} \qquad (15.17)$$

This equation expresses the condition that the mass parameters, $\overset{k}{\mu}$, be constants; for the integral $\oint (\text{grad}\ \psi\cdot d\mathbf{S})$ is simply the "number of lines of force" originating in the interior of S, and, therefore, proportional to $\overset{k}{\mu}$. Eq. (15.17) is equivalent to the condition

$$\frac{d\overset{k}{\mu}}{d\xi^4} = 0, \qquad k = 1 \cdots N. \qquad (15.18)$$

The last step in the first approximation consists of solving the differential equations G_{4s} with respect to the $\underset{2}{\gamma}_{4s}$,

$$\left.\begin{aligned} \underset{2}{\gamma}_{4t,ts} - \underset{2}{\gamma}_{4s,tt} &= \psi_{,4s}, \\ \psi &= -\sum_k \frac{\overset{k}{\mu}}{\overset{k}{r}}, \end{aligned}\right\} \qquad (15.19)$$

where the $\overset{k}{\mu}$ are now constants. Eqs. (15.6) and (15.7) suggest that these equations may be solved by the expressions

$$\underset{2}{\gamma}_{4r} = +\sum_{k=1}^{N} \dot{\overset{k}{y}}{}^r \frac{\overset{k}{\mu}}{\overset{k}{r}} \equiv \psi_r, \qquad (15.20)$$

where the dot denotes differentiation with respect to ξ^4; substitution into eqs. (15.19) shows that this is really the case.

To obtain the equations of motion, we must now turn to the second approximation.

The second approximation and the equations of motion. For the second approximation, the following contravariant components of the metric tensor are required:

$$\left.\begin{aligned} \underset{1}{h}^{44} &= -\underset{1}{h}_{44} = -\tfrac{1}{2}\psi, \\ \underset{1}{h}^{4s} &= \underset{2}{h}_{4s} = \psi_s, \\ \underset{0}{h}^{rs} &= -\underset{2}{h}_{rs} = -\tfrac{1}{2}\delta_{rs}\psi. \end{aligned}\right\} \tag{15.21}$$

With their help, we shall form these Christoffel symbols:

$$\left.\begin{aligned} \underset{2}{[44, 4]} &= \tfrac{1}{2}\underset{2}{h}_{44,4}, \\ \underset{2}{[44, t]} &= -\tfrac{1}{2}\underset{2}{h}_{44,t} + \psi_{t,4}, \\ \underset{2}{[4r, 4]} &= \tfrac{1}{2}\underset{2}{h}_{44,r}, \\ \underset{3}{[4r, t]} &= \tfrac{1}{2}(\underset{3}{h}_{4t,r} - \underset{3}{h}_{4r,t} + \underset{3}{h}_{rt,4}), \\ \underset{3}{[rs, t]} &= \tfrac{1}{2}(\underset{3}{h}_{rt,s} + \underset{3}{h}_{st,r} - \underset{3}{h}_{rs,t}), \end{aligned}\right\} \tag{15.22}$$

and

$$\left.\begin{aligned} \underset{2}{\begin{Bmatrix} 4 \\ 44 \end{Bmatrix}} &= \underset{2}{[44, 4]} + \underset{1}{h}^{44}\underset{1}{[44, 4]} + \underset{1}{h}^{4n}\underset{1}{[44, n]} \\ &= \tfrac{1}{2}\underset{2}{h}_{44,4} - \tfrac{1}{8}\psi\psi_{,4} - \tfrac{1}{4}\psi_n\psi_{,n}, \\ \underset{1}{\begin{Bmatrix} t \\ 44 \end{Bmatrix}} &= -\underset{2}{[44, t]} + \underset{0}{h}^{tn}\underset{1}{[44, n]} = \tfrac{1}{2}\underset{2}{h}_{44,t} - \psi_{t,4} + \tfrac{1}{8}\psi\psi_{,t}, \\ \underset{2}{\begin{Bmatrix} 4 \\ 4r \end{Bmatrix}} &= \underset{2}{[4r, 4]} + \underset{1}{h}^{44}\underset{1}{[4r, 4]} = \tfrac{1}{2}\underset{2}{h}_{44,r} - \tfrac{1}{8}\psi\psi_{,r}, \\ \underset{2}{\begin{Bmatrix} t \\ 4r \end{Bmatrix}} &= -\underset{3}{[4r, t]} + \underset{0}{h}^{tn}\underset{2}{[4r, n]} + \underset{1}{h}^{t4}\underset{1}{[4r, 4]} \\ &= \tfrac{1}{2}(\underset{3}{h}_{4r,t} - \underset{3}{h}_{4t,r} - \underset{3}{h}_{rt,4}) + \tfrac{1}{4}\psi(\psi_{r,t} - \psi_{t,r}) \\ &\qquad - \tfrac{1}{8}\delta_{rt}\psi\psi_{,4} + \tfrac{1}{4}\psi_t\psi_{,r}, \\ \underset{2}{\begin{Bmatrix} t \\ rs \end{Bmatrix}} &= -\underset{3}{[rs, t]} + \underset{0}{h}^{tn}\underset{2}{[rs, n]} = \tfrac{1}{2}(\underset{3}{h}_{rs,t} - \underset{3}{h}_{rt,s} - \underset{3}{h}_{st,r}) \\ &\qquad + \tfrac{1}{8}\psi(\delta_{rs}\psi_{,t} - \delta_{rt}\psi_{,s} - \delta_{st}\psi_{,r}). \end{aligned}\right\} \tag{15.23}$$

The components of the contracted curvature tensor in this approximation are somewhat lengthy expressions. If the equations of the first approximation are satisfied, the $R_{\mu\nu}$ of the second approximation are

$$
\left.\begin{aligned}
\underset{1}{R_{44}} &= \underset{1}{\left\{ {\rho \atop 4\rho} \right\}}_{,4} - \underset{1}{\left\{ {4 \atop 44} \right\}}_{,4} - \underset{1}{\left\{ {t \atop 44} \right\}}_{,t} - \underset{0}{\left\{ {t \atop 44} \right\}} \underset{1}{\left\{ {\rho \atop t\rho} \right\}} + 2 \underset{0}{\left\{ {t \atop 44} \right\}} \underset{1}{\left\{ {4 \atop 4t} \right\}} \\
&= -\tfrac{1}{2} \underset{2}{h_{44,tt}} - \tfrac{3}{4}\psi_{,44} + \psi_{t,t4} + \tfrac{1}{8}\psi_{,t}\psi_{,t}, \\
\underset{2}{R_{4s}} &= \underset{2}{\left\{ {\rho \atop s\rho} \right\}}_{,4} - \underset{2}{\left\{ {4 \atop 4s} \right\}}_{,4} - \underset{2}{\left\{ {t \atop 4s} \right\}}_{,t} - \underset{1}{\left\{ {4 \atop 4s} \right\}} \underset{1}{\left\{ {\rho \atop 4\rho} \right\}} - \underset{1}{\left\{ {t \atop 4s} \right\}} \underset{1}{\left\{ {\rho \atop t\rho} \right\}} \\
&\qquad + \underset{1}{\left\{ {4 \atop 44} \right\}} \underset{1}{\left\{ {4 \atop s4} \right\}} + \underset{0}{\left\{ {t \atop 44} \right\}} \underset{2}{\left\{ {4 \atop st} \right\}} + \underset{1}{\left\{ {4 \atop 4t} \right\}} \underset{1}{\left\{ {t \atop s4} \right\}} + \underset{1}{\left\{ {n \atop 4m} \right\}} \underset{1}{\left\{ {m \atop sn} \right\}} \\
&= \tfrac{1}{2}\big(\underset{3}{h_{4t,st}} - \underset{3}{h_{4s,tt}} + \underset{3}{h_{st,4t}} - \underset{3}{h_{tt,s4}}\big) + \tfrac{1}{4}[\psi(\psi_{t,s} - \psi_{s,t})]_{,t} \\
&\qquad - \tfrac{1}{4}(\psi\psi_{,s})_{,4} - \tfrac{1}{4}(\psi_t\psi_{,s})_{,t} + \tfrac{1}{8}\psi_{,4}\psi_{,s} + \tfrac{1}{4}\psi_{,t}\psi_{s,t}, \\
\underset{2}{R_{rs}} &= \underset{2}{\left\{ {\rho \atop r\rho} \right\}}_{,s} - \underset{2}{\left\{ {4 \atop rs} \right\}}_{,4} - \underset{2}{\left\{ {t \atop rs} \right\}}_{,t} - \underset{1}{\left\{ {t \atop rs} \right\}} \underset{1}{\left\{ {\rho \atop t\rho} \right\}} \\
&\qquad + \underset{1}{\left\{ {4 \atop r4} \right\}} \underset{1}{\left\{ {4 \atop s4} \right\}} + \underset{1}{\left\{ {n \atop rm} \right\}} \underset{1}{\left\{ {m \atop sn} \right\}} \\
&= \tfrac{1}{2}\big(\underset{3}{h_{rt,st}} + \underset{3}{h_{st,rt}} - \underset{3}{h_{rs,tt}} + \underset{2}{h_{44,rs}} - \underset{3}{h_{tt,rs}}\big) + \tfrac{1}{4}\delta_{rs}\psi_{,44} \\
&\qquad - \tfrac{1}{2}(\psi_{r,s4} + \psi_{s,r4}) - \tfrac{1}{8}\psi_{,r}\psi_{,s} - \tfrac{1}{4}\psi\psi_{,rs} - \tfrac{1}{8}\delta_{rs}\psi_{,t}\psi_{,t}, \\
\underset{1}{R} &= \underset{3}{h_{mm,nn}} - \underset{2}{h_{44,tt}} - \underset{3}{h_{mn,mn}} - \tfrac{3}{2}\psi_{,44} + 2\psi_{t,t4} + \tfrac{5}{8}\psi_{,t}\psi_{,t}.
\end{aligned}\right\} \quad (15.24)
$$

To set up the equations of motion, we need the expressions $\underset{2}{G_{rs}}$. They are

$$
\left.\begin{aligned}
\underset{2}{G_{rs}} &= \tfrac{1}{2}\big[\underset{3}{h_{rt,st}} + \underset{3}{h_{st,rt}} - \underset{3}{h_{rs,tt}} + \underset{2}{h_{44,rs}} - \underset{3}{h_{tt,rs}} \\
&\qquad + \delta_{rs}\big(\underset{3}{h_{mm,nn}} - \underset{2}{h_{44,tt}} - \underset{3}{h_{mn,mn}}\big)\big] \\
&\qquad - \tfrac{1}{2}\delta_{rs}\psi_{,44} - \tfrac{1}{2}(\psi_{r,s4} + \psi_{s,r4}) + \delta_{rs}\psi_{t,t4} - \tfrac{1}{8}\psi_{,r}\psi_{,s} \\
&\qquad - \tfrac{1}{4}\psi\psi_{,rs} + \tfrac{3}{16}\delta_{rs}\psi_{,t}\psi_{,t} \\
&= \tfrac{1}{2}\big[\underset{3}{\gamma_{rt,st}} + \underset{3}{\gamma_{st,rt}} - \underset{3}{\gamma_{rs,tt}} - \delta_{rs}\underset{3}{\gamma_{mn,mn}}\big] - \tfrac{1}{2}(\psi_{r,s4} + \psi_{s,r4}) \\
&\qquad + \delta_{rs}\psi_{t,t4} - \tfrac{1}{2}\delta_{rs}\psi_{,44} - \tfrac{1}{8}\psi_{,r}\psi_{,s} - \tfrac{1}{4}\psi\psi_{,rs} \\
&\qquad + \tfrac{3}{16}\delta_{rs}\psi_{,t}\psi_{,t}.
\end{aligned}\right\} \quad (15.25)
$$

Some of these terms can be written as divergences of skewsymmetric expressions. One is the square bracket,

$$\left.\begin{aligned}&\underset{3}{\gamma}_{rt,st} + \underset{3}{\gamma}_{st,rt} - \underset{3}{\gamma}_{rs,tt} - \delta_{rs}\underset{3}{\gamma}_{mn,mn} \\ &\qquad = (\underset{3}{\gamma}_{rt,s} - \underset{3}{\gamma}_{rs,t})_{,t} + (\delta_{rt}\underset{3}{\gamma}_{sn,n} - \delta_{rs}\underset{3}{\gamma}_{tn,n})_{,t}\,;\end{aligned}\right\} \tag{15.26}$$

the other is

$$-\tfrac{1}{2}\psi_{s,r4} + \tfrac{1}{2}\delta_{rs}\psi_{t,t4} = \tfrac{1}{2}(\delta_{rs}\psi_{t,4} - \delta_{rt}\psi_{s,4})_{,t}\,. \tag{15.27}$$

From now on, the argument proceeds as in the case of the mass conservation law. If we form the integrals of $\underset{2}{G}_{rs}\cos(s, n)$ over a closed surface S, which does not go through any singularity, then these integrals must vanish. Those expressions which are equivalent to curls make no contribution. Therefore, the integral of the remaining term must vanish by itself. *These remaining terms contain only the components of the first approximation. The condition that the integral of the remaining terms vanish is an integrability condition which the first approximation must satisfy if the second approximation is to exist.* The three integrals which must vanish are

$$\left.\begin{aligned}&\oint_S \{\tfrac{1}{2}(\delta_{rs}\psi_{t,t4} - \psi_{r,s4}) - \tfrac{1}{2}\delta_{rs}\psi_{,44} - \tfrac{1}{8}\psi_{,r}\psi_{,s} \\ &\qquad - \tfrac{1}{4}\psi\psi_{,rs} + \tfrac{3}{16}\delta_{rs}\psi_{,t}\psi_{,t}\}\cos(s, n)\, dS = 0.\end{aligned}\right\} \tag{15.28}$$

Unless the surface S encloses singularities, the conditions (15.28) are empty. For if the interior of S is regular, we can apply Gauss' law and transform the surface integrals into volume integrals, and we shall find that the integrands of these volume integrals vanish identically. If we call the integrands of the surface integrals K_{rs}, then we have:

$$\oint_S K_{rs}\cos(s, n)\, dS \equiv \oint_S \underset{2}{G}_{rs}\cos(s, n)\, dS = \int_V \underset{2}{G}_{rs,s}\, d\xi^1\, d\xi^2\, d\xi^3. \tag{15.29}$$

With the help of the Bianchi identities, we shall prove that the integrands $\underset{2}{G}_{rs,s}$ vanish. Let us denote the expressions $G_\mu{}^\rho{}_{;\rho}$ by B_μ. If we expand the B_μ into power series with respect to c^{-2}, we obtain for $\underset{1}{B}_r$

$$\left.\begin{aligned}\underset{1}{B}_r &= -\underset{2}{G}_{rs,s} + \underset{1}{G}_{r4,4} + \underset{0}{h}^{mn}\underset{1}{G}_{rm,n} \\ &\qquad - \underset{1}{\begin{Bmatrix}4\\r4\end{Bmatrix}}\underset{0}{G}_{44} + \underset{1}{\begin{Bmatrix}m\\rn\end{Bmatrix}}\underset{1}{G}_{mn} + \underset{1}{\begin{Bmatrix}s\\mm\end{Bmatrix}}\underset{1}{G}_{rs} - \underset{0}{\begin{Bmatrix}s\\44\end{Bmatrix}}\underset{1}{G}_{rs}\,.\end{aligned}\right\} \tag{15.30}$$

Since the equations of the first approximation, $\underset{0}{G_{44}}$, $\underset{1}{G_{4r}}$, $\underset{1}{G_{rs}}$, are already satisfied throughout the domain V, it turns out that the first term, $-\underset{2}{G_{rs,s}}$, vanishes by itself, even when the equations of the second and higher approximations are not satisfied.

The conditions (15.28) are significant only if the surface S encloses singularities. The value of the integrals (15.28) depends on the singularities enclosed by S, but is otherwise independent of the shape and size of S. We shall choose for S a small, spherical surface with the radius R, the center of which is the pth singularity.

It is now in order to compute explicitly the three integrals (15.28). For the expressions $\frac{1}{R}(\xi^s - \overset{p}{y}{}^s)$, which are the cosines of the angles (s, n), we shall introduce the abbreviations η_s. The derivatives of η_s with respect to ξ^r and ξ^4 are

$$\left.\begin{aligned} \eta_{s,r} &= \frac{1}{R}(\delta_{sr} - \eta_s\eta_r), \\ \eta_{s,4} &= \frac{1}{R}(\eta_s\eta_r - \delta_{sr})\dot{\overset{p}{y}}{}^r. \end{aligned}\right\} \tag{15.31}$$

The necessary computations will be simplified if the integrands of eqs. (15.28) are expanded into power series with respect to the radius R. It is, therefore, advantageous to introduce, instead of three coördinates ξ^s, the radius R and the directional cosines η_s, which satisfy the relationship

$$\eta_s\eta_s = 1. \tag{15.32}$$

Since the values of the integrals do not depend on the shape or size of S, they are independent of R. If the integrands K_{rs} are expanded into power series with respect to R (power series which contain both negative and positive powers), then all those terms which contain R in any power other than R^{-2} cannot contribute to the integrals. For the "area" of the surface S is proportional to R^2, and only those terms of the integrands which are proportional to R^{-2} will contribute to the integrals amounts which are independent of R. We shall, therefore, reduce the necessary computations by expanding all terms in K_{rs} into power series and by neglecting all expressions which are not multiplied by R^{-2}.

Furthermore, if $\overset{p}{\mu}$ were to vanish, that is, if the pth singularity were nonexistent, then the integrals (15.28) would vanish identically. We are, therefore, assured that all those terms which do not depend on $\overset{p}{\kappa}$

will not contribute to the integrals, and, consequently, we shall neglect them. We shall split up ψ and ψ_s into two parts each,

$$\left.\begin{aligned} \psi &= \iota + \lambda, \qquad \iota = \frac{\overset{p}{\mu}}{R}, \qquad \lambda = \sum_{k=1}^{N}{}' \frac{\overset{k}{\mu}}{\overset{k}{r}}, \\ \psi_s &= \iota_s + \lambda_s, \qquad \iota_s = \dot{\overset{p}{y}}{}^s \frac{\overset{p}{\mu}}{R}, \qquad \lambda_s = \sum_{k=1}^{N}{}' \dot{\overset{k}{y}}{}^s \frac{\overset{k}{\mu}}{\overset{k}{r}}, \end{aligned}\right\} \tag{15.33}$$

where the summation $\sum'$ is to be extended over all values of k except $k = p$.

K_{rs} contains terms which are linear in the ψ, ψ_s and quadratic terms. In the linear terms, we need to consider only the expressions depending on ι and ι_s. We have:

$$\left.\begin{aligned} \iota_{t,t4} &= -\ddot{\overset{p}{y}}{}^t \eta_t \frac{\overset{p}{\mu}}{R^2} + \text{terms which depend on } R^{-3}, \\ \iota_{r,s4} &= -\ddot{\overset{p}{y}}{}^r \eta_s \frac{\overset{p}{\mu}}{R^2} + \cdots, \\ \iota_{,44} &= -\ddot{\overset{p}{y}}{}^t \eta_t \frac{\overset{p}{\mu}}{R^2} + \cdots; \end{aligned}\right\} \tag{15.34}$$

the linear terms which contribute to the integral, multiplied by

$$\cos(s, n) = \eta_s,$$

are

$$\frac{1}{2} \frac{\overset{p}{\mu}}{R^2} \ddot{\overset{p}{y}}{}^r.$$

Integrated over S, they contribute the amount

$$2\pi \overset{p}{\mu} \ddot{\overset{p}{y}}{}^r \equiv L_r. \tag{15.35}$$

Let us now turn to the quadratic terms. ι and ι_s are multiplied by R^{-1}, their first derivatives with respect to spatial coördinates by R^{-2}, and their second derivatives with respect to spatial coördinates by R^{-3}. It is apparent that all terms which are quadratic in $\overset{p}{\mu}$ are multiplied by R^{-4} and, therefore, do not contribute to the integrals. We need to consider only terms which are bilinear in ι- and λ-expressions. Of ι-expressions, we shall require $\iota_{,r}$ and $\iota_{,rs}$. These are:

$$\left.\begin{aligned} \iota_{,r} &= +\frac{\overset{p}{\mu}}{R^2}\eta_r , \\ \iota_{,rs} &= \frac{\overset{p}{\mu}}{R^3}(\delta_{rs} - 3\eta_r\eta_s). \end{aligned}\right\} \tag{15.36}$$

These ι-expressions are multiplied by λ and its derivatives. λ itself is multiplied by a second derivative of ι, that is, by an expression which is proportional to R^{-3}. Therefore, only the term in the power expansion of λ which is proportional to R^{+1} will contribute to the integrals. For similar reasons, of the first derivatives of λ only those terms which are multiplied by R^0 contribute to the integrals (15.28); these terms are the first derivatives of the contributing term in the power expansion of λ. Finally, of the second derivatives, we should require the terms multiplied by R^{-1}; but as λ is regular at the center of S, these terms vanish.

To obtain the power expansion of λ, let us first expand the expressions $\overset{k}{r}$. We write $\overset{k}{r}$ in the form

$$\left.\begin{aligned} \overset{k}{r} &= [(\xi^t - \overset{k}{y}{}^t)(\xi^t - \overset{k}{y}{}^t)]^{1/2} = [(\overset{p,k}{y}{}^t + R\eta_t)(\overset{p,k}{y}{}^t + R\eta_t)]^{1/2}, \\ \overset{p,k}{y}{}^t &= \overset{p}{y}{}^t - \overset{k}{y}{}^t. \end{aligned}\right\} \tag{15.37}$$

Let us call the "coördinate distance" between the pth and the kth mass points, $[(\overset{p}{y}{}^t - \overset{k}{y}{}^t)(\overset{p}{y}{}^t - \overset{k}{y}{}^t)]^{1/2}$, $r_{p,k}$; and we obtain for $\overset{k}{r}$ this power expansion:

$$\left.\begin{aligned} \overset{k}{r} &= r_{p,k}\left[1 + 2\,\frac{\overset{p,k}{y}{}^t\eta_t}{(r_{p,k})^2}R + \frac{R^2}{(r_{p,k})^2}\right]^{1/2} \\ &= r_{p,k}\left[1 + \frac{\overset{p,k}{y}{}^t\eta_t}{(r_{p,k})^2}R + \cdots\right]. \end{aligned}\right\} \tag{15.38}$$

The power expansion of λ takes the form

$$\lambda = -\sum_{k=1}^{N}{}' \frac{\overset{k}{\mu}}{r_{p,k}}\left(1 - \frac{\overset{p,k}{y}{}^t\eta_t}{(r_{p,k})^2}R + \cdots\right). \tag{15.39}$$

Only the second term of this expansion,

$$\lambda = \cdots + \sum_{k=1}^{N}{}' \frac{\overset{k}{\mu}}{(r_{p,k})^3}\overset{p,k}{y}{}^t\eta_t R + \cdots , \tag{15.40}$$

contributes to the integrals (15.28). Its derivatives are

$$\lambda_{,r} = \cdots \sum_{k=1}^{N}{}' \frac{\overset{k}{\mu}}{(r_{p,k})^3} \overset{p,k}{y^r} + \cdots. \tag{15.41}$$

The nonlinear terms in the integrals (15.28) are, then,

$$-\tfrac{1}{8}(\iota_{,r}\lambda_{,s} + \iota_{,s}\lambda_{,r}) - \tfrac{1}{4}\lambda\iota_{,rs} + \tfrac{3}{8}\delta_{rs}\iota_{,t}\,\lambda_{,t} \equiv N_{rs}\,. \tag{15.42}$$

By substituting the expressions (15.36), (15.40), and (15.41) in eq. (15.42), we obtain for N_{rs}

$$N_{rs} = \sum_{k=1}^{N}{}' \frac{\overset{p}{\mu}\overset{k}{\mu}}{(r_{p,k})^3} \frac{1}{R^2} [-\tfrac{1}{8}(\eta_r \overset{p,k}{y^s} + \eta_s \overset{p,k}{y^r}) + \eta_t \overset{p,k}{y^t}(\tfrac{1}{8}\delta_{rs} + \tfrac{3}{4}\eta_r\eta_s)], \tag{15.43}$$

and the product of N_{rs} by cos (s, n) is

$$N_{rs}\eta_s = \sum_{k=1}^{N}{}' \frac{\overset{p}{\mu}\overset{k}{\mu}}{(r_{p,k})^3} \frac{1}{R^2} [-\tfrac{1}{8}\overset{p,k}{y^r} + \tfrac{3}{4}\eta_t \overset{p,k}{y^t}\eta_r]. \tag{15.44}$$

This expression is to be integrated over the surface S. To do this, we must consider the integral

$$\oint_S \eta_t \overset{p,k}{y^t} \eta_r \, dS.$$

We must compute separately the two integrals

$$\oint_S \overset{p,k}{y^r}(\eta_r)^2 \, dS \quad \text{(do not sum over the index } r!) \tag{15.45}$$

and

$$\oint_S \eta_t \overset{p,k}{y^t} \eta_r \, dS, \qquad t \neq r. \tag{15.46}$$

The second integral vanishes. For we can divide S into two hemispheres: one on which η_r is positive, the other on which η_r is negative. On either hemisphere, η_t takes positive and negative values on one quartersphere each, and the contributions of these two quarterspheres cancel. To compute the first integral, (15.45), we shall introduce polar coördinates, with the poles at the two points $\eta_r = \pm 1$. The integral becomes, then,

$$\oint_S \overset{p,k}{y^r}(\eta_r)^2 \, dS = 2\pi \oint_{\theta=-\pi/2}^{+\pi/2} \overset{p,k}{y^r} \cdot \sin^2\theta \cdot R^2 \cdot \cos\theta \, d\theta = \tfrac{4}{3}\pi R^2 \overset{p,k}{y^r},$$

and the integral of the expression (15.44) is

$$\oint_S N_{rs}\eta_s \, dS = \frac{\pi}{2} \sum_{k=1}^{N} \frac{\overset{p}{\mu}\overset{k}{\mu} \cdot \overset{p,k}{y^r}}{(r_{p,k})^3} \equiv Q_r\,. \tag{15.47}$$

Our condition states that the sum of L_r, eq. (15.35), and Q_r must vanish,

$$4\overset{p}{\mu}\overset{p}{\ddot{y}}{}^{r} + \sum_{k=1}^{N}{}' \frac{\overset{p}{\mu}\overset{k}{\mu}}{(r_{p,k})^3} \overset{p,k}{y}{}^{r} = 0. \tag{15.48}$$

Dividing this equation by $\overset{p}{\mu}$, and substituting for $\overset{k}{\mu}$ its value, eq. (15.14), we obtain the equation of motion of Newtonian physics,

$$\overset{p}{\ddot{y}}{}^{r} = -\sum_{k=1}^{N}{}' \frac{\kappa \overset{k}{M}}{(r_{p,k})^3} \overset{p,k}{y}{}^{r}. \tag{15.49}$$

The method of Einstein, Infeld, and Hoffmann is also applicable to a combination of gravitational and electromagnetic fields. Again, the field equations of the second approximation have solutions only if the first approximation satisfies certain integrability conditions. These conditions are identical with eq. (15.48), except for an additional term which represents the Coulomb force.

Conclusion. It is important to examine why in one field theory (the theory of gravitation) the equations of motion follow from the field equations and why in another field theory (Maxwell's theory) they do not. The main reason is that the field equations of gravitation satisfy four identities, while Maxwell's equations satisfy only one. The significance of the identities for the equations of motion is revealed both in Weyl's investigation of the rigorous solutions with rotational symmetry and in the approximation method of Einstein, Infeld, and Hoffmann.

Because of the identities, the field equations of the general theory of relativity are, so to speak, at once "overdetermined" and "underdetermined." The field equations are linear with respect to the second derivatives. However, it is impossible to solve them in terms of the second derivatives with respect to a single coördinate, such as ξ^4, because four of the ten field equations, G_{4s} and G_{44}, contain only first derivatives with respect to ξ^4. It is, therefore, not permissible to choose all the field variables and their first derivatives with respect to ξ^4 freely on some hypersurface $\xi^4 =$ const.; these quantities must satisfy four conditions on any such hypersurface. In this respect, the equations are overdetermined.

But once the variables and their first derivatives with respect to ξ^4 have been chosen on a hypersurface $\xi^4 =$ const., subject to the four equations G_{4s} and G_{44}, there are only six equations which restrict the continuation of the ten variables in the ξ^4-direction; the second derivatives of four variables with respect to ξ^4 remain arbitrary. In this respect, the equations are underdetermined. However, it can be shown that, because of the four contracted Bianchi identities, the four equations G_{4s} and G_{44} remain satisfied outside the initial hypersurface if they are satisfied on the hypersurface and if the six equations G_{rs} are satisfied everywhere.

If, on an initial hypersurface $\xi^4 = \text{const.}$, there are isolated regions in which the equations are not satisfied, then these regions can be enclosed in (spatial) surfaces, and the identities cause four integrals over these surfaces to vanish. These four integrals yield the equations of motion. The one identity which is satisfied by Maxwell's equations furnishes one condition for isolated singular regions, which is that the charges be constant (see Chapter XII, p. 194). However, the number of identities is not sufficient to yield equations of motion.

The field equations of gravitation differ in still another respect from Maxwell's equations of the electromagnetic field: they are nonlinear equations. It is impossible, therefore, to obtain new solutions of the field equations merely by combining several solutions linearly. If solutions were obtained by linear combination, mass points would not interact with each other; the approximation method of Einstein, Infeld, and Hoffmann shows that even the classical interaction of mass points is brought about by the nonlinear terms in the field equations.

A field theory can, it appears, lead to reasonable laws of motion only if the field equations satisfy at least four identities, and only if they are nonlinear. In the general theory of relativity, the nonlinearity of the field equations is brought about by their covariance with respect to general coördinate transformations. The interaction of mass points is, thus, an additional argument in favor of the principle of general covariance.

Before concluding our discussion of the general theory of relativity, we shall point out some of the problems it raises. The general theory of relativity has furnished us with a logically unified theory of gravitation. But the electromagnetic field remains logically independent of the gravitational field; the general theory of relativity has not succeeded in establishing a conceptual relationship between these two kinds of field. Several attempts have been made to create a "unified" field theory, in which the electromagnetic field as well as the gravitational field would appear as parts of the geometrical structure of space. Some of these attempts will be discussed in the third part of this book.

Furthermore, the general theory of relativity has failed to furnish us with a satisfactory theory of matter. In nature, all particles belong to some definite type of elementary particle which has a characteristic mass, a characteristic charge, and so forth. The theory of relativity, however, in which particles are merely singularities of the field, cannot explain why this is the case.

Quantum phenomena are completely outside the scopeof the general theory of relativity. Wave mechanics is not relativistic, because it cannot dispense with the concept of action at a distance Dirac's original theory of the electron describes in a Lorentz-covariant manner the action of the electromagnetic field on *one* electron; but since the field itself is not quantized, this theory does not account for the dependence of the field on the electron, and, therefore, cannot treat problems which involve the interaction of several charged particles.

Attempts to obtain a relativistic quantum theory by quantizing the electromagnetic field itself have been fairly successful. But all "quantum field theories" suffer from the mathematical inadequacy that their solutions are always divergent. Thus, there is at present no really satisfactory quantum theory which is Lorentz-covariant or covariant with respect to general coördinate transformations.

Both the theory of relativity and quantum theory have made tremendous advances beyond nineteenth-century physics. But these advances have brought us face to face with new problems to which there are no solutions at present. It is hard to predict how these difficulties will be overcome, but future theories will undoubtedly include the best of quantum theory and of relativity.

PROBLEM

Derive the equations of motion of electrically charged mass points (Coulomb's law).

PART III

Unified Field Theories

CHAPTER XVI

Weyl's Gauge-Invariant Geometry

In the general theory of relativity, the gravitational field forms the basic geometric structure of a metric space, while the electromagnetic field is unrelated to the geometry of the space. Many attempts have been made to base a new theory of gravitation and electromagnetism on a modified geometry in which there would be room and need for another geometric object besides the metric tensor. One of the most ingenious of these attempts is undoubtedly H. Weyl's gauge-invariant geometry.[1] However, in spite of the beauty of the geometrical conception, this geometry has not led to a successful theory.

In this chapter, we shall report on Weyl's geometry and the field theories based on it. The formalism by means of which Weyl's geometry is presented here differs from Weyl's own, but is equivalent to it as far as the formation of covariants goes. All covariant formations of Weyl's formalism correspond to covariants in the representation chosen here, and vice versa. Our formalism, however, eliminates one non-invariant feature of Weyl's formalism, the gauge parameter, and the corresponding transformations, the gauge transformations.

The geometry. In a metric space, the space-time interval between two infinitesimally near world points, $d\tau$, is an invariant. To each world point belongs an invariant "cone" of directions along which $d\tau$ vanishes, the "light cone." Weyl's idea was to modify the geometry so that the invariance of the light cone would be maintained, while $d\tau$ would lose its invariant character. Whether this proposal has a basis in physical experience is open to debate. There is little doubt that the possible directions of light rays are an invariant property of physical space. But there are also "atomic clocks" which provide us with universal standards for a unit of proper time. Of course, it may be that in reality the frequency of a standard spectral line is subject to small variations, so that, strictly speaking, there are no

[1] H. Weyl, *Sitzungsber. d. Preuss. Akad. d. Wiss.*, 1918, p. 465; *Ann. d. Physik*, **59**, 101 (1919).

rigorous standards for proper time measurements. Be that as it may, we shall tentatively accept Weyl's basic proposition and develop a geometry which satisfies his requirements.

The zero directions are completely determined by the *ratios* between the various components of the metric tensor. Weyl assumed, therefore, in addition to coördinate transformations, the existence of "*gauge transformations*," which multiply all components of the metric tensor by a factor which is an arbitrary function of the coördinates. The line element $d\tau$ is multiplied by the same factor and is, therefore, not "gauge invariant." It is possible to construct a geometry which is covariant with respect to both coördinate and gauge transformations.

Instead of introducing gauge transformations, we shall introduce a "normalizing condition" for the arbitrary factor of the line element, and require that the determinant of the $g_{\mu\nu}$ be equal to -1,

$$| g_{\mu\nu} | = -1. \tag{16.1}$$

This normalizing condition would not be invariant if we should preserve the transformation character of the $g_{\mu\nu}$. We shall, therefore, assume in this chapter that the $g_{\mu\nu}$ transform as the components of a tensor density of weight $-\frac{1}{2}$,

$$\overset{*}{g}_{\mu\nu} = \left| \frac{\partial \xi^{*\alpha}}{\partial \xi^{\beta}} \right|^{1/2} \frac{\partial \xi^{\iota}}{\partial \xi^{*\mu}} \frac{\partial \xi^{\kappa}}{\partial \xi^{*\nu}} g_{\iota\kappa} . \tag{16.2}$$

The line element $d\tau$,

$$d\tau^2 = g_{\mu\nu} d\xi^{\mu} d\xi^{\nu},$$

thus, loses its invariant character. The determinant can be represented as the multiple product of the metric tensor and the Levi-Civita tensor density,

$$| g_{\mu\nu} | \equiv \tfrac{1}{24}\, \delta^{\alpha_1 \alpha_2 \alpha_3 \alpha_4} \cdot \delta^{\beta_1 \beta_2 \beta_3 \beta_4} \cdot g_{\alpha_1 \beta_1} \cdot g_{\alpha_2 \beta_2} \cdot g_{\alpha_3 \beta_3} \cdot g_{\alpha_4 \beta_4} . \tag{16.3}$$

The Levi-Civita densities each have the weight $+1$; the determinant is, therefore, a scalar, and eq. (16.1) is an invariant condition. Because of the normalizing conditions (16.1), the metric tensor density has only nine algebraically independent components.

Analysis in gauge-invariant geometry. In order to formulate field equations in Weyl's geometry, we must again introduce an affine connection and a curvature. Since tensor densities play a major role in this geometry, we must extend the concept of covariant differentiation to tensor densities. To do this, we shall introduce a set of variables with one index, φ_μ, which enters the definition of covariant differentiation like this:

$$\mathfrak{A}^{\iota\cdots}{}_{\kappa\cdots;\sigma} = \mathfrak{A}^{\iota\cdots}{}_{\kappa\cdots,\sigma} + \Gamma_{\rho}{}^{\iota}{}_{\sigma} \mathfrak{A}^{\rho\cdots}{}_{\kappa\cdots} - \Gamma_{\kappa}{}^{\rho}{}_{\sigma} \mathfrak{A}^{\iota\cdots}{}_{\rho\cdots} - n \mathfrak{A}^{\iota\cdots}{}_{\kappa\cdots} \varphi_{\sigma}, \tag{16.4}$$

where n is the weight of the tensor density $\mathfrak{A}^{\iota\cdots}{}_{\kappa\cdots}$. To derive the transformation law of φ_μ, it is sufficient to consider the covariant derivatives of a scalar density D of weight n. These derivatives form a vector density,

$$D^*_{;\sigma^*} = \left|\frac{\partial \xi^\alpha}{\partial \xi^{*\beta}}\right|^n \frac{\partial \xi^\iota}{\partial \xi^{*\sigma}} D_{;\iota}.$$

Thus, we have

$$\left(\left|\frac{\partial \xi^\alpha}{\partial \xi^{*\beta}}\right|^n D\right)_{,\iota} \frac{\partial \xi^\iota}{\partial \xi^{*\sigma}} - n\overset{*}{\varphi}_\sigma \left|\frac{\partial \xi^\alpha}{\partial \xi^{*\beta}}\right|^n D = \left|\frac{\partial \xi^\alpha}{\partial \xi^{*\beta}}\right|^n (D_{,\iota} - n\varphi_\iota D) \frac{\partial \xi^\iota}{\partial \xi^{*\sigma}};$$

and, therefore,

$$n\left[\left(\log\left|\frac{\partial \xi^\alpha}{\partial \xi^{*\beta}}\right|\right)_{,\iota} \frac{\partial \xi^\iota}{\partial \xi^{*\sigma}} - \overset{*}{\varphi}_\sigma\right] D = -n\varphi_\iota \frac{\partial \xi^\iota}{\partial \xi^{*\sigma}} D. \tag{16.5}$$

The logarithmic derivative of the determinant can be written in a simplified form,

$$\left(\log\left|\frac{\partial \xi^\alpha}{\partial \xi^{*\beta}}\right|\right)_{,\iota} = \frac{\partial \xi^{*\beta}}{\partial \xi^\alpha} \frac{\partial \xi^{*\sigma}}{\partial \xi^\iota} \frac{\partial^2 \xi^\alpha}{\partial \xi^{*\beta} \partial \xi^{*\sigma}} = -\frac{\partial \xi^\alpha}{\partial \xi^{*\beta}} \frac{\partial^2 \xi^{*\beta}}{\partial \xi^\alpha \partial \xi^\iota}. \tag{16.6}$$

So we obtain finally the transformation law

$$\overset{*}{\varphi}_\sigma = \frac{\partial \xi^\iota}{\partial \xi^{*\sigma}} \varphi_\iota + \frac{\partial \xi^{*\beta}}{\partial \xi^\alpha} \frac{\partial^2 \xi^\alpha}{\partial \xi^{*\beta} \partial \xi^{*\sigma}} = \frac{\partial \xi^\iota}{\partial \xi^{*\sigma}} \left(\varphi_\iota - \frac{\partial \xi^\alpha}{\partial \xi^{*\beta}} \frac{\partial^2 \xi^{*\beta}}{\partial \xi^\alpha \partial \xi^\iota}\right). \tag{16.7}$$

In Riemannian geometry, the requirement that the covariant derivatives of the Kronecker tensor, δ^κ_ι, vanish leads to the conclusion that the $\overset{\text{I}}{\Gamma}$ and the $\overset{\text{II}}{\Gamma}$ (see Chapter V, p. 71) are equal. Once the concept of covariant differentiation is extended to densities, it is reasonable to postulate that the covariant derivatives of the Levi-Civita tensor density vanish. This condition establishes a relationship between the "pseudo-vector" φ_μ and the components of the affine connection,

$$\left.\begin{aligned} &\delta^{\rho\iota_2\iota_3\iota_4}\Gamma^{\iota_1}_{\rho\sigma} + \delta^{\iota_1\rho\iota_3\iota_4}\Gamma^{\iota_2}_{\rho\sigma} + \delta^{\iota_1\iota_2\rho\iota_4}\Gamma^{\iota_3}_{\rho\sigma} \\ &\qquad\qquad + \delta^{\iota_1\iota_2\iota_3\rho}\Gamma^{\iota_4}_{\rho\sigma} - \delta^{\iota_1\iota_2\iota_3\iota_4}\varphi_\sigma = 0. \end{aligned}\right\} \tag{16.8}$$

The left-hand side is skewsymmetric with respect to the four indices $\iota_1 \cdots \iota_4$, as can be shown by straightforward computation. Only those equations (16.8) are not empty in which these four indices all have different values. In the significant equations, the summation over ρ in the first term reduces to the one value $\rho = \iota_1$; likewise, in the second term, ρ equals ι_2; and so forth. The first four terms in eq. (16.8) are equal to

$$\delta^{\iota_1\iota_2\iota_3\iota_4}\Gamma_{\rho}{}^{\rho}{}_{\sigma},$$

and the conditions (16.8) reduce to the equations

$$\varphi_\sigma = \Gamma_{\rho}{}^{\rho}{}_{\sigma}\,. \tag{16.9}$$

The metric tensor density in Weyl's geometry plays a role similar to that of the metric tensor in Riemannian geometry. We shall, therefore, assume that its covariant derivatives vanish. If we assume, as before, that the components of the affine connection are symmetric in their subscripts, we obtain the equations

$$g_{\mu\nu;\rho} \equiv g_{\mu\nu,\rho} - g_{\mu\sigma}\Gamma_{\nu}{}^{\sigma}{}_{\rho} - g_{\sigma\nu}\Gamma_{\mu}{}^{\sigma}{}_{\rho} + \tfrac{1}{2}g_{\mu\nu}\varphi_\rho = 0. \tag{16.10}$$

We can solve these equations if we introduce the "contravariant metric tensor density" of weight $(+\frac{1}{2})$, $g^{\mu\nu}$,

$$g_{\iota\mu}g^{\mu\nu} = \delta_\iota^\nu\,. \tag{16.11}$$

The solutions of eq. (16.10) then take the form

$$\Gamma_{\iota\kappa}^{\lambda} = \tfrac{1}{2}g^{\lambda\sigma}(g_{\iota\sigma,\kappa} + g_{\kappa\sigma,\iota} - g_{\iota\kappa,\sigma}) + \tfrac{1}{4}g^{\lambda\sigma}(g_{\iota\sigma}\varphi_\kappa + g_{\kappa\sigma}\varphi_\iota - g_{\iota\kappa}\varphi_\sigma) \equiv \binom{\lambda}{\iota\kappa}. \tag{16.12}$$

If we form the contracted affine connection, $\binom{\iota}{\iota\kappa}$, we obtain again eq. (16.9). The pseudovector φ_ι and the metric tensor density $g_{\mu\nu}$ are independent of each other and are both needed to form the components of the affine connection, $\binom{\lambda}{\iota\kappa}$.

The curvature tensor, $R_{\iota\kappa\lambda\cdot}{}^{\nu}$, is skewsymmetric in ι and κ, has the cyclical symmetry expressed by eq. (11.29), and satisfies the Bianchi identities, eq. (11.35), for these identities hold for *any* curvature tensor which is formed from symmetric $\Gamma_{\iota\kappa}^{\lambda}$. However, it does not have the other symmetry properties of the curvature tensor of Riemannian geometry. Neither is the contracted tensor $R_{\iota\kappa\lambda\cdot}{}^{\iota}$ symmetric in the indices κ and λ. In forming the contracted Bianchi identities, we must, therefore, proceed somewhat cautiously.

Let us first contract the identities,

$$R_{\iota\kappa\sigma\cdot}{}^{\nu}{}_{;\lambda} + R_{\kappa\lambda\sigma\cdot}{}^{\nu}{}_{;\iota} + R_{\lambda\iota\sigma\cdot}{}^{\nu}{}_{;\kappa} \equiv 0 \tag{16.13}$$

with respect to the indices ι and ν. Thus, we obtain the equations

$$R_{\rho\kappa\sigma\cdot}{}^{\rho}{}_{;\lambda} + R_{\kappa\lambda\sigma\cdot}{}^{\rho}{}_{;\rho} + R_{\lambda\rho\sigma\cdot}{}^{\rho}{}_{;\kappa} = R_{\rho\kappa\sigma\cdot}{}^{\rho}{}_{;\lambda} + R_{\kappa\lambda\sigma\cdot}{}^{\rho}{}_{;\rho} - R_{\rho\lambda\sigma\cdot}{}^{\rho}{}_{;\kappa} \equiv 0. \tag{16.14}$$

If we now raise the index σ and contract the equations with respect to κ and σ, we obtain the equations

$$R_{\rho\sigma\cdot\cdot}{}^{\sigma\rho}{}_{;\lambda} + R_{\sigma\lambda\cdot\cdot}{}^{\sigma\rho}{}_{;\rho} - R_{\rho\lambda\cdot\cdot}{}^{\sigma\rho}{}_{;\sigma} \equiv 0, \tag{16.15}$$

in which the R-quantities are *densities*.

In Riemannian geometry, the curvature tensor is skewsymmetric in its last two indices. With the help of this skewsymmetry, it is shown that the second term and the third term of eq. (16.15) are equal. In Weyl's geometry, we cannot apply the same reasoning. To obtain the symmetry relationship of the last index pair in this geometry, we shall form the covariant curvature tensor density, $R_{\iota\kappa\lambda\mu}$. We shall introduce the symbols

$$(\iota\kappa, \lambda) = g_{\lambda\sigma}\begin{pmatrix} \sigma \\ \iota\kappa \end{pmatrix} = \tfrac{1}{2}(g_{\iota\lambda,\kappa} + g_{\kappa\lambda,\iota} - g_{\iota\kappa,\lambda}) + \tfrac{1}{4}(g_{\iota\lambda}\varphi_\kappa + g_{\kappa\lambda}\varphi_\iota - g_{\iota\kappa}\varphi_\lambda), \tag{16.16}$$

in terms of which the derivatives of the metric tensor are

$$g_{\iota\kappa,\lambda} = (\kappa\lambda, \iota) + (\iota\lambda, \kappa) - \tfrac{1}{2}g_{\iota\kappa}\varphi_\lambda . \tag{16.17}$$

With the help of the $(\iota\kappa, \lambda)$, the covariant curvature tensor density can be written in the form

$$\left.\begin{aligned} R_{\iota\kappa\lambda\mu} &= (\lambda\iota, \mu)_{,\kappa} + \tfrac{1}{2}(\lambda\iota, \mu)\varphi_\kappa - (\lambda\kappa, \mu)_{,\iota} - \tfrac{1}{2}(\lambda\kappa, \mu)\varphi_\iota \\ &\qquad + g^{\rho\sigma}[(\lambda\kappa, \rho)(\mu\iota, \sigma) - (\lambda\iota, \rho)(\mu\kappa, \sigma)] \\ &= \bar{R}_{\iota\kappa\lambda\mu} + \tfrac{1}{4}[g_{\iota\mu}\varphi_{\lambda,\kappa} - g_{\kappa\mu}\varphi_{\lambda,\iota} + g_{\lambda\mu}(\varphi_{\iota,\kappa} - \varphi_{\kappa,\iota}) \\ &\qquad + g_{\kappa\lambda}\varphi_{\mu,\iota} - g_{\iota\lambda}\varphi_{\mu,\kappa}]. \end{aligned}\right\} \tag{16.18}$$

On the right-hand side, $\bar{R}_{\iota\kappa\lambda\mu}$ stands for a number of terms which have all the algebraic symmetry properties of the Riemannian curvature tensor, while the remaining terms do not. These remaining terms have only the symmetry properties (11.28) and (11.29). If we form the expression $R_{\iota\kappa\lambda\mu} + R_{\iota\kappa\mu\lambda}$, which vanishes in Riemannian geometry, we obtain here

$$R_{\iota\kappa\lambda\mu} = -R_{\iota\kappa\mu\lambda} + \tfrac{1}{2}g_{\lambda\mu}(\varphi_{\iota,\kappa} - \varphi_{\kappa,\iota}). \tag{16.19}$$

Although φ_ι is not a vector, its skewsymmetric derivatives,

$$\varphi_{\iota\kappa} = \varphi_{\iota,\kappa} - \varphi_{\kappa,\iota} , \tag{16.20}$$

form a tensor, which can be verified with the help of eq. (16.7).

If we apply eq. (16.19) to the second term in eq. (16.15), we obtain the equation

$$R_{\rho\sigma\cdot\cdot;\lambda}^{\cdot\cdot\sigma\rho} - 2R_{\rho\lambda\cdot\cdot;\sigma}^{\cdot\cdot\sigma\rho} + \tfrac{1}{2}g^{\rho\sigma}\varphi_{\sigma\lambda;\rho} \equiv 0. \tag{16.21}$$

However, we can introduce into eq. (16.21) the symmetric part of $R_{\rho\kappa\lambda\cdot}^{\cdot\cdot\cdot\rho}$,

$$R_{\kappa\lambda} = \tfrac{1}{2}(R_{\rho\kappa\lambda\cdot}^{\cdot\cdot\cdot\rho} + R_{\rho\lambda\kappa\cdot}^{\cdot\cdot\cdot\rho}). \tag{16.22}$$

With the help of eq. (16.18), we find that $R_{\kappa\lambda}$ and $R_{\rho\kappa\lambda\cdot}{}^{\rho}$ are connected by the equation

$$R_{\rho\kappa\lambda\cdot}{}^{\rho} = R_{\kappa\lambda} + \tfrac{1}{2}\varphi_{\lambda\kappa}\,. \tag{16.23}$$

By substituting this expression in eq. (16.21), we obtain the contracted Bianchi identities in the form

$$(R^{\lambda\sigma} - \tfrac{1}{2}g^{\lambda\sigma}R)_{;\sigma} - \tfrac{1}{4}\varphi^{\lambda\sigma}{}_{;\sigma} = 0. \tag{16.24}$$

$\varphi^{\lambda\sigma}$, by the way, is a tensor density of weight $+1$, and $\varphi^{\lambda\sigma}{}_{;\sigma}$ is, therefore, the *ordinary* divergence of $\varphi^{\lambda\sigma}$, or

$$\varphi^{\lambda\sigma}{}_{;\sigma} \equiv \varphi^{\lambda\sigma}{}_{,\sigma}\,. \tag{16.25}$$

Physical interpretation of Weyl's geometry. In Weyl's geometry, the geometric structure of space is characterized by the symmetric tensor density $g_{\mu\nu}$ and the "pseudovector" φ_{μ}. It appears reasonable to assume that the $g_{\mu\nu}$ represent the gravitational field, and that the φ_{μ} are the components of the world vector potential. In Weyl's original formalism, the φ_{μ} transform as a vector with respect to coördinate transformations, but are changed by a gradient when a gauge transformation is carried out. This is the historical reason for calling the addition of a gradient to the electromagnetic world vector potential a gauge transformation.

We shall now attempt to set up field equations for the $g_{\mu\nu}$ and φ_{μ}.

Weyl's variational principle. Weyl considered it desirable to derive the field equations as the Euler-Lagrange equations of a variational principle. We shall show that the Euler-Lagrange equations of a variational principle which is invariant with respect to coördinate transformations always satisfy the necessary number of identities.

Let us consider a variational principle of the form

$$\delta I \equiv \delta \int_V \mathfrak{L}(y_A\,, y_{A,\rho}\,, \cdots)\, d\xi = 0, \tag{16.26}$$

where the indices $A, B \cdots$ are used to number the field variables y. If we vary the y_A so that the variations and their derivatives up to the order needed vanish on the boundary of the domain V, the variation of the integral (16.26) can be brought into the form

$$\delta I = \int_V \sum_A \mathfrak{L}^A\, \delta y_A\, d\xi, \tag{16.27}$$

and the equations

$$\mathfrak{L}^A = 0 \tag{16.28}$$

are the Euler-Lagrange equations of the variational principle (16.26) Let us now vary the variables y_A by amounts which correspond to an infinitesimal coördinate transformation,

$$\xi^{*\alpha} = \xi^\alpha + \delta\xi^\alpha(\xi^\rho). \tag{16.29}$$

If the $\delta\xi^\alpha$ and their derivatives vanish on the boundary of V, the value of I is not changed by the variation, regardless of whether eqs. (16.28) are satisfied or not.

Because of the coördinate transformation (16.29), the functions y_A undergo infinitesimal changes δy_A, which depend on the $\delta\xi^\alpha$ and on the particular type of transformation law of the y_A. Let us consider the case for which some of the y_A are the components of a contravariant vector, Y^ρ. If we carry out a coördinate transformation from ξ^α to $\xi^{*\alpha}$, the transformed vector components will be new functions of the ξ^α; in addition, the arguments of these functions, ξ^α, have to be replaced by $\xi^{*\alpha}$. If we call the *transformed* vector field components, as functions of the *original* coördinates, $\bar{Y}^\rho(\xi^\alpha)$, we have:

$$\bar{Y}^\rho(\xi^\alpha) = Y^\rho(\xi^\alpha) + (\delta\xi^\rho)_{,\sigma}Y^\sigma(\xi^\alpha). \tag{16.30}$$

At each world point, the quantities $\bar{Y}^\rho(\xi^\alpha)$ equal the $Y^{*\rho}(\xi^{*\alpha})$, or

$$\bar{Y}^\rho(\xi^\alpha) = Y^{*\rho}(\xi^{*\alpha}), \tag{16.31}$$

where ξ^α and $\xi^{*\alpha}$ are the original and the new coördinates, respectively, of the same world point. To obtain the difference between the functions $\bar{Y}^\rho$ and $Y^{*\rho}$ for the same values of their respective arguments, we must compare $\bar{Y}^\rho$ at the world point having the coördinates $\xi^\alpha = \overset{0}{\xi^\alpha}$ with $Y^{*\rho}$ at the world point having the coördinates $\xi^{*\alpha} = \overset{0}{\xi^\alpha}$. We have:

$$Y^{*\rho}(\overset{0}{\xi^\alpha}) = \bar{Y}^\rho(\overset{0}{\xi^\alpha} - \delta\xi^\alpha) = \bar{Y}^\rho(\overset{0}{\xi^\alpha}) - Y^\rho_{,\sigma}\delta\xi^\sigma; \tag{16.32}$$

and the infinitesimal changes in the functions Y^ρ are, therefore,

$$\delta Y^\rho = (\delta\xi^\rho)_{,\sigma}Y^\sigma - Y^\rho_{,\sigma}\delta\xi^\sigma. \tag{16.33}$$

Generally speaking, tensor and tensor density components transform according to laws of the type

$$\delta y_A = -y_{A,\sigma}\delta\xi^\sigma + \sum_B F^{B\rho}_{A\sigma} y_B(\delta\xi^\sigma)_{,\rho}, \tag{16.34}$$

where the constants $F^{B\rho}_{A\sigma}$ depend on the particular transformation laws. The transformation law of the φ_μ contains one term which is not of the type (16.34). To account for this term, we shall assume that the field variables considered transform according to a law

$$\delta y_A = -y_{A,\sigma}\delta\xi^\sigma + \sum_B F^{B\rho}_{A\sigma} y_B(\delta\xi^\sigma)_{,\rho} + G^{\rho\tau}_{A\sigma}(\delta\xi^\sigma)_{,\rho\tau}, \tag{16.35}$$

where $G^{\rho\tau}_{A\sigma}$ are additional constants.

To express the condition that I is invariant with respect to general coördinate transformations, we have to replace δy_A in eq. (16.27) by the expression (16.35). Then δI must vanish for arbitrary $\delta\xi^\alpha$, if $\delta\xi^\alpha$ and its first and second derivatives vanish on the boundary of the domain V. We have

$$\left.\begin{aligned}\delta I = \int_V \sum_A \mathfrak{L}^A \{-y_{A,\sigma}\delta\xi^\sigma + \sum_B F^{B\rho}_{A\sigma} y_B(\delta\xi^\sigma)_{,\rho}& \\ + G^{\rho\tau}_{A\sigma}(\delta\xi^\sigma)_{,\rho\tau}\}\, d\xi \equiv 0.&\end{aligned}\right\} \tag{16.36}$$

By carrying out a number of integrations by parts, we can obtain eq. (16.36) in the form

$$\int_V \sum_A \{G^{\rho\tau}_{A\sigma}\mathfrak{L}^A{}_{,\rho\tau} - \sum_B F^{B\rho}_{A\sigma}(y_B\mathfrak{L}^A)_{,\rho} - y_{A,\sigma}\mathfrak{L}^A\}\delta\xi^\sigma\, d\xi \equiv 0;$$

and as the $\delta\xi^\sigma$ are arbitrary in the interior of V, we find that the Euler-Lagrange equations satisfy the four differential identities

$$\sum_A \{G^{\rho\tau}_{A\sigma}\mathfrak{L}^A{}_{,\rho\tau} - \sum_B F^{B\rho}_{A\sigma}(y_B\mathfrak{L}^A)_{,\rho} - y_{A,\sigma}\mathfrak{L}^A\} \equiv 0. \tag{16.37}$$

Now let us proceed to the construction of a variational principle in Weyl's gauge-invariant theory. The integrand of an invariant integral must be a scalar density of the weight $+1$. The curvature scalar density,

$$R = g^{\kappa\lambda}R_{\kappa\lambda}, \tag{16.38}$$

has the weight $+\frac{1}{2}$. The Lagrangian of a variational principle must be quadratic with respect to the curvature. There are several scalar densities of weight $+1$ which can be constructed from the components of the curvature tensor. They are:

$$R^2; \qquad R_{\kappa\lambda}R^{\kappa\lambda}; \qquad R_{\iota\kappa\lambda\mu}R^{\iota\kappa\lambda\mu}; \qquad \varphi_{\kappa\lambda}\varphi^{\kappa\lambda}. \tag{16.39}$$

The first three are of the fourth differential order with respect to the $g_{\mu\nu}$. The most general Lagrangian of this differential order would be a linear combination of the four expressions (16.39). The variations of the ten $g_{\mu\nu}$ are not independent of one another, but are restricted by the condition that the determinant g remain normalized,

$$g_{\mu\nu}\delta g^{\mu\nu} = 0. \tag{16.40}$$

This auxiliary condition can be taken into account with the help of the method of Lagrangian factors. The additional parameter thus entering the equations can, however, be eliminated, and we are finally left with thirteen differential equations for the thirteen variables $g_{\mu\nu}$, φ_{μ} [the $g_{\mu\nu}$ satisfy the condition (16.1)]. The thirteen equations satisfy the four differential identities (16.37).

There are two main objections to field equations of this type. First, the equations are of the fourth differential order in the $g_{\mu\nu}$. There is every reason to believe that equations of this high order have many more solutions than field equations of the second order, and it becomes very difficult to explain why the solutions of these hypothetical equations of the fourth order are so closely approximated in nature by the solutions of equations of the second order. Second, the Lagrangian of the variational principle is not uniquely determined, for any linear combination of the expressions (16.39) is suitable. The desired unification of the field is actually not accomplished, for there remains a distinctly "electromagnetic" scalar density (the last one), which can be introduced into the Lagrangian with an arbitrary coefficient.

The equations $G_{\mu\nu} = 0$. Einstein and the author attempted to set up field equations of the *second* differential order which would be similar in structure to the equations of the general theory of relativity.[2] They investigated in particular the differential equations

$$G_{\mu\nu} = 0, \tag{16.41}$$

which are ten equations for thirteen variables; they satisfy no differential identities, but are completely independent of one another. According to eq. (16.24), Maxwell's equations,

$$\varphi^{\mu\sigma}{}_{,\sigma} = 0, \tag{16.42}$$

are a consequence of these field equations. It turned out that these equations have a static solution with spherical symmetry and with two arbitrary parameters, which can be interpreted as mass and charge, respectively. When the φ_{μ} vanish, the equations (16.41) for the $g_{\mu\nu}$ are formally identical with the field equations of the general theory of relativity; the solution with vanishing charge is Schwarzschild's solution, the determinant of which happens to be equal to (-1). But if the "electric charge" does not vanish, the boundary conditions at infinity cannot be satisfied, and the deviation of the metric tensor from $\epsilon_{\rho\sigma}$ increases with increasing r. This result is convincing proof that the field equations (16.41) also fail to meet the requirements of our physical experience.

[2] This work has not been published.

CHAPTER XVII

Kaluza's Five Dimensional Theory and the Projective Field Theories

Kaluza's theory. Kaluza also attempted to create a geometry in which the gravitational and the electromagnetic potentials together would determine the structure of space.[1] While Weyl constructed a non-Riemannian geometry, Kaluza increased the number of components of the metric tensor by changing the number of dimensions. He assumed that, in addition to the four dimensions of physical space, there was a fifth dimension, which had no direct physical significance.

The number of components of a symmetric tensor of rank 2 in n dimensions is

$$N = \tfrac{1}{2}n(n + 1). \tag{17.1}$$

In a five dimensional space, the metric tensor has, therefore, fifteen components. To account for the four dimensional character of the physical world, Kaluza assumed that, with a suitable choice of coördinates, the components of the metric tensor were independent of the fifth coördinate. Finally, to reduce the number of variables by one, Kaluza assumed that in those coördinate systems in which the field variables did not depend on ξ^5, the component of the metric tensor with two indices 5 was constant and equal to unity. With these assumptions, Kaluza showed that, at least in the first approximation, the fourteen differential equations

$$G_{\mu\nu} = 0 \quad (\mu,\, \nu = 1 \cdots 5, \text{ exclude } G_{55} = 0) \tag{17.2}$$

were equivalent to the fourteen field equations (12.55), which determine the gravitational and the electromagnetic field, if the components of the metric tensor with one index 5 were assumed to be the electromagnetic potentials.

It was soon shown that this equivalence was not only approximate, but rigorous, if the proper combinations of metric components were accepted as the gravitational potentials.

[1] Th. Kaluza, *Sitzungsber. d. Preuss. Akad. d. Wiss.*, 1921, p. 966.

To facilitate the comparison of Kaluza's theory with other theories, we shall first develop a general formalism which is applicable to several different theories, and then return to a rigorous formulation of Kaluza's proposal.

A four dimensional formalism in a five dimensional space. Let us consider a five dimensional space with the coördinates ξ^α ($\alpha = 1, \cdots, 5$) and a metric tensor $\gamma_{\mu\nu}$. (In the remainder of this chapter and in Chapter XVIII, Greek indices shall assume the values $1 \cdots 5$, while Latin indices shall run from 1 to 4.) In this space, we shall introduce four *parameters* x^a ($a = 1 \cdots 4$), which are functions of the coördinates ξ^α. The derivatives of these four parameters with respect to the coördinates, $x^a{}_{,\alpha}$, shall be everywhere linearly independent,

$$\left.\begin{aligned} \delta_{a_1 a_2 a_3 a_4} \cdot \delta^{\alpha_1 \alpha_2 \alpha_3 \alpha_4 \beta} \cdot x^{a_1}{}_{,\alpha_1} \cdot x^{a_2}{}_{,\alpha_2} \cdot x^{a_3}{}_{,\alpha_3} \cdot x^{a_4}{}_{,\alpha_4} \neq 0, \\ \text{for at least one value of } \beta. \end{aligned}\right\} \tag{17.3}$$

These four parameters define a set of curves, $x^a =$ const., in the five dimensional space. Through every point of the five dimensional space one of these curves passes. We shall consider the curves a new invariant structure in the five dimensional space. This structure remains unchanged if we carry out a "parameter transformation,"

$$x^{*a} = f^a(x^b). \tag{17.4}$$

We shall find that it is possible to introduce quantities which transform with respect to the parameter transformations as four dimensional tensors. In the physical applications which we shall consider later, the manifold which is characterized by the parameters, x^a, is assumed to be physical space; we shall, therefore, investigate how far the "geometry" of the "space" of the x^a resembles the geometry of an ordinary four dimensional, Riemannian space.

We shall call a "*p-tensor*" (short for "parameter tensor") a set of functions of the ξ^α which are characterized by indices running from 1 to 4 and which transform with respect to parameter transformations like four dimensional tensors. The derivatives of the x^a with respect to the coördinates,

$$\gamma^a_\alpha = x^a{}_{,\alpha}, \tag{17.5}$$

are contravariant p-vectors and covariant (ordinary) vectors. With the help of these quantities, we can introduce a vector field A^α, which consists of the tangential vectors of the curves x^a = const. and which satisfies the equations

$$\left.\begin{aligned} \gamma^{a}_{\alpha} A^{\alpha} &= 0, \\ \gamma_{\alpha\beta} A^{\alpha} A^{\beta} &= 1. \end{aligned}\right\} \tag{17.6}$$

These five conditions determine the **A**-field completely.

With the help of both the γ^{a}_{α} and the A^{α}, we can define a field of "reciprocal γ," γ^{α}_{a}, which satisfy these conditions

$$\left.\begin{aligned} \gamma^{a}_{\alpha} \gamma^{\alpha}_{b} &= \delta^{a}_{b}, \\ A_{\alpha} \gamma^{\alpha}_{a} &= 0. \end{aligned}\right\} \tag{17.7}$$

These twenty conditions determine the γ^{α}_{a} completely.

With each (ordinary) vector V^{α} or W_{β}, we can associate a scalar and a p-vector,

$$\left.\begin{aligned} V^{a} &= \gamma^{a}_{\alpha} V^{\alpha}, \\ V &= A_{\alpha} V^{\alpha}, \end{aligned}\right\} \tag{17.8}$$

and

$$\left.\begin{aligned} W_{b} &= \gamma^{\beta}_{b} W_{\beta}, \\ W &= A^{\beta} W_{\beta}. \end{aligned}\right\} \tag{17.9}$$

The scalar is the part of **V** or **W** which is parallel to **A**, while the p-vector represents the part which is normal to **A**. Conversely, we can associate with a p-vector, U^{a}, an ordinary vector,

$$U^{\alpha} = \gamma^{\alpha}_{a} U^{a} \tag{17.10}$$

(the summation convention is to be extended to p-indices); this vector is orthogonal to **A**, regardless of the choice of U^{a} [because of the last four eqs. (17.7)].

With the metric tensor $\gamma_{\alpha\beta}$ we can associate a covariant p-tensor,

$$g_{ab} = \gamma^{\alpha}_{a} \gamma^{\beta}_{b} \gamma_{\alpha\beta}, \tag{17.11}$$

which we shall call the metric p-tensor or the p-metric.

The mixed γ-quantities of both kinds and the vector A^{α} enable us to split up every five dimensional relationship into "four dimensional" and scalar relations. The mixed γ-quantities "project" the five dimensional space on our four dimensional one. If their product is contracted with respect to the p-indices, we obtain a typical projection tensor,

$$\left.\begin{aligned} &\gamma^{a}_{\alpha} \gamma^{\beta}_{a} = \epsilon^{\beta}_{\alpha}, \\ &\epsilon^{\beta}_{\alpha} A_{\beta} = 0, \qquad \epsilon^{\beta}_{\alpha} A^{\alpha} = 0, \qquad \epsilon^{\beta}_{\alpha} \epsilon^{\gamma}_{\beta} = \epsilon^{\gamma}_{\alpha}. \end{aligned}\right\} \tag{17.12}$$

Any vector multiplied by ϵ_α^β becomes orthogonal to **A**. It can easily be shown that ϵ_α^β can be represented by the Kronecker symbol and the vector **A**,

$$\epsilon_\alpha^\beta = \delta_\alpha^\beta - A_\alpha A^\beta. \tag{17.13}$$

The proof can be carried out by multiplying the expression $(\delta_\alpha^\beta - A_\alpha A^\beta - \epsilon_\alpha^\beta)$ by the five linearly independent sets of five components, γ_a^α, A^α. All five products,

$$(\delta_\alpha^\beta - A_\alpha A^\beta - \epsilon_\alpha^\beta)\gamma_a^\alpha = 0,$$

$$(\delta_\alpha^\beta - A_\alpha A^\beta - \epsilon_\alpha^\beta)A^\alpha = 0$$

vanish. If the products of a five-rowed matrix by five independent "vectors" vanish, the matrix is the zero matrix.

A vector can be represented by its associated p-vector and its associated scalar as follows:

$$U^\alpha = \gamma_a^\alpha U^a + A^\alpha U. \tag{17.14}$$

For, if we replace U^a and U by the expressions (17.8), we obtain for the right-hand side of eq. (17.14)

$$\gamma_a^\alpha \gamma_\beta^a U^\beta + A^\alpha A_\beta U^\beta = (\epsilon_\beta^\alpha + A^\alpha A_\beta)U^\beta = \delta_\beta^\alpha U^\beta = U^\alpha.$$

Analysis in the p-formalism. In addition to the differential covariants of a Riemannian five dimensional space, there are differential covariants peculiar to the formalism developed on the preceding pages. We shall discuss them one by one.

The inner product of the (ordinary) derivative of a p-tensor by A^α,

$$t^{i\cdots}{}_{k\cdots,\alpha}A^\alpha$$

is a p-tensor of the same type. The proof can be carried out by straightforward computation. We shall call this type of covariant differentiation A-differentiation. If the A-derivative of a p-tensor vanishes, we call the p-tensor field cylindrical with respect to the **A**-field, or, for short, A-cylindrical.

We shall now consider a few of these A-derivatives. Let us consider first the A-derivative of the metric p-tensor, $g_{mn,\alpha}A^\alpha$. We shall show that it can be represented by covariant derivatives of the **A**-field. We can replace $g_{mn,\alpha}A^\alpha$ by the expression

$$g_{mn,\alpha}A^\alpha = \gamma_m^\rho \gamma_n^\sigma \cdot \gamma_\rho^r \gamma_\sigma^s g_{rs,\alpha}A^\alpha,$$

and then work on those factors to the right of the dot. We have:

$$\begin{aligned}
\gamma^r_\rho\gamma^s_\sigma g_{rs,\alpha}A^\alpha &= [(\gamma^r_\rho\gamma^s_\sigma g_{rs})_{,\alpha} - (\gamma^r_\rho\gamma^s_\sigma)_{,\alpha}g_{rs}]A^\alpha \\
&= [(\gamma_{\rho\sigma} - A_\rho A_\sigma)_{,\alpha} - (\gamma^r_{\alpha,\rho}\gamma^s_\sigma + \gamma^r_\rho\gamma^s_{\alpha,\sigma})g_{rs}]A^\alpha \\
&= [\gamma_{\rho\sigma,\alpha} - A_{\rho,\alpha}A_\sigma - A_{\sigma,\alpha}A_\rho]A^\alpha + (\gamma^r_\alpha\gamma^s_\sigma A^\alpha{}_{,\rho} + \gamma^r_\rho\gamma^s_\alpha A^\alpha{}_{,\sigma})g_{rs} \\
&= (\gamma_{\rho\sigma,\alpha} - A_{\rho,\alpha}A_\sigma - A_{\sigma,\alpha}A_\rho)A^\alpha + (\gamma_{\alpha\sigma} - A_\alpha A_\sigma)A^\alpha{}_{,\rho} \\
&\qquad + (\gamma_{\alpha\rho} - A_\alpha A_\rho)A^\alpha{}_{,\sigma} \\
&= (\gamma_{\rho\sigma,\alpha} - A_{\rho,\alpha}A_\sigma - A_{\sigma,\alpha}A_\rho)A^\alpha + (A_{\sigma,\rho} + A_{\rho,\sigma}) \\
&\qquad - A^\alpha(\gamma_{\alpha\sigma,\rho} + \gamma_{\alpha\rho,\sigma}) + A^\alpha(A_\sigma A_{\alpha,\rho} + A_\rho A_{\alpha,\sigma}).
\end{aligned}$$

At this point, we can combine a number of terms so that we obtain a (five dimensional) tensor expression. We shall rewrite the terms in this order:

$$\left.\begin{aligned}
\gamma^r_\rho\gamma^s_\sigma g_{rs,\alpha}A^\alpha &= (A_{\rho,\sigma} + A_{\sigma,\rho} - 2[\rho\sigma,\alpha]A^\alpha) \\
&\qquad - (A_{\rho\alpha}A_\sigma + A_{\sigma\alpha}A_\rho)A^\alpha \\
&= A_{\rho;\sigma} + A_{\sigma;\rho} - (B_\rho A_\sigma + B_\sigma A_\rho),
\end{aligned}\right\} \tag{17.15}$$

where $A_{\rho\sigma}$ and B_ρ are a tensor and a vector, respectively, which denote the expressions

$$A_{\rho\sigma} = A_{\rho,\sigma} - A_{\sigma,\rho}\,, \tag{17.16}$$

$$B_\rho = A_{\rho\sigma}A^\sigma. \tag{17.17}$$

For the A-derivative of the metric p-tensor, we find the expression

$$g_{mn,\alpha}A^\alpha = \gamma^\rho_m\gamma^\sigma_n(A_{\rho;\sigma} + A_{\sigma;\rho}). \tag{17.18}$$

The square of the line element,

$$d\tau^2 = \gamma_{\alpha\beta}\,d\xi^\alpha\,d\xi^\beta,$$

equals the sum of the squares of the p-line element,

$$ds^2 = g_{ab}\gamma^a_\alpha\gamma^b_\beta\,d\xi^\alpha\,d\xi^\beta, \tag{17.19}$$

and of the component of the coördinate differential vector in the A-direction,

$$d\tau_A^2 = (A_\alpha\,d\xi^\alpha)^2.$$

Let us consider an arbitrary curve in the five dimensional space, having two end points, P_1 and P_2. If we displace the whole curve by letting each of its points slide along an A-curve for the same distance, the p-length of the curve,

$$s_{1,2} = \int_{P_1}^{P_2} ds, \tag{17.20}$$

will remain the same if the p-metric is A-cylindric. The A-length of the curve remains the same if the vector B_ρ vanishes,

$$B_\rho = 0; \tag{17.21}$$

and the (five dimensional) length of the curve will remain unchanged if the **A**-field satisfies Killing's equation,

$$A_{\rho;\sigma} + A_{\sigma;\rho} = 0. \tag{17.22}$$

For the p-metric to be cylindric, it is sufficient that the right-hand side of eq. (17.15) vanish. This condition is weaker than Killing's equation, for the product of the right-hand side of eq. (17.15) and A^ρ vanishes identically; in other words, there are only ten algebraically different components of eq. (17.15), while Killing's equation has fifteen components. It is obvious that the cylindricity of the p-metric and the cylindricity of the "A-metric" together imply cylindricity of the (five dimensional) metric, and vice versa.

Next, we shall consider the A-derivative of the skewsymmetric p-tensor φ_{rs},

$$\varphi_{rs} = \gamma^{\rho}_{r}\gamma^{\sigma}_{s}A_{\rho\sigma}. \tag{17.23}$$

This tensor φ_{rs} will eventually be interpreted as the electromagnetic field tensor. Its A-derivative is

$$\begin{aligned}
\varphi_{rs,\alpha}A^\alpha &= \gamma^{\rho}_{r}\gamma^{\sigma}_{s}\cdot\gamma^{m}_{\rho}\gamma^{n}_{\sigma}\varphi_{mn,\alpha}A^\alpha \\
&= \gamma^{\rho}_{r}\gamma^{\sigma}_{s}\cdot[\varphi_{\rho\sigma,\alpha} - \varphi_{mn}(\gamma^{m}_{\rho}\gamma^{n}_{\sigma})_{,\alpha}]A^\alpha, \\
\varphi_{\rho\sigma} &= \gamma^{r}_{\rho}\gamma^{s}_{\sigma}\varphi_{rs}.
\end{aligned}$$

In the square bracket, we shall eliminate the γ^{m}_{ρ} and γ^{n}_{σ}. This can be done by the following transformation:

$$\begin{aligned}
\varphi_{mn}(\gamma^{m}_{\rho}\gamma^{n}_{\sigma})_{,\alpha}A^\alpha &= \varphi_{mn}(\gamma^{m}_{\alpha,\rho}\gamma^{n}_{\sigma} + \gamma^{m}_{\rho}\gamma^{n}_{\alpha,\sigma})A^\alpha \\
&= -\varphi_{mn}(\gamma^{m}_{\alpha}\gamma^{n}_{\sigma}A^\alpha{}_{,\rho} + \gamma^{m}_{\rho}\gamma^{n}_{\alpha}A^\alpha{}_{,\sigma}) \\
&= \varphi_{\sigma\alpha}A^\alpha{}_{,\rho} - \varphi_{\rho\alpha}A^\alpha{}_{,\sigma}.
\end{aligned}$$

We have, therefore,

$$\varphi_{rs,\alpha}A^\alpha = \gamma^{\rho}_{r}\gamma^{\sigma}_{s}[\varphi_{\rho\sigma,\alpha}A^\alpha + \varphi_{\rho\alpha}A^\alpha{}_{,\sigma} - \varphi_{\sigma\alpha}A^\alpha{}_{,\rho}].$$

Since the product of $\varphi_{\rho\alpha}$ and A^{α} vanishes, we may write:

$$\begin{aligned}\varphi_{rs,\alpha}A^{\alpha} &= \gamma_r^{\rho}\gamma_s^{\sigma}(\varphi_{\rho\sigma,\alpha} + \varphi_{\sigma\alpha,\rho} + \varphi_{\alpha\rho,\sigma})A^{\alpha} \\ &= \gamma_r^{\rho}\gamma_s^{\sigma}[(\epsilon_\rho^{\mu}\epsilon_\sigma^{\nu}A_{\mu\nu})_{,\alpha} + (\epsilon_\sigma^{\mu}\epsilon_\alpha^{\nu}A_{\mu\nu})_{,\rho} + (\epsilon_\alpha^{\mu}\epsilon_\rho^{\nu}A_{\mu\nu})_{,\sigma}]A^{\alpha}.\end{aligned}$$

To compute the square bracket, let us keep in mind that the cyclic derivatives of $A_{\mu\nu}$ vanish and that both indices of $A_{\mu\nu}$ cannot be multiplied by the same vector. Only those terms will contribute in which one of the two factors ϵ_ρ^{μ} has been replaced by δ_ρ^{μ} and the other by $(-A^{\mu}A_{\rho})$. We obtain, therefore,

$$\begin{aligned}\varphi_{rs,\alpha}A^{\alpha} = &-\gamma_r^{\rho}\gamma_s^{\sigma}[(A_{\mu\sigma}A^{\mu}A_{\rho} + A_{\rho\mu}A^{\mu}A_{\sigma})_{,\alpha} + (A_{\alpha\mu}A^{\mu}A_{\rho} + A_{\mu\rho}A^{\mu}A_{\alpha})_{,\sigma} \\ &\qquad + (A_{\sigma\mu}A^{\mu}A_{\alpha} + A_{\mu\alpha}A^{\mu}A_{\sigma})_{,\rho}]A^{\alpha} \\ = &\ \gamma_r^{\rho}\gamma_s^{\sigma}[(B_{\sigma}A_{\rho} - B_{\rho}A_{\sigma})_{,\alpha} + (B_{\alpha}A_{\rho} - B_{\rho}A_{\alpha})_{,\sigma} \\ &\qquad + (B_{\sigma}A_{\alpha} - B_{\alpha}A_{\sigma})_{,\rho}]A^{\alpha}.\end{aligned}$$

If the differentiations are carried out, most terms cancel, either because $B_{\alpha}A^{\alpha}$ vanishes, or because undifferentiated A_{ρ} are multiplied by γ_r^{ρ}. The final result is

$$\varphi_{rs,\alpha}A^{\alpha} = \gamma_r^{\rho}\gamma_s^{\sigma}(B_{\rho,\sigma} - B_{\sigma,\rho}). \tag{17.24}$$

Having considered the "A-differentiation," we shall consider another differential operation, the "*p-differentiation.*" We shall call the expression

$$V_{|a} \equiv V_{,\alpha}\gamma_a^{\alpha} \tag{17.25}$$

the p-derivative of V with respect to x^a. If the function happens to be A-cylindrical, the p-derivative is the ordinary derivative of V with respect to the argument x^a. (An A-cylindrical function can be considered as a function of the parameters x^a only.)

The p-derivatives of a scalar form a p-vector.

The p-differentiations do not, in general, commute. Their commutation law is

$$V_{|ab} - V_{|ba} = V_{,\rho}(\gamma_{a|b}^{\rho} - \gamma_{b|a}^{\rho}). \tag{17.26}$$

If V is A-cylindrical, the p-differentiations, being ordinary differentiations, must commute, and the right-hand side of eq. (17.26) vanishes. The bracket is, therefore, proportional to A^{ρ},

$$\gamma_{a|b}^{\rho} - \gamma_{b|a}^{\rho} = A^{\rho}Q_{ab}\,. \tag{17.27}$$

We can find Q_{ab} by multiplying eq. (17.27) by A_{ρ},

$$Q_{ab} = A_{\rho}(\gamma_{a|b}^{\rho} - \gamma_{b|a}^{\rho}) = \gamma_b^{\rho}A_{\rho|a} - \gamma_a^{\rho}A_{\rho|b} = \gamma_a^{\alpha}\gamma_b^{\beta}A_{\beta\alpha} = \varphi_{ba}\,.$$

We have, therefore,

$$\gamma^{\rho}_{a|b} - \gamma^{\rho}_{b|a} = A^{\rho}\varphi_{ba} , \tag{17.28}$$

and

$$V_{|ab} - V_{|ba} = V_{,\rho}A^{\rho}\varphi_{ba} . \tag{17.26a}$$

The p-derivatives of a p-tensor are not, in general, covariant. But the skewsymmetric p-derivatives of a p-vector form a p-tensor, and the cyclic p-derivatives of a skewsymmetric p-tensor of rank 2 also have p-tensor character.

Of particular interest is the p-tensor $\varphi_{rs|t} + \varphi_{st|r} + \varphi_{tr|s}$. We can express this tensor in terms of the skewsymmetric derivatives of A_ρ,

$$\varphi_{rs|t} + \varphi_{st|r} + \varphi_{tr|s} = (\gamma^{\rho}_{r}\gamma^{\sigma}_{s}A_{\rho\sigma})_{,\tau}\gamma^{\tau}_{t} + (\gamma^{\sigma}_{s}\gamma^{\tau}_{t}A_{\sigma\tau})_{,\rho}\gamma^{\rho}_{r} + (\gamma^{\tau}_{t}\gamma^{\rho}_{r}A_{\tau\rho})_{,\sigma}\gamma^{\sigma}_{s} .$$

As the cyclic derivatives of $A_{\rho\sigma}$ vanish, we have left only the terms

$$\varphi_{rs|t} + \varphi_{st|r} + \varphi_{tr|s} = A_{\mu\nu}[\gamma^{\nu}_{r}(\gamma^{\mu}_{t|s} - \gamma^{\mu}_{s|t}) + \gamma^{\nu}_{s}(\gamma^{\mu}_{r|t} - \gamma^{\mu}_{t|r})$$
$$+ \gamma^{\nu}_{t}(\gamma^{\mu}_{s|r} - \gamma^{\mu}_{r|s})].$$

In accordance with eq. (17.28), the skewsymmetric p-derivatives of the γ^{μ}_{t} can be eliminated, and our final result is

$$\varphi_{rs|t} + \varphi_{st|r} + \varphi_{tr|s} = \gamma^{\rho}_{r}\gamma^{\sigma}_{s}\gamma^{\tau}_{t}(B_{\rho}A_{\tau\sigma} + B_{\sigma}A_{\rho\tau} + B_{\tau}A_{\sigma\rho}). \tag{17.29}$$

We shall now define the *covariant differentiation of p-tensors.* If V^a is a *p-vector*, then it can be easily shown that the differential expressions

$$V^{a}{}_{,\rho} + \Gamma^{a}_{s\rho}V^{s}$$

form a mixed tensor (p-vector with respect to a, vector with respect to ρ) if the coefficients $\Gamma^{a}_{s\rho}$ transform according to the law

$$\Gamma^{*a}_{s\rho} = \frac{\partial\xi^{\sigma}}{\partial\xi^{*\rho}}\frac{\partial x^{k}}{\partial x^{*s}}\left(\frac{\partial x^{*a}}{\partial x^{i}}\Gamma^{i}_{k\sigma} - \gamma^{l}_{\sigma}\frac{\partial^{2}x^{*a}}{\partial x^{l}\partial x^{k}}\right). \tag{17.30}$$

There are $5 \times 4 \times 4$ quantities $\Gamma^{a}_{b\rho}$. To determine them, we must formulate an equal number of conditions. The conditions

$$\gamma^{\alpha}_{a;\rho} = 0 \tag{17.31}$$

would be too restrictive, while

$$g_{ab;\rho} = 0 \tag{17.32}$$

would be too loose. Suitable conditions must depend on three indices, one coördinate index and two parameter indices. The conditions

$$\gamma^{a}_{\alpha}\gamma^{\alpha}_{s;\rho} = 0 \tag{17.33}$$

are of this type. By straightforward computation we find that these conditions are satisfied if the $\Gamma^{b}_{a\rho}$ take the values

$$\Gamma^{b}_{a\rho} = \gamma^{b}_{\beta}\gamma^{\alpha}_{a}\begin{Bmatrix}\beta\\ \alpha\rho\end{Bmatrix} - \gamma^{b}_{\rho|a} \equiv \begin{Bmatrix}b\\ a\rho\end{Bmatrix}. \tag{17.34}$$

It can be shown that by this choice for the $\Gamma^{b}_{a\rho}$, eqs. (17.32) are satisfied, too. However, eqs. (17.31) are not satisfied, and we have, instead,

$$\gamma^{\alpha}_{a;\rho} = A^{\alpha}A_{\kappa}\left(\gamma^{\kappa}_{a,\rho} + \begin{Bmatrix}\kappa\\ \sigma\rho\end{Bmatrix}\gamma^{\sigma}_{a}\right) = A^{\alpha}A_{\kappa}\gamma^{\kappa}_{a;\rho} = -\gamma^{\kappa}_{a}A^{\alpha}A_{\kappa;\rho}. \tag{17.35}$$

The covariant p-derivative of a tensor or p-tensor with respect to x^{a} shall be defined as the covariant derivative with respect to ξ^{α}, multiplied by γ^{α}_{a},

$$V^{\alpha}{}_{,b} = V^{\alpha}{}_{|b} + \begin{Bmatrix}\alpha\\ \rho b\end{Bmatrix}V^{\rho}, \qquad \begin{Bmatrix}\alpha\\ \rho b\end{Bmatrix} = \begin{Bmatrix}\alpha\\ \rho\beta\end{Bmatrix}\gamma^{\beta}_{b}, \tag{17.36}$$

and

$$V^{a}{}_{;b} = V^{a}{}_{|b} + \begin{Bmatrix}a\\ sb\end{Bmatrix}V^{s}, \qquad \begin{Bmatrix}a\\ sb\end{Bmatrix} = \begin{Bmatrix}a\\ s\beta\end{Bmatrix}\gamma^{\beta}_{b}. \tag{17.37}$$

The $\begin{Bmatrix}a\\ sb\end{Bmatrix}$ are symmetric in s and b, and we know that the expressions $g_{ab;s}$ vanish if they are formed with the help of the expressions (17.34). It follows that the $\begin{Bmatrix}s\\ ab\end{Bmatrix}$ have the values

$$\begin{Bmatrix}s\\ ab\end{Bmatrix} = \tfrac{1}{2}g^{sr}(g_{ar|b} + g_{br|a} - g_{ab|r}). \tag{17.38}$$

We shall now turn to the various curvature tensors which we can form with the help of the different types of affine connection. The easiest method of obtaining them is to formulate the various commutation laws. We have, of course, the Riemannian commutation law,

$$V^{\nu}{}_{;\iota\kappa} - V^{\nu}{}_{;\kappa\iota} = R_{\iota\kappa\lambda}{}^{\nu}V^{\lambda}. \tag{17.39}$$

To obtain the commutation law for the differentiation of a p-vector, we need the covariant derivatives of γ^{a}_{α}. They are

$$\gamma^{a}_{\alpha;\rho} = A_{\alpha}A^{\kappa}\left(\gamma^{a}_{\kappa,\rho} - \begin{Bmatrix}\sigma\\ \kappa\rho\end{Bmatrix}\gamma^{a}_{\sigma}\right) = A_{\alpha}A^{\kappa}\gamma^{a}_{\kappa;\rho} = -A_{\alpha}\gamma^{a}_{\kappa}A^{\kappa}{}_{;\rho}. \tag{17.40}$$

Thus, the commutation law is

$$\left.\begin{aligned} V^{n}{}_{;\iota\kappa} - V^{n}{}_{;\kappa\iota} &= (\gamma^{n}_{\nu}V^{\nu})_{;\iota\kappa} - (\gamma^{n}_{\nu}V^{\nu})_{;\kappa\iota} \\ &= \gamma^{n}_{\nu}(V^{\nu}{}_{;\iota\kappa} - V^{\nu}{}_{;\kappa\iota}) + V^{\nu}(\gamma^{n}_{\nu;\iota\kappa} - \gamma^{n}_{\nu;\kappa\iota}) \\ &= V^{\lambda}\gamma^{n}_{\nu}R_{\iota\kappa\lambda}{}^{\nu} + (\gamma^{n}_{\nu;\iota\kappa} - \gamma^{n}_{\nu;\kappa\iota})V^{\nu}. \end{aligned}\right\} \tag{17.41}$$

The expression $\gamma^{n}_{\nu;\iota\kappa} - \gamma^{n}_{\nu;\kappa\iota}$ can be computed as follows:

$$\left.\begin{aligned}\gamma^{n}_{\nu;\iota\kappa} - \gamma^{n}_{\nu;\kappa\iota} &= (\gamma^{n}_{\sigma} A^{\sigma}{}_{;\kappa} A_{\nu})_{;\iota} - (\gamma^{n}_{\sigma} A^{\sigma}{}_{;\iota} A_{\nu})_{;\kappa} \\ &= (\gamma^{n}_{\sigma;\iota} A^{\sigma}{}_{;\kappa} - \gamma^{n}_{\sigma;\kappa} A^{\sigma}{}_{;\iota})A_{\nu} + \gamma^{n}_{\sigma} A_{\nu}(A^{\sigma}{}_{;\kappa\iota} - A^{\sigma}{}_{;\iota\kappa}) \\ &\qquad + \gamma^{n}_{\sigma}(A^{\sigma}{}_{;\kappa} A_{\nu;\iota} - A^{\sigma}{}_{;\iota} A_{\nu;\kappa}) \\ &= \gamma^{n}_{\sigma}[A^{\sigma}{}_{;\kappa} A_{\nu;\iota} - A^{\sigma}{}_{;\iota} A_{\nu;\kappa} - R_{\iota\kappa\rho\cdot}{}^{\sigma} A^{\rho} A_{\nu}],\end{aligned}\right\} \tag{17.42}$$

for the first parenthesis in the second line vanishes. If we substitute this expression in eq. (17.41), we obtain, after a few changes,

$$\left.\begin{aligned}V^{n}{}_{;\iota\kappa} - V^{n}{}_{;\kappa\iota} &= R_{\iota\kappa l\cdot}{}^{n} V^{l}, \\ R_{\iota\kappa l\cdot}{}^{n} &= \gamma^{n}_{\nu}\gamma^{\lambda}_{l}(R_{\iota\kappa\lambda\cdot}{}^{\nu} + A^{\nu}{}_{;\kappa} A_{\lambda;\iota} - A^{\nu}{}_{;\iota} A_{\lambda;\kappa}).\end{aligned}\right\} \tag{17.43}$$

The "mixed" curvature tensor $R_{\iota\kappa l\cdot}{}^{n}$ can also be represented by means of the $\left\{\begin{matrix} b \\ a\sigma \end{matrix}\right\}$,

$$R_{\iota\kappa l\cdot}{}^{n} = \left\{\begin{matrix} n \\ l\iota \end{matrix}\right\}_{,\kappa} - \left\{\begin{matrix} n \\ l\kappa \end{matrix}\right\}_{,\iota} - \left\{\begin{matrix} n \\ s\iota \end{matrix}\right\}\left\{\begin{matrix} s \\ l\kappa \end{matrix}\right\} + \left\{\begin{matrix} n \\ s\kappa \end{matrix}\right\}\left\{\begin{matrix} s \\ l\iota \end{matrix}\right\}. \tag{17.44}$$

Next we shall set up the commutation law for the covariant p-derivatives of an (ordinary) vector:

$$\left.\begin{aligned}V^{\nu}{}_{;ik} - V^{\nu}{}_{;ki} &= \gamma^{\kappa}_{k}(\gamma^{\iota}_{i} V^{\nu}{}_{;\iota})_{;\kappa} - \gamma^{\iota}_{i}(\gamma^{\kappa}_{k} V^{\nu}{}_{;\kappa})_{;\iota} \\ &= \gamma^{\iota}_{i}\gamma^{\kappa}_{k}(V^{\nu}{}_{;\iota\kappa} - V^{\nu}{}_{;\kappa\iota}) + V^{\nu}{}_{;\rho}(\gamma^{\sigma}_{k}\gamma^{\rho}_{i;\sigma} - \gamma^{\sigma}_{i}\gamma^{\rho}_{k;\sigma}) \\ &= \gamma^{\iota}_{i}\gamma^{\kappa}_{k}[R_{\iota\kappa\lambda\cdot}{}^{\nu} V^{\lambda} + A_{\kappa\iota} A^{\rho} V^{\nu}{}_{;\rho}].\end{aligned}\right\} \tag{17.45}$$

A similar calculation furnishes us with the commutation law for the p-derivatives of a p-vector:

$$\left.\begin{aligned}V^{n}{}_{;ik} - V^{n}{}_{;ki} &= (\gamma^{n}_{\nu}{}_{;ik} - \gamma^{n}_{\nu}{}_{;ki})V^{\nu} + \gamma^{n}_{\nu}(V^{\nu}{}_{;ik} - V^{\nu}{}_{;ki}) \\ &= \gamma^{\iota}_{i}\gamma^{\kappa}_{k}(R_{\iota\kappa l\cdot}{}^{n} V^{l} + A_{\kappa\iota} A^{\rho} V^{n}{}_{;\rho}) \\ &= \gamma^{\iota}_{i}\gamma^{\kappa}_{k}[(R_{\iota\kappa l\cdot}{}^{n} + A_{\kappa\iota}\gamma^{n}_{\nu}\gamma^{\lambda}_{l} A^{\nu}{}_{;\lambda})V^{l} + A_{\kappa\iota} A^{\rho} V^{n}{}_{,\rho}].\end{aligned}\right\} \tag{17.46}$$

On the other hand, we can express the same commutation law by means of the $\left\{\begin{matrix} l \\ ik \end{matrix}\right\}$,

$$\left.\begin{aligned}V^{n}{}_{;ik} - V^{n}{}_{;ki} &= R_{ikl\cdot}{}^{n} V^{l} + V^{n}{}_{,\rho} A^{\rho}\varphi_{ki}, \\ R_{ikl\cdot}{}^{n} &= \left\{\begin{matrix} n \\ li \end{matrix}\right\}_{|k} - \left\{\begin{matrix} n \\ lk \end{matrix}\right\}_{|i} - \left\{\begin{matrix} n \\ si \end{matrix}\right\}\left\{\begin{matrix} s \\ lk \end{matrix}\right\} + \left\{\begin{matrix} n \\ sk \end{matrix}\right\}\left\{\begin{matrix} s \\ li \end{matrix}\right\}.\end{aligned}\right\} \tag{17.47}$$

Comparison of the right-hand sides of eqs. (17.47), (17.46), and (17.43) furnishes us with a relation between $R_{ikl\cdot}{}^{n}$ and $R_{\iota\kappa\lambda\cdot}{}^{\nu}$:

$$R_{ikl\cdot}{}^{n} = \gamma_i^{\iota}\gamma_k^{\kappa}\gamma_l^{\lambda}\gamma_\nu^{n}(R_{\iota\kappa\lambda\cdot}{}^{\nu} + A^{\nu}{}_{;\lambda}A_{\kappa\iota} + A^{\nu}{}_{;\kappa}A_{\lambda;\iota} - A^{\nu}{}_{;\iota}A_{\lambda;\kappa}). \qquad (17.48)$$

Of particular importance for the application of our formalism to Kaluza's theory is the double contraction of this equation,

$$\left.\begin{aligned}
&\delta_n^i\, g^{kl}\, R_{ikl\cdot}{}^{n} \\
&\quad = \epsilon_\nu^{\iota}(\gamma^{\kappa\lambda} - A^{\kappa}A^{\lambda})[R_{\iota\kappa\lambda\cdot}{}^{\nu} + A^{\nu}{}_{;\lambda}A_{\kappa\iota} + A^{\nu}{}_{;\kappa}A_{\lambda;\iota} - A^{\nu}{}_{;\iota}A_{\lambda;\kappa}] \\
&\quad = (\delta_\nu^{\iota} - A^{\iota}A_\nu)(\gamma^{\kappa\lambda} - A^{\kappa}A^{\lambda})R_{\iota\kappa\lambda\cdot}{}^{\nu} \\
&\qquad\qquad + (\gamma^{\kappa\lambda} - A^{\kappa}A^{\lambda})[A^{\rho}{}_{;\lambda}A_{\kappa\rho} + A^{\rho}{}_{;\kappa}A_{\lambda;\rho} - A^{\rho}{}_{;\rho}A_{\lambda;\kappa}] \\
&\quad = R - 2A^{\kappa}A^{\lambda}R_{\kappa\lambda} + \gamma^{\kappa\lambda}[A^{\rho}{}_{;\lambda}A_{\kappa\rho} + A^{\rho}{}_{;\kappa}A_{\lambda;\rho} - A^{\rho}{}_{;\rho}A_{\lambda;\kappa}] \\
&\qquad\qquad + B_\rho B^{\rho} \\
&\quad = R - 2A^{\kappa}A^{\lambda}R_{\kappa\lambda} + 2A^{\rho}{}_{;\sigma}A^{\sigma}{}_{;\rho} - \gamma^{\rho\sigma}A^{\tau}{}_{;\rho}A_{\tau;\sigma} - (A^{\rho}{}_{;\rho})^2 \\
&\qquad\qquad + B_\rho B^{\rho}.
\end{aligned}\right\} \qquad (17.49)$$

In this expression, we shall put the term $(-2A^{\kappa}A^{\lambda}R_{\kappa\lambda})$ into a somewhat different form, which is

$$A^{\kappa}A^{\lambda}R_{\iota\kappa\lambda\cdot}{}^{\iota} = A^{\kappa}(A^{\iota}{}_{;\iota\kappa} - A^{\iota}{}_{;\kappa\iota}) = A^{\kappa}(A^{\iota}{}_{;\iota})_{,\kappa} - B^{\iota}{}_{;\iota} + A^{\kappa}{}_{;\iota}A^{\iota}{}_{;\kappa}. \qquad (17.50)$$

Substituting this expression in eq. (17.49), we find that the p-curvature scalar is related to the (ordinary) curvature scalar by the equation

$$\delta_n^i g^{kl}R_{ikl\cdot}{}^{n} = R - \gamma^{\rho\sigma}A^{\tau}{}_{;\rho}A_{\tau;\sigma} - (A^{\rho}{}_{;\rho})^2 + B_\rho B^{\rho} - 2A^{\kappa}(A^{\iota}{}_{;\iota})_{,\kappa} + 2B^{\iota}{}_{;\iota} \qquad (17.51)$$

It should be emphasized that, from the point of view of the theory of invariants, the formalism which has been developed here is nothing but the theory of a unit vector field in a metric space. The value of this presentation is that in those unified field theories which employ a five dimensional space as the background of the physical world, the four parameters x^a are assumed to represent the four coördinates of our physical experience.

A special type of coördinate system. Apart from the covariant assumptions regarding the field **A**, which characterizes a given unified field theory, most authors have restricted their presentation to a special type of coördinate system, which is characterized by the condition that the first four coördinates $\xi^1 \cdots \xi^4$ be identical with the parameters $x^1 \cdots x^4$, while the choice of the fifth coördinate is restricted by the condition that the component A^5 of **A** shall be equal to unity. The other four components, $A^1 \cdots A^4$, vanish. Coördinate systems which satisfy these conditions will be referred to as "special coördinate systems." The only coördinate transformations which lead from one "special coördinate system" to another "special coördinate system" are of the type

$$\left.\begin{aligned} \xi^{*a} &= f^a(\xi^s), \text{ coupled with } x^{*a} = f^a(x^s) \\ \xi^{*5} &= \xi^5 + f^5(\xi^s) \end{aligned}\right\} \tag{17.52}$$

We shall call these transformations "special coördinate transformations."

In a special coördinate system, the metric tensor has the components

$$\gamma_{\alpha\beta} = \begin{Bmatrix} g_{ab} + \varphi_a\varphi_b\,, & \varphi_a \\ \varphi_b \quad , & 1 \end{Bmatrix}, \qquad \gamma^{\alpha\beta} = \begin{Bmatrix} g^{ab} \quad , & -g^{as}\varphi_s \\ -g^{bs}\varphi_s\,, & 1 + g^{rs}\varphi_r\varphi_s \end{Bmatrix}, \tag{17.53}$$

where φ_a are the first four covariant components of $\mathbf{A}$,

$$A_\rho = (\varphi_r\,, 1). \tag{17.54}$$

The transformation law of g_{ab} is that of a p-tensor and, as such, independent of the function f^5, eq. (17.52). The φ_r, on the other hand, transform according to the law,

$$\overset{*}{\varphi_r} = \frac{\partial \xi^s}{\partial \xi^{*r}} (\varphi_s - f^5{}_{,s}). \tag{17.55}$$

These are the transformation laws of gravitational and electromagnetic potentials with respect to "coördinate" and "gauge" transformations. (The expression "gauge transformation" is not to be understood as a Weyl gauge transformation!)

The various differential operations which we have discussed in the preceding section assume special forms in the special coördinate systems. The A-differentiation turns into the differentiation with respect to ξ^5,

$$V^{i\cdots}{}_{k\cdots,\alpha}A^\alpha = V^{i\cdots}{}_{k\cdots,5}\,. \tag{17.56}$$

To obtain the expression for p-differentiation, we must first know the values of γ^a_α and γ^α_a in the special coördinate system. They are

$$\gamma^a_\alpha = \left\{\begin{array}{c|c} \alpha = 1\cdots 4 & \alpha = 5 \\ \delta^a_\alpha & 0 \end{array}\right\}, \qquad \gamma^\alpha_a = \left\{\begin{array}{c|c} \alpha = 1\cdots 4 & \alpha = 5 \\ \delta^\alpha_a & -\varphi_a \end{array}\right\}. \tag{17.57}$$

For the p-differentiation, we obtain the expression

$$V_{|a} = V_{,a} - \varphi_a V_{,5}\,. \tag{17.58}$$

The p-tensor φ_{rs} takes the form

$$\varphi_{rs} = \gamma_r^\rho \gamma_s^\sigma A_{\rho\sigma} = \varphi_{r,s} - \varphi_{s,r} - \varphi_s \varphi_{r,5} + \varphi_r \varphi_{s,5} . \tag{17.59}$$

The vector B_ρ has the components

$$B_\rho = (\varphi_{\rho,5} , 0), \tag{17.60}$$

and the scalar $A^\rho{}_{;\rho}$ becomes

$$A^\rho{}_{;\rho} = A^\rho{}_{,\rho} + \begin{Bmatrix} \rho \\ \sigma\rho \end{Bmatrix} A^\sigma = \begin{Bmatrix} \rho \\ 5\rho \end{Bmatrix} = \tfrac{1}{2}\gamma^{\rho\sigma}\gamma_{\rho\sigma,5} = \tfrac{1}{2}(\log |\gamma_{\alpha\beta}|)_{,5} .$$

The determinant $|\gamma_{\alpha\beta}|$ can be expressed by the g_{rs} only. If we multiply the last column of the components of $\gamma_{\alpha\beta}$, eq. (17.53), by each of the four quantities φ_b in turn and subtract the resulting columns from the first four, we obtain

$$|\gamma_{\alpha\beta}| = \begin{vmatrix} g_{ab} + \varphi_a \varphi_b , & \varphi_a \\ \varphi_b \quad , & 1 \end{vmatrix} = \begin{vmatrix} g_{ab} , & \varphi_a \\ 0 , & 1 \end{vmatrix} = |g_{ab}|. \tag{17.61}$$

We find, therefore, for $A^\rho{}_{;\rho}$

$$A^\rho{}_{;\rho} = \tfrac{1}{2}(\log |g_{rs}|)_{,5} = \tfrac{1}{2}g^{rs}g_{rs,5} . \tag{17.62}$$

In a special coördinate system, the derivatives of an A-cylindric p-tensor with respect to ξ^5 vanish. *If the derivatives of an (ordinary) tensor with respect to ξ^5 vanish in a special coördinate system, they vanish in every special coördinate system, and we shall call such a five dimensional tensor A-cylindric.*

It turns out that the (ordinary) derivatives of a tensor with respect to ξ^5 form a tensor of the same type. In terms of general coördinate systems, these differential formations are

$$\left.\begin{aligned} V^{\iota\cdots}{}_{\kappa\cdots,5} &= V^{\iota\cdots}{}_{\kappa\cdots,\rho}A^\rho - A^\iota{}_{,\rho}V^{\rho\cdots}{}_{\kappa\cdots} - \cdots \\ &\qquad + A^\rho{}_{,\kappa}V^{\iota\cdots}{}_{\rho\cdots} + \cdots \\ &= V^{\iota\cdots}{}_{\kappa\cdots;\rho}A^\rho - A^\iota{}_{;\rho}V^{\rho\cdots}{}_{\kappa\cdots} - \cdots \\ &\qquad + A^\rho{}_{;\kappa}V^{\iota\cdots}{}_{\rho\cdots} + \cdots . \end{aligned}\right\} \tag{17.63}$$

If this tensor vanishes, we call the tensor $V^{\iota\cdots}{}_{\kappa\cdots}$ A-cylindrical. As previously, we find that the metric tensor $\gamma_{\alpha\beta}$ is A-cylindrical if Killing's equation, (17.22), is satisfied. According to the extended definition of A-cylindricity, the γ_α^a are A-cylindrical, while the γ_a^α in general are not.

Covariant formulation of Kaluza's theory. We obtain Kaluza's restrictions of the five dimensional metric by assuming that the (five dimensional) metric is A-cylindrical, that is, that A_ρ satisfies Killing's equation, (17.22). As a consequence, the vector B_ρ vanishes, and the A-curves are geodesics. From eq. (17.18), it follows that the p-metric is also A-cylindrical.

Furthermore, we find that φ_{rs} is A-cylindrical, because of eq. (17.24). In the p-tensor $(\varphi_{rs|t} + \varphi_{st|r} + \varphi_{tr|s})$, the strokes can be replaced by commas—ordinary differentiation with respect to the parameter x^a—and this p-tensor vanishes, because of eq. (17.29),

$$\varphi_{rs,t} + \varphi_{st,r} + \varphi_{tr,s} = 0. \tag{17.64}$$

This set of equations is just the condition which must be satisfied if the φ_{rs} are to be the skewsymmetric derivatives of four functions, Φ_r. These four functions are determined by the φ_{rs} except for an arbitrary additive gradient.

If we employ a "special coördinate system," we find that the quantities φ_r are independent of ξ^5, because of eq. (17.60), and that the φ_{rs} are their skewsymmetric derivatives,

$$\varphi_{rs} = \varphi_{r,s} - \varphi_{s,r}, \tag{17.65}$$

because of eq. (17.59).

We find that in Kaluza's theory, if we use a special coördinate system, both the g_{rs} and the φ_r are independent of ξ^5, and that the skewsymmetric derivatives of the φ_r form a p-tensor. If we do not use a special coördinate system, the quantities φ_r are not defined, but the p-tensor φ_{rs}, nevertheless, satisfies the second set of Maxwell's equations, (17.64), and both the g_{rs} and the φ_{rs} are functions of the *four* arguments x^a only. g_{rs} and φ_{rs} have, therefore, all the properties of gravitational potentials and electromagnetic field intensities, respectively, in the general theory of relativity.

To obtain field equations, Kaluza assumed that the five dimensional curvature scalar R, multiplied by the square root of the determinant $|\gamma_{\alpha\beta}|$, was the Lagrangian of a variational principle. The five dimensional curvature scalar is connected with the p-curvature scalar by eq. (17.51). Since B_ρ vanishes in Kaluza's theory, and since only the skewsymmetric parts of the covariant derivatives of A_ρ do not vanish, eq. (17.51) reduces to the equation

$$R = \delta^i_n g^{kl} R_{ikl\cdot}^{\ \ \ \ n} + \tfrac{1}{4} A_{\rho\sigma} A^{\rho\sigma} = \delta^i_n g^{kl} R_{ikl\cdot}^{\ \ \ \ n} + \tfrac{1}{4}\varphi_{rs}\varphi^{rs}. \tag{17.66}$$

If we use a special coördinate system, the determinant $|\gamma_{\alpha\beta}|$ may be replaced by $|g_{ab}|$, because of eq. (17.61); and since the Lagrangian is A-cylindrical, the integral may be taken either over a five dimensional domain of the coördinates ξ^α or over a four dimensional domain of the parameters x^a. Kaluza's variational principle is, therefore,

$$\delta \int (\delta^i_n g^{kl} R_{ikl\cdot}^{\cdot\cdot\cdot n} + \tfrac{1}{4}\varphi_{rs}\varphi^{rs})\sqrt{-g}\,\mathbf{dx} = 0, \tag{17.67}$$

where the variation is subject to the conditions

$$(\delta g^{rs})_{,5} = 0, \qquad (\delta\varphi_r)_{,5} = 0. \tag{17.68}$$

This variational principle is identical with eq. (12.56), and the resulting field equations are those of the general theory of relativity (with an electromagnetic field).

Projective field theories. Kaluza introduced the fifth dimension solely for the purpose of increasing the number of components of the metric tensor, and assigned to it no real significance. A similar procedure is employed in the so-called projective geometries, which represent an n dimensional space by means of $(n + 1)$ homogeneous coördinates. In projective geometry, all those "projective points," the $(n + 1)$ homogeneous coördinate values of which have the same ratios, are considered to be "the same" point. Several authors, particularly Veblen and Hoffmann[2] and Pauli,[3] applied the same principle in their unified field theories. Our general formalism is applicable to their theories, but the geometric interpretation is different. The five dimensional coördinates are to be considered as "projective coördinates," while the real space is the four dimensional space of the parameters x^a. Each "projective" A-curve is only one point in the real space. The metric, therefore, is assumed to be A-cylindrical as a matter of course. From the point of view of our general formalism, there is no difference between the theories of Kaluza, Veblen and Hoffmann, and Pauli. The field equations are the same in all three theories. But each of these three theories was presented by its authors in a different type of coördinate system.

We have already studied Kaluza's special coördinate systems. To identify his coördinates, we shall denote them by $x^1 \cdots x^4$, $x^0 (= \xi^5)$. Veblen and Hoffmann chose a type of coördinate system which is more frequently encountered in projective geometry. It is related to Kaluza's coördinates by the equations

$$\left.\begin{aligned} \xi^s &= x^s, \\ \xi^5 &= e^{x^0}. \end{aligned}\right\} \tag{17.69}$$

[2] O. Veblen, *Projektive Relativitätstheorie*, Berlin, Springer, 1933, part of "Ergebnisse d. Mathematik u. ihrer Grenzgebiete." Contains bibliography.
[3] W. Pauli, *Ann. d. Physik*, **18**, 305 (1933); **18**, 337 (1933).

In such a coördinate system, the components of the metric tensor become

$$\left.\begin{aligned} \gamma_{55} &= e^{-2x^0}, \\ \gamma_{5a} &= \varphi_a e^{-x^0}, \\ \gamma_{ab} &= (g_{ab} + \varphi_a \varphi_b). \end{aligned}\right\} \quad (17.70)$$

Apart from transformations of the first four coördinates among themselves, their choice of coördinates permits transformations of the type

$$\xi^{*5} = F(x^a) \cdot \xi^5. \quad (17.71)$$

With respect to these transformations, the components of the metric tensor transform according to the equations

$$\left.\begin{aligned} \overset{*}{\gamma}_{55} &= \frac{1}{F^2} \gamma_{55} = \frac{1}{F^2} e^{-2x^0}, \\ \overset{*}{\gamma}_{5a} &= \frac{1}{F} [A_a - (\log F)_{,a}] e^{-x^0}, \\ \overset{*}{g}_{ab} &= g_{ab} \,. \end{aligned}\right\} \quad (17.72)$$

Pauli chose a type of coördinate system which can properly be called homogeneous. His coördinates are related to Kaluza's by the equations

$$X^\alpha = f^\alpha(x^s) e^{x^0}, \quad (17.73)$$

$$\left.\begin{aligned} x^a &= \overset{(0)}{h}{}^a(X^\rho), \\ x^0 &= \log \overset{(1)}{h}(X^\rho), \end{aligned}\right\} \quad (17.74)$$

where $\overset{(0)}{h}{}^a$ are four *homogeneous* functions of the zeroth degree,

$$\overset{(0)}{h}{}^a(\alpha X^\rho) = \overset{(0)}{h}{}^a(X^\rho), \quad (17.75)$$

and $\overset{(1)}{h}(X^\rho)$ is a homogeneous function of the first degree,

$$\overset{(1)}{h}(\alpha X^\rho) = \alpha \overset{(1)}{h}(X^\rho). \quad (17.76)$$

Transitions from one homogeneous coördinate system to another are accomplished by means of homogeneous transformation equations of the first degree,

$$X^{*\rho} = \overset{(1)}{H}{}^\rho(X^\sigma). \quad (17.77)$$

The contravariant components of the vector **A** in homogeneous coördinates are

$$A^\rho = \frac{\partial X^\rho}{\partial x^0} = X^\rho. \tag{17.78}$$

In Pauli's formalism, the coördinates themselves have vector character and are identical with the components A^ρ.

The A-cylindricity of tensors finds a very peculiar expression in Pauli's coördinate system. In terms of Kaluza's coördinates, an A-cylindrical tensor is independent of the coördinate x^0. When we go over to Pauli's coördinates, a tensor $V_{\alpha\cdots}{}^{\beta\cdots}$ transforms according to the law

$$V^*_{\alpha\cdots}{}^{\beta\cdots} = \frac{\partial X^\beta}{\partial x^\sigma} \cdots \frac{\partial x^\rho}{\partial X^\alpha} \cdots V_{\rho\cdots}{}^{\sigma\cdots}.$$

$V_{\rho\cdots}{}^{\sigma\cdots}$ are functions of $x^1 \cdots x^4$ only and, therefore, homogeneous functions of the zeroth degree of the X^ρ. Each coefficient $\frac{\partial X^\beta}{\partial x^\sigma}$ is a function of $x^1 \cdots x^4$, multiplied by e^{x^0}, and, therefore, homogeneous of the first degree in the X^ρ. Conversely, the coefficients $\frac{\partial x^\rho}{\partial X^\alpha}$ are each homogeneous of the (-1)st degree in the X^ρ. *An A-cylindrical tensor is homogeneous, and the degree of its homogeneity equals the difference between the number of its contravariant and of its covariant indices.*

CHAPTER XVIII

A Generalization of Kaluza's Theory

Possible generalizations of Kaluza's theory. Kaluza's theory appeared to be a very attractive starting point for a modification of the general theory of relativity, though it did not lead to new field equations nor solve any of the unsolved problems of theoretical physics. But it suggested several avenues of approach to a new theory which would make gravitational and electromagnetic fields parts of one unified field.

One generalization of Kaluza's theory was attempted by Einstein and Mayer.[1] They assumed, as did Kaluza, that physical space was four dimensional, but they introduced five dimensional tensor calculus without, at the same time, introducing a five dimensional space and a five dimensional coördinate system. They assumed that there were tensors, with indices running from 1 to 5, the components of which were functions of four coördinates only. Apart from four dimensional coördinate transformations, there were to be transformations of the "five dimensional tensors," with transformation matrices $M_\alpha{}^\beta$ and $M^\alpha{}_\beta$, so that, for instance, "five dimensional" vectors transformed according to the equations

$$\left.\begin{aligned} V^{*\alpha} &= M^\alpha{}_\beta V^\beta \\ W^*_\alpha &= M_\alpha{}^\beta W_\beta , \qquad M^\alpha{}_\beta M_\gamma{}^\beta = \delta^\alpha_\gamma , \end{aligned}\right\} \tag{18.1}$$

where the $M^\alpha{}_\beta$ were *arbitrary* functions of the four coördinates. If the existence of a five dimensional coördinate system had been assumed, the matrices $M^\alpha{}_\beta$ would be partial derivatives of the new coördinates with respect to the old ones,

$$M^\alpha{}_\beta = \frac{\partial \xi^{*\alpha}}{\partial \xi^\beta} ,$$

and as such subject to differential identities,

$$M^\alpha{}_{\rho,\sigma} - M^\alpha{}_{\sigma,\rho} = M^\alpha{}_{\rho,s}\gamma^s_\sigma - M^\alpha{}_{\sigma,r}\gamma^r_\rho = 0.$$

[1] *Berl. Ber.*, 1931, p. 541; *Berl. Ber.*, 1932, p. 130.

Nothing of this kind is assumed in the Einstein-Mayer theory. A number of the differential covariants of the Kaluza theory, therefore, have no counterpart in the Einstein-Mayer theory. While they also introduce quantities γ^a_α and γ^α_a to lead from five dimensional tensors to four dimensional ones, and vice versa, the expressions

$$\gamma^a_{\alpha,\beta} - \gamma^a_{\beta,\alpha} = \gamma^a_{\alpha,b}\gamma^b_\beta - \gamma^a_{\beta,b}\gamma^b_\alpha,$$

which vanish in the Kaluza theory, do not in general vanish in the Einstein-Mayer theory and are not even covariant. The expressions

$$\gamma^\alpha_{a,b} - \gamma^\alpha_{b,a},$$

which are covariant in the Kaluza theory, are not covariant in the Einstein-Mayer theory. On the other hand, in the Einstein-Mayer theory there are a number of differential covariants the counterparts of which in the Kaluza theory vanish identically. In fact, there are many more of them than can be interpreted in a physical theory.

It is also possible to generalize Kaluza's theory by relaxing the condition of A-cylindricity. This was done by Einstein and Bergmann, and later by Einstein, Bargmann, and Bergmann, in two papers which will be reported on briefly in this chapter.[2] †

The geometry of the closed, five dimensional world. While the projective theories, and particularly the Einstein-Mayer theory, consider space as four dimensional, and introduce the fifth dimension only as a means of constructing a new type of tensor calculus, the theory presented in the two papers mentioned above was designed to give the fifth dimension a stronger physical significance.

Physical considerations motivated the development of this theory. It appeared impossible for an ironclad four dimensional field theory ever to account for the results of quantum theory, in particular, for Heisenberg's indeterminacy relation. Since the description of a five dimensional world in terms of a four dimensional formalism would be incomplete, it was hoped that the indeterminacy of "four dimensional" laws would account for the indeterminacy relation and that quantum phenomena would, after all, be explained by a field theory. It appears fairly certain today that these high hopes were unjustified. Whether any part of the five dimensional approach will stand the test of time remains to be seen.

---

[2] A. Einstein and P. Bergmann, *Ann. of Math.*, **39**, 683 (1938). A. Einstein, V. Bargmann, and P. G. Bergmann, *Theodore von Kármán Anniversary Volume*, Pasadena, 1941, p. 212.

†See Appendix B.

The macroscopic world, at any rate, is four dimensional, and the five dimensional world must be, at least approximately, cylindrical with respect to the fifth dimension. Einstein and his collaborators assumed, therefore, that the world was *closed* with respect to the fifth dimension and formed what might be called a tube.

If we cut out of a five dimensional continuum a thin slice of infinite extension and identify the two open (four dimensional) faces of the slice, we shall have a model of such a closed five dimensional space. All field functions are, of course, supposed to be continuous across the "seam"; and, therefore, if the tube is sufficiently narrow (that is, if the slice is sufficiently thin), the variation of a field quantity around the tube will be small compared with its variation along the tube.

The closed five dimensional space is assumed to have Riemannian geometry. In addition, it is subject to another restriction which reduces the number of field variables from 15 to 14. In treating Kaluza's theory, we called the metric of his theory A-cylindrical. In other words, Kaluza's space is cylindrical not only with respect to *some* vector field, but to a *unit* vector field, **A**. Because of this, the A-curves of Kaluza turn out to be geodesics, and in a special coördinate system, γ_{55} equals unity. In the geometry which we are discussing at present, Kaluza's cylindricity condition is replaced by the condition that the five dimensional space be closed. In addition, it is assumed that those geodesics which connect a point with itself around the tube intersect themselves at an angle zero, in other words, that they are continuous, closed lines. This assumption takes the place of Kaluza's condition that his space be cylindrical with respect to a *unit* vector field.

Exactly one of the closed geodesic lines goes through each point of the five dimensional space. The length of such a line, taken once around the tube and back to the starting point, shall be called the "*circumference*" of the space in the fifth dimension. We shall now prove that *this circumference is everywhere the same.*

Let us consider a closed geodesic line passing through the point P. Its length S equals

$$S = \int_P^P \sqrt{\gamma_{\iota\kappa} \frac{d\xi^\iota}{dp} \frac{d\xi^\kappa}{dp}}\, dp = \int_P^P \sqrt{\gamma_{\iota\kappa} \xi^{\iota\prime} \xi^{\kappa\prime}}\, dp, \qquad (18.2)$$

where p is an arbitrary function of the coördinates; the path of integration is a closed geodesic line. Let us now vary the coördinates of every point along the closed geodesic line by infinitesimal amounts $\delta\xi^\iota$, so that we obtain a new line. The "end point" P is not to be kept fixed. The difference between the length of the new and the old lines is

$$\left.\begin{aligned} \delta S &= \int_P^P \frac{\frac{1}{2}\gamma_{\iota\kappa,\rho}\xi^{\iota\prime}\xi^{\kappa\prime}\,\delta\xi^\rho + \gamma_{\iota\rho}\xi^{\iota\prime}\,\delta\xi^{\rho\prime}}{\sqrt{\gamma_{\iota\kappa}\xi^{\iota\prime}\xi^{\kappa\prime}}}\,dp \\ &= \int_P^P \{\tfrac{1}{2}\gamma_{\iota\kappa,\rho}\dot{\xi}^\iota\dot{\xi}^\kappa\,\delta\xi^\rho + \gamma_{\iota\rho}\dot{\xi}^\iota\,\delta\dot{\xi}^\rho\}\,d\tau, \end{aligned}\right\} \tag{18.3}$$

where the dots denote differentiation with respect to τ, the arc length from P. By integrating by parts, we obtain

$$\left.\begin{aligned} \delta S &= \int_P^P \{\tfrac{1}{2}\gamma_{\iota\kappa,\rho}\dot{\xi}^\iota\dot{\xi}^\kappa - \gamma_{\iota\rho,\kappa}\dot{\xi}^\iota\dot{\xi}^\kappa\,\delta\xi^\rho - \gamma_{\iota\rho}\ddot{\xi}^\iota\,\delta\xi^\rho\}\,d\tau + \mathop{[\gamma_{\iota\rho}\dot{\xi}^\iota\,\delta\xi^\rho]}_P^P \\ &= -\int_P^P \gamma_{\iota\rho}\left(\ddot{\xi}^\iota + \left\{ {\iota \atop \kappa\lambda} \right\}\dot{\xi}^\kappa\dot{\xi}^\lambda\right)\delta\xi^\rho\,d\tau + \mathop{[\gamma_{\iota\rho}\dot{\xi}^\iota\,\delta\xi^\rho]}_P^P. \end{aligned}\right\} \tag{18.4}$$

The contents of the parenthesis vanish, because the original line is geodesic. The variation of S depends, therefore, only on the coördinate variations at the double end point,

$$\delta S = \mathop{[\gamma_{\iota\rho}\dot{\xi}^\iota\,\delta\xi^\rho]}_P^P. \tag{18.5}$$

The expression in the square bracket is the scalar product of the two vectors $\dot{\xi}^\iota$ and $\delta\xi^\rho$, and thus invariant. Since the $\delta\xi^\rho$ at the two boundaries actually represent the same infinitesimal displacement, and the $\dot{\xi}^\iota$ the same direction, the contributions of the two boundaries cancel, and δS vanishes. If we proceed, by successive variations, from one closed geodesic line to another, so that all the intermediate lines are also geodesics, S will remain constant during this process.

The circumference of a closed space, in which all self-intersecting geodesics are closed lines without discontinuities of direction, is thus a characteristic constant of that space.

The tangential vectors of the closed geodesics, $\frac{d\xi^\alpha}{d\tau}$, form a field of unit vectors. We shall call this unit vector field **A** and apply to the closed five dimensional continuum the formalism which was developed in the preceding chapter.

Since the **A**-field consists of the tangential vectors of geodesics, it satisfies the differential equations

$$A^\alpha{}_{;\rho}A^\rho = 0, \tag{18.6}$$

which means, according to eqs. (17.17) and (17.21), that the A-metric,

$$d\tau_A = A_\alpha \, d\xi^\alpha, \tag{18.7}$$

is A-cylindrical. Eq. (18.6) takes a particularly simple form in a special coördinate system.

Introduction of the special coördinate system. In a special coördinate system, **A** has the contravariant components (0, 0, 0, 0, 1) and the covariant components $(\gamma_{5s}, 1)$. Eq. (18.6), therefore, takes the form

$$A^\alpha{}_{,5} + \begin{Bmatrix} \alpha \\ 55 \end{Bmatrix} = 0 \tag{18.8}$$

or

$$\left. \begin{aligned} &\begin{Bmatrix} \alpha \\ 55 \end{Bmatrix} = 0, \qquad [55, \alpha] = 0, \\ &\gamma_{\alpha 5,5} - \tfrac{1}{2}\gamma_{55,\alpha} = 0. \end{aligned} \right\} \tag{18.8a}$$

Since, in a special coördinate system, γ_{55} is constant and equal to unity, eq. (18.6) becomes finally

$$\gamma_{a5,5} = \varphi_{a,5} = 0; \tag{18.9}$$

in other words, *the quantities* φ_s *do not depend on* ξ^5.

The remaining components of the metric tensor are periodic in ξ^5, for the A-curves are closed curves, and return to a point through which they have passed once. The coördinate distance of a point from itself (around the tube) equals the metric distance, because γ_{55} equals unity; it is, thus, the same for every point in the space considered. *All* fields which are uniquely defined in our closed space are periodic functions of ξ^5 with the period S, eq. (18.2).

In a special coördinate system, the five dimensional metric thus decomposes into a set of ten functions g_{mn},

$$g_{mn} = \gamma_{mn} - \varphi_m \varphi_n, \tag{18.10}$$

which are periodic in ξ^5 with the period S, and into four functions φ_m, which depend only on $\xi^1 \cdots \xi^4$.

Because of eq. (18.9), the vector B_ρ, eqs. (17.17) and (17.60), vanishes, and the only tensors of the first differential order are

$$\varphi_{rs} = \varphi_{r,s} - \varphi_{s,r} \tag{18.11}$$

[because of eq. (17.59)] and

$$g_{mn,5} = A_{m;n} + A_{n;m} \tag{18.12}$$

[because of eq. (17.18)].

The derivation of field equations from a variational principle. Einstein and his collaborators formulated two different sets of field equations which are based on the geometry of a closed five dimensional space with closed coördinates. In this section, we shall discuss the first of these two sets.

We can set up field equations which are the Euler-Lagrange equations of a variational principle. There are four different scalars of the second differential order, each of which makes a different contribution to the field equations. They are:

$$\left.\begin{array}{l} \delta^i_n\, g^{kl} R_{ikl\cdot}{}^n; \qquad \varphi_{rs}\varphi^{rs}; \qquad (A_{\mu;\nu} + A_{\nu;\mu})(A_{\rho;\sigma} + A_{\sigma;\rho})\gamma^{\mu\nu}\gamma^{\rho\sigma}; \\ (A_{\mu;\nu} + A_{\nu;\mu})(A_{\rho;\sigma} + A_{\sigma;\rho})\gamma^{\mu\rho}\gamma^{\nu\sigma}. \end{array}\right\} \tag{18.13}$$

All other scalars of the second differential order differ from a linear combination of these four only by divergences, which do not contribute to the Euler-Lagrange equations. A linear combination of these four scalars, multiplied by the square root of the negative determinant $\sqrt{-|\gamma_{\rho\sigma}|}$ (or, in the case of a special coördinate system, by $\sqrt{-g}$) and integrated over a five dimensional domain of the coördinates $\xi^1 \cdots \xi^5$, is an invariant.

The variation of such an integral,

$$I = \int H\sqrt{-g}\, \mathbf{d}\xi, \tag{18.14}$$

must preserve the typical geometrical properties of the closed space; that is, if we use a special coördinate system, the variations of the φ_r must be independent of ξ^5, and the variations of the g_{mn} must be periodic in ξ^5 with the period S. We cannot require, therefore, that the variations of the φ_r and g^{rs} vanish everywhere on the boundary of an arbitrary domain of integration. However, the variation on the boundary will not contribute to the variation of the integral if the domain extends exactly once around the tube, or, in other words, over one period of ξ^5, and if δg^{rs} and $\delta\varphi_s$ vanish on that part of the boundary which is generated by A-curves. The variation will, thus, have the form

$$\left.\begin{array}{l} \delta \int H\sqrt{-g}\, d\xi^1 \cdots d\xi^5 \\ \qquad = \int \{Q_{rs}\,\delta g^{rs} + J^s\,\delta\varphi_s\}\sqrt{-g}\, d\xi^1 \cdots d\xi^5, \end{array}\right\} \tag{18.15}$$

provided that the domain of integration extends over one period (or an integral number of periods) of ξ^5, and provided that the δg^{rs} and $\delta\varphi_s$

vanish on the significant part of the boundary of the domain. The $\delta\varphi_s$ are independent of ξ^5; the δg^{rs} are periodic functions of ξ^5, but locally arbitrary. The integral I will be stationary if these equations are satisfied:

$$\left.\begin{aligned} Q_{rs} &= 0, \\ \int_{\xi^5=0}^{S} J^s \sqrt{-g}\, d\xi^5 &= 0. \end{aligned}\right\} \tag{18.16}$$

We cannot require that the expressions J^s vanish everywhere, for if the integral (18.16), which is to be extended once around an A-curve, vanishes, the variations of φ_s, which are constant along each A-curve, will not contribute to δI.

The integro-differential equations (18.16) satisfy five integro-differential identities. If we carry out an infinitesimal special coördinate transformation,

$$\left.\begin{aligned} \xi^{*a} &= \xi^a + \delta\xi^a(\xi^s), \\ \xi^{*5} &= \xi^5 + \delta\xi^5(\xi^s), \end{aligned}\right\} \tag{18.17}$$

the field variables g^{rs} and φ_s transform as follows:

$$\begin{aligned} \delta g^{rs} &= g^{rt}(\delta\xi^s)_{,t} + g^{ts}(\delta\xi^r)_{,t} - \delta\xi^t g^{rs}{}_{,t} - \delta\xi^5 g^{rs}{}_{,5}, \\ \delta\varphi_s &= -\varphi_t(\delta\xi^t)_{,s} - (\delta\xi^5)_{,s} - \delta\xi^t\varphi_{s,t}, \end{aligned}$$

or

$$\left.\begin{aligned} \delta g^{rs} &= g^{rt}(\delta\xi^s)_{;t} + g^{st}(\delta\xi^r)_{;t} - g^{rs}{}_{,5}(A_\rho\,\delta\xi^\rho), \\ \delta\varphi_s &= -\varphi_{st}\,\delta\xi^t - (A_\rho\,\delta\xi^\rho)_{,s}, \\ A_\rho\,\delta\xi^\rho &= \varphi_t\,\delta\xi^t + \delta\xi^5. \end{aligned}\right\} \tag{18.18}$$

If we choose a set of $\delta\xi^\alpha$ which vanish on the significant part of the boundary of the domain of integration, the variation of I on account of this infinitesimal transformation must vanish, even if eqs. (18.16) are not satisfied,

$$\left.\begin{aligned} 0 &\equiv \int \{Q_{rs}[g^{rt}(\delta\xi^s)_{;t} + g^{st}(\delta\xi^r)_{;t} - g^{rs}{}_{,5}(A_\rho\,\delta\xi^\rho)] \\ &\qquad\qquad + J^s[-(A_\rho\,\delta\xi^\rho)_{,s} - \varphi_{st}\,\delta\xi^t]\}\sqrt{-g}\,d\xi \\ &= \int \{[-2Q_{s\cdot;t}^{\cdot t} + \varphi_{ts}J^t]\delta\xi^s \\ &\qquad\qquad + [-Q_{rs}g^{rs}{}_{,5} - J^s{}_{;s}](A_\rho\,\delta\xi^\rho)\}\sqrt{-g}\,d\xi. \end{aligned}\right\} \tag{18.19}$$

In carrying out the partial integrations involved in the derivation of this equation, we have omitted all the terms which are covariant divergences. The covariant divergence of a p-vector density is a linear combination of *ordinary* derivatives,

$$\mathfrak{V}^{s}{}_{;s} = \mathfrak{V}^{s}{}_{,s} - (\varphi_s \mathfrak{V}^{s})_{,5}\,, \tag{18.20}$$

and vanishes, therefore, when it is integrated over a domain on the boundary of which the components $\mathfrak{V}^s$ vanish.

Since the $\delta\xi^\alpha$ in eq. (18.19) are independent of ξ^5, but otherwise arbitrary in the interior of the domain of integration, we conclude that eqs. (18.16) satisfy the identities

$$\left.\begin{aligned} \int_{\xi^5=0}^{S} (2Q_{s\cdot;t}^{\ t} + \varphi_{ts}J^t)\sqrt{-g}\,d\xi^5 &\equiv 0, \\ \int_{\xi^5=0}^{S} (J^s{}_{;s} + Q_{rs}g^{rs}{}_{,5})\sqrt{-g}\,d\xi^5 &\equiv 0. \end{aligned}\right\} \tag{18.21}$$

As for the form of the expressions J^s and Q_{rs}, Q_{rs} contains the same terms as the field equations of the general theory of relativity, except that all derivatives are replaced by p-derivatives,

$$g_{rs|t} = g_{rs,t} - g_{rs,5}\varphi_t\,,$$

and so forth; in addition, the Q_{rs} contain terms in which the p-metric is differentiated with respect to ξ^5. The expressions J^s contain the Maxwell term, $\varphi^{rs}{}_{;s}$, and, in addition, terms which are products of p-derivatives by A-derivatives of the g_{rs}. In other words, the world current density does not vanish in this theory.

Differential field equations. There are two objections to the field equations which can be derived from a variational principle. First, these equations are not uniquely determined; any linear combination of the scalars (18.13), multiplied by $\sqrt{-g}$, is suitable as a Lagrangian. Second, the field equations are not pure differential equations.

If the field equations were a set of pure differential equations, they would have to satisfy stronger identities in order to have significant solutions. Actually, there is a set of fourteen differential equations which satisfy four differential identities and one integro-differential identity; moreover, this set is uniquely determined.

Let us consider the fifteen expressions $G_{\rho\sigma}$ which are formed from the contracted five dimensional curvature tensor,

$$G_{\rho\sigma} = R_{\rho\sigma} - \tfrac{1}{2}\gamma_{\rho\sigma}R. \tag{18.22}$$

These fifteen quantities satisfy the five relationships

$$(\sqrt{\gamma}\, G^{\rho\sigma})_{,\sigma} + \sqrt{\gamma}\, G^{\tau\sigma}\begin{Bmatrix}\rho\\ \tau\sigma\end{Bmatrix} \equiv 0. \tag{18.23}$$

If $\sqrt{\gamma}G^{\rho\sigma}$ is denoted by $\mathfrak{G}^{\rho\sigma}$, and if special coördinates are employed, eq. (18.23) takes the form

$$\left.\begin{aligned} \mathfrak{G}^{rs}{}_{,s} + \mathfrak{G}^{r5}{}_{,5} + \mathfrak{G}^{ts}\begin{Bmatrix}r\\ ts\end{Bmatrix} + 2\mathfrak{G}^{t5}\begin{Bmatrix}r\\ t5\end{Bmatrix} &= 0, \\ \mathfrak{G}^{5s}{}_{,s} + \mathfrak{G}^{ts}\begin{Bmatrix}5\\ ts\end{Bmatrix} + 2\mathfrak{G}^{5s}\begin{Bmatrix}5\\ 5s\end{Bmatrix} &= -\mathfrak{G}^{55}{}_{,5}, \end{aligned}\right\} \tag{18.24}$$

for the Christoffel symbols $\left\{ {\alpha \atop 55} \right\}$ vanish because of eq. (18.8a). We find that the *fourteen* expressions G^{rs}, G^{r5} satisfy four differential identities; in addition, an expression of the first differential order in the G^{rs} and G^{r5} is identically equal to a derivative with respect to ξ^5, and its integral over one period of an A-curve, therefore, vanishes. The fourteen equations

$$\left.\begin{aligned} G^{rs} &= 0, \\ I^s &= G^{s5} + \varphi_t G^{st} = 0, \end{aligned}\right\} \tag{18.25}$$

thus, satisfy the necessary identities. They are also covariant with respect to special coördinate transformations, for G^{rs} and I^s are p-tensors,

$$G^{rs} = \gamma^r_\rho \gamma^s_\sigma G^{\rho\sigma},$$

$$I^s = \gamma^s_\sigma A_\rho G^{\rho\sigma}.$$

A lengthy but straightforward calculation shows that there is no other set of fourteen differential equations of the second order which are covariant with respect to special coördinate transformations and which satisfy identities similar to those of eqs. (18.25). The form of the equations is the same as the form of the equations which are derived from a suitable variational principle, except that all fourteen equations are differential equations. Again the world current density does not vanish.

APPENDIX A

Ponderomotive Theory by Surface Integrals

In Chapter XV the laws of motion of ponderable bodies under the influence of gravitation were obtained from the field equations, more or less by the methods pioneered by A. Einstein, L. Infeld, and B. Hoffmann.[1] Their procedure, by successive approximations, was refined in a number of subsequent papers.[2] It has been presented in a monograph by L. Infeld and J. Plebański.[3] This original approach, usually referred to as the EIH theory, demonstrates that ponderable bodies must move in the gravitational field according to definite laws in order to permit the existence of solutions of the vacuum field equations in the surrounding space.

The same result was obtained by a different route, and independently, by V. A. Fock.[4] Whereas EIH treated the ponderable body as a point mass, Fock worked with an extended body, whose internal equation of state was assumed known. Both approaches are equally suitable for the treatment of planetary systems in celestial mechanics. They lend themselves to the calculation of post-Newtonian effects that can be observed, and even radiative effects.

But considering that the relationship between field equations and ponderomotive laws is fundamental, and unique, to general-relativistic physics, one would like to obtain this connection independently of any approximation procedures. One would like the connection elucidated independently of the choice of coordinates. Finally, one would like to be able to describe those properties of the ponderable masses that influence their motions in an invariant manner; a typical

---

[1] Cf. footnote on p. 225.

[2] L. Infeld and P. R. Wallace, *Phys. Rev.*, **57**, 797 (1940). L. Infeld and A. Schild, *Rev. Modern Phys.*, **21**, 408 (1949). A. Einstein and L. Infeld, *Can. J. Math.*, **1**, 209 (1949). Many additional important papers are listed in the bibliography of the monograph referred to below.

[3] L. Infeld and J. Plebański, *Motion and Relativity*, Państwowe Wydawnictwo Naukowe, Warszawa, and Pergamon Press, London and New York, 1960.

[4] V. A. Fock, *J. Phys.* (Moscou), **1**, 81 (1939); *JETP*, **9**, 375 (1939).

characteristic of this kind would be "absence of mass multipole moments." The last one of these objectives has not as yet been achieved, but the approaches to be described in this appendix go a long way toward meeting the first two. To this extent, the theory has been greatly improved.

Essentially the new method involves formulating the laws of motion as statements concerning the time dependence of certain two-dimensional closed-surface integrals, which are interpreted as specific characteristics of the particles inside, such as their mass or their linear momentum. The rates of change of these integrals with time are found to equal other closed-surface integrals, whose integrands represent the fluxes of mass density, linear momentum density, and so forth. Whereas the first type of integrals is taken over expressions that are linear in the first derivatives of the metric, the flux expressions are quadratic in the first derivatives of the metric. Both are free of any higher-order derivatives.

There is a close analogy to this procedure in electromagnetic theory. The electric charge inside a closed two-dimensional surface can be represented by an integral on that surface over the electric displacement, $\oint \mathbf{D} \cdot \mathbf{dS}$. If Maxwell's equations (without charge and without current) hold on that surface, the time derivative of the integrand is proportional to the curl of the magnetic field strength; the time derivative of the electric charge is found to vanish, as a precondition that the field equations can be satisfied on the enclosing surface, though nothing has been assumed concerning the interior. Isolated electric charges cannot change their values in the course of time.

Integral relations of this type can be obtained in any theory whose field equations are derivable from a stationary-action principle that satisfies the *principle of general covariance* (cf. p. 159). They are corollaries of *Noether's theorem*, which systematizes most conservation laws of physics. Noether's theorem in turn involves the notion of *invariance group*. That concept will be our point of departure.

Invariance groups. The principle of relativity asserts that the laws of physics take the same form if expressed in terms of the space and time coordinates of any inertial frame of reference. Likewise, the principle of general covariance asserts the formal equivalence of all curvilinear four-dimensional coordinate systems. In both examples a given physical situation may be described in formally different ways, which however have in common that the form of the dynamical laws is the same in all of them. The transition from one description to another is governed by the transformation laws of the physical variables involved.

The importance of the transformation laws lies in the circumstance that an inertial frame of reference in the first example, or a curvilinear coordinate system in the second, cannot be identified as such, but only its relationship to other frames, or to other coordinate systems. It is the nature of these relationships that distinguishes special relativity from Newtonian-Galilean physics. In the latter theory two inertial frames of reference are related to each other by a Galilean transformation, such as Eqs. (1.3) on p. 5, whereas in special relativity they will be connected with each other by a Lorentz (or Poincaré) transformation. Whatever the physical theory of concern, the transformations that interconnect the equivalent descriptions, and with respect to which the physical laws remain formally unchanged, are referred to as *invariance transformations*. Galilean transformations in Newtonian mechanics, Lorentz (or Poincaré) transformations in special relativity, curvilinear transformations in general relativity, and gauge transformations in electrodynamics are all invariance transformations.

Generally, invariance transformations form a *transformation group*. We call a set of transformations a transformation group if it satisfies these requirements:

1. If two transformations belonging to the set are performed one after the other, the resulting transformation (called their *product*) also belongs to the set;
2. The identity transformation belongs to the set; and
3. Together with every transformation in the set, its inverse belongs to the set.

The notion of product of two transformations is borrowed from algebra. If two transformations are denoted by the symbols A and B, the symbol BA usually indicates that the transformation A is performed first, B afterwards. (This convention may be assumed to be obeyed in any paper or book whose author does not expressly adopt the opposite convention.) In general, AB and BA are different transformations. If $AB = BA$, then B and A are said to *commute* with each other. If all members of a transformation group commute with each other, the group is called *commutative* or *Abelian*. Except for the group of gauge transformations, none of the transformation groups mentioned in this section are commutative; even two ordinary spatial rotations do not commute with each other unless they are rotations about the same axis.

All the transformation groups that we shall consider are *continuous*. In addition to satisfying the algebraic requirements above, they permit the construction of one-parametric sets of transformations that depend on one parameter, λ, so that in the equation

$$\bar{\xi}^{\mu} = f^{\mu}(\xi^{\rho}, \lambda) \tag{A.1}$$

the derivative $(\partial\bar{\xi}^{\mu}/\partial\lambda)$ exists. For the spatial rotations, for instance, consider rotations about a fixed axis, with the parameter λ standing for the angle through which one rotates. For Lorentz transformations, λ might denote the relative velocity of two Lorentz frames, with the direction of that velocity held fixed.

Consider now one-parametric subsets of a transformation group that include the identity transformation for $\lambda = 0$. The functions

$$\left.\frac{\partial\bar{\xi}^{\mu}}{\partial\lambda}\right|_{\lambda=0} \equiv \delta\xi^{\mu}(\xi^{\rho}) \tag{A.2}$$

are then called *infinitesimal transformations*. Two different one-parametric sets may lead to the same infinitesimal transformation. All infinitesimal transformations together form the *infinitesimal transformation group*, which is a much simpler structure than the (finite) transformation group. It contains much of the structural information on the original group. The members of the infinitesimal transformation group possess two kinds of composition. They are called, respectively, *linear combination* and *commutation*. Both correspond to specific operations within the finite group.

Consider two one-parametric sets of transformations within the group, $A(\lambda)$ and $B(\lambda)$. From them we can form a third one-parametric set, $C(\lambda)$,

$$C(\lambda) = B(\beta\lambda)A(\alpha\lambda)\,. \tag{A.3}$$

α and β denote two arbitrary real nonvanishing numbers. If we denote the corresponding infinitesimal transformations by a, b, and c, respectively, we find the interrelation

$$c = \alpha a + \beta b \tag{A.4}$$

between them. Hence c is a linear combination of a and b.

This relationship can be visualized in terms of the underlying coordinate transformations if these transformations are interpreted as mappings. For a fixed point, Eq. (A.1) represents a one-parametric set of coordinate assignments, or alternatively, a one-parametric set of possible maps, which together form a curve passing through the point ξ^{ρ}. This curve is parametrized by λ, and $\delta\xi^{\mu}$ is a vector tangent to the curve at the point ξ^ρ. Thus, the infinitesimal transformation a gives rise to a vector at each world point, or a *vector field*. Likewise, b represents a different vector field, and so does c. The coefficients α, β are arbitrary, but constant; they cannot change from point to point.

Consider again the two sets A and B. If A and B do not commute, i.e., if $AB \neq BA$, then AB multiplied by the inverse of BA, denoted by $(BA)^{-1}$, is not the identity transformation. The product $AB(BA)^{-1}$ is known as the *commutator* of A and B. Its infinitesimal analog can, for instance, be obtained by forming the infinitesimal transformation corresponding to the commutator of $A(\sqrt{\lambda})$ and $B(\sqrt{\lambda})$. If the two vector fields corresponding to a and b are denoted by $\delta_a\xi^\mu$ and $\delta_b\xi^\mu$, respectively, then the vector field belonging to the commutator is given by the expression:

$$u^\mu = (\delta_a\xi^\mu)_{,\rho}\,\delta_b\xi^\rho - (\delta_b\xi^\mu)_{,\rho}\,\delta_a\xi^\rho. \tag{A.5}$$

This expression is also known as the *Lie derivative* of the vector field $\delta_a\xi^\mu$ with respect to the field $\delta_b\xi^\mu$. In terms of the infinitesimal transformations themselves, the relationship between a, b, and their commutator u is often symbolized by a so-called commutator bracket,

$$u = [a, b] \equiv -[b, a]. \tag{A.6}$$

Commutator brackets satisfy the *Jacobi identities*,

$$[[a, b], c] + [[b, c], a] + [[c, a], b] \equiv 0. \tag{A.7}$$

Why is one concerned with the structure of transformation groups, rather than merely with statements concerning equivalent frames, and with transformations leading from one to another? Given a physical theory, with its dynamical variables, there are transformation laws not only for the space-time coordinates but also for the dynamical variables, which may be vectors, tensors, affine connections, and even more exotic objects. Suppose that three transformations belonging to the invariance group have the property $ABC = 1$ (where "1" stands for the identity transformation). Then the transformation laws for the variables occurring in the theory must be such that if we apply to any particular choice of the values of the variables first the transformation engendered by C, then the one corresponding to B, and finally the one belonging to A, we must recover the original values. That is to say, any proposed transformation law for field variables or other relevant quantities must reflect the algebraic properties of the invariance group.

The concept of transformation group is not confined to transformations of coordinates. Gauge transformations of the electromagnetic potentials, discussed in Chapter VII, p. 115, form a transformation group, in that two sets of transformations of the type (7.27), one performed after the other, together are a transformation of the same type. Here the sequence of transformations is immaterial, and the gauge group is commutative. The gauge group is an invariance group: if a set of electrodynamical potential fields satisfies Maxwell's equations, then any new set obtained from the original potentials by a gauge transformation will again satisfy these laws. As the field intensities are unaffected by the gauge transformation, two sets of potentials obtained from each other by a gauge transformation are considered to describe the same electromagnetic field.

In analytical mechanics the point-to-point transformations of generalized coordinates q_k spanning configuration space also form a group. Most of these transformations change the form of the Lagrangian as a function of its arguments q_k, $\dot{q}_k$ (and possibly t). Those that do not are called *invariant transformations*. They form a group by themselves, a *subgroup* of the point-to-point transformations. (The invariant transformations are a subset of the point-to-point transformations, as well as a group in their own right.)

Similarly the canonical transformations in phase space form a group. Among them those canonical transformations that leave unchanged the form of the Hamiltonian as a function of its arguments, $H(q_k, p_k, t)$, again called invariant transformations, form a subgroup. The invariant canonical transformations include the invariant transformations that are point-to-point in configuration space; thus the latter are a subgroup of the former. The invariant canonical transformations are used in one of the formulations of Noether's theorem.

Noether's theorem. In classical mechanics Noether's theorem relates the invariant transformations to the constants of the motion. When applied to field theories it connects the invariant transformations to equations of continuity, which in turn result in certain integrals being constants of the motion. We shall obtain the theorem first within Hamilton's canonical formalism.

Hamilton's equations of motion may be obtained from a variational problem formulated in phase space:

$$\delta S = 0, \qquad S = \int_{t_1}^{t_2} \left(\sum_k p_k \dot{q}_k - H \right) dt. \tag{A.8}$$

The Hamiltonian H is a given function of the canonical coordinates

q_k, p_k (and possibly of the time t), which incorporates the assumed dynamical law of the mechanical system. We shall now consider an infinitesimal transformation of the canonical coordinates, to be denoted by the symbols δq_k, δp_k, and simultaneously a change in the functional dependence of H on its arguments, to be denoted by $\bar{\delta} H$,

$$\bar{\delta} H = \delta H - \sum_k \left(\frac{\partial H}{\partial q_k} \delta q_k + \frac{\partial H}{\partial p_k} \delta p_k \right). \tag{A.9}$$

δH stands for the change in value of H at a fixed point of phase space. The resulting change in the form of the action integral will be:

$$\bar{\delta} S = \int_{t_1}^{t_2} \left[\sum_k (\dot{p}_k \delta q_k - \dot{q}_k \delta p_k) - \delta H \right] dt - \left[\sum_k p_k \delta q_k \right]_{t_1}^{t_2}. \tag{A.10}$$

The change in its value will equal the integral over $-\bar{\delta} H$ from t_1 to t_2.

The extremal curves corresponding to the altered action will be identical with those belonging to the original action if the change of value depends only on the end points,

$$\delta S = -\int_{t_1}^{t_2} \bar{\delta} H \, dt = Q_2 - Q_1, \qquad \bar{\delta} H = -\frac{dQ}{dt}, \tag{A.11}$$

where Q is to be some function of the canonical coordinates and the time. If this expression for $\bar{\delta} H$ is substituted into Eq. (A.9), and the latter into Eq. (A.10), the resulting change in the form of the action integral will be:

$$\begin{aligned} \bar{\delta} S = \int \Bigg[\sum_k \dot{p}_k \left(\delta q_k - \frac{\partial C}{\partial p_k} \right) - \sum_k \dot{q}_k \left(\delta p_k + \frac{\partial C}{\partial q_k} \right) \\ + \sum_k \left(\frac{\partial H}{\partial q_k} \delta q_t + \frac{\partial H}{\partial p_k} \delta p_k \right) - \frac{\partial C}{\partial t} \Bigg] dt, \end{aligned} \tag{A.12}$$

$$C \equiv \sum_k p_k \delta q_k - Q.$$

The new equations of motion will again be Hamiltonian if the coefficients of $\dot{q}_k$ and of $\dot{p}_k$ vanish (and that means that the infinitesimal coordinate transformation is canonical, and generated by C). The change in the dependence of the Hamiltonian on its arguments will then be given by the expression:

$$\bar{\delta} H = \frac{\partial C}{\partial t} + [C, H] \equiv \frac{dC}{dt}. \tag{A.13}$$

The symbol [,] denotes the Poisson bracket. The right-hand side equals the time rate of change in the value of C along a trajectory obeying Hamilton's equations of motion.

The infinitesimal canonical transformation generated by C will leave the form of the Hamiltonian unchanged—it will be an *infinitesimal invariant transformation*—if (dC/dt) vanishes, that is to say, if C is a *constant of the motion*. This is one form of Noether's theorem, the one adapted to the Hamiltonian formulation of classical mechanics. The theorem may be employed in either direction. Given some invariant transformations, one can construct the generating constants of the motion. Conversely, if one knows a constant of the motion, one can obtain the corresponding infinitesimal invariant transformation.

In the Lagrangian formalism (in configuration space) one may derive Noether's theorem by examining the change in the form of the Lagrangian as a function of its arguments q_k, $\dot{q}_k$, t resulting from an infinitesimal change $\delta q_k(q, \dot{q}, t)$. Permitting again the addition to the Lagrangian of an appropriate exact time derivative, $\dot{Q}$, one finds:

$$\begin{aligned} \delta L &= -\sum_k \delta q_k \left[\frac{\partial L}{\partial q_k} - \frac{d}{dt}\left(\frac{\partial L}{\partial \dot{q}_k}\right)\right] - \frac{dC}{dt}, \\ C &\equiv \sum_k \frac{\partial L}{\partial \dot{q}_k}\, \delta q_k - Q. \end{aligned} \tag{A.14}$$

L will retain its form only if the generator C and the infinitesimal changes in the configuration variables, δq_k, are adjusted to each other so that the time derivative of C along a trajectory equals identically the linear combination of Lagrangian equations of motion that are indicated in Eq. (A.14). If it does, then the time derivative of C vanishes along a trajectory that obeys the equations of motion; C is a constant of the motion.

There are some applications of Noether's theorem to specific dynamical systems, in which all invariant transformations are related to all constants of the motion. More commonly the theorem is exploited with respect to an invariance group, that is to say a group of transformations with respect to which a whole class of dynamical interactions is invariant. For instance, all special-relativistic theories have in common their invariance with respect to the Poincaré group, the transformation group that consists of Lorentz transformations, and rigid displacements. With respect to this group, the relevant constants of the motion are the (total) linear momentum, which generates spatial displacements, the (total) mass/energy, which generates a displacement along the time axis, the components of the angular momentum, which generate rotations of the three spatial coordinates about spatial axes, and the components of the so-called *boost*, the generators of Lorentz transformations proper, those that involve relative motions of the inertial frames of reference. The three components of the boost are:

$$\mathbf{B} = M\mathbf{X} - t\mathbf{P}. \tag{A.15}$$

M denotes the (total) relativistic mass, $\mathbf{X}$ the coordinates of the center of mass, and $\mathbf{P}$ the (total) linear momentum. The corresponding nonrelativistic constants of the motion, incidentally, generate Galilean transformations.

Next we shall proceed to field theories, where Noether's theorem relates invariant transformations to equations of continuity, of the form Eq. (12.59), p. 194. We shall denote all the components of the field by symbols φ^A, $A = 1, \ldots, N$, where the superscript A is used to number the field components consecutively in some arbitrary manner, which need not be related to their transformation properties. Suppose the field equations are derivable from an action principle of the form

$$\delta S = 0, \qquad S = \int L(\varphi^A, \varphi^A{}_{,\rho})\, d^4x, \tag{A.16}$$

with the usual understanding that the variation extends only over the interior of the domain of integration D. The field equations will have the form:

$$\frac{\delta L}{\delta \varphi^A} \equiv \frac{\partial L}{\partial \varphi^A} - \left(\frac{\partial L}{\partial \varphi^A{}_{,\rho}}\right)_{,\rho} = 0. \tag{A.17}$$

We shall now consider infinitesimal changes in the field variables, $\delta\varphi^A$, with the property that they carry solutions of the field equations over into solutions. They will do so if as a result of the changes contemplated the action will change only by a surface integral,

$$\delta S = \int_D d^4x \left(\frac{\partial L}{\partial \varphi^A}\, \delta\varphi^A + \frac{\partial L}{\partial \varphi^A{}_{,\rho}}\, \delta\varphi^A{}_{,\rho}\right) \equiv \oint Q^\nu\, d^3\Sigma_\nu,$$
$$d^3\Sigma_\nu \equiv \delta_{\nu\alpha\beta\gamma} \frac{\partial x^\alpha}{\partial u^1} \frac{\partial x^\beta}{\partial u^2} \frac{\partial x^\gamma}{\partial u^3}\, d^3u. \tag{A.18}$$

The symbol $\delta_{\nu\alpha\beta\gamma}$ denotes the tensor density of Levi-Civita, defined on p. 60, and the three parameters u^1, u^2, u^3 serve to parametrize the three-dimensional hypersurface that bounds the four-dimensional domain D. The import of the equality (A.18) is to specify that the change in the value of S does not depend on the values of the field variables φ^A in the interior of the domain D, but only on changes on the bounding hypersurface.

If this condition is satisfied, then the left-hand side of Eq. (A.18) can be transformed by an integration by parts, with the result:

$$\int_D d^4x \frac{\delta L}{\delta \varphi^A} \delta\varphi^A + \oint C^\nu \, d^3\Sigma_\nu \equiv 0\,, \qquad C^\nu \equiv \frac{\partial L}{\partial \varphi^A{}_{,\nu}} \delta\varphi^A - Q^\nu. \tag{A.19}$$

The surface integral $\oint C^\nu \, d^3\Sigma_\nu$ will vanish whenever the surface of integration bounds a domain throughout which the field equations (A.17) hold. This is possible only if throughout that domain we have:

$$C^\nu{}_{,\nu} = 0\,. \tag{A.20}$$

If the infinitesimal transformation of variables $\delta\varphi^A$ carries all solutions of the field equations into new solutions, then the new Lagrangian density is the same function of its arguments as the original Lagrangian density; $\delta\varphi^A$ is an invariant infinitesimal transformation. C^ν is called the generating density of that transformation, and the three-dimensional integral $\int C^0 \, d^3x$ the generator (or generating integral). With appropriate boundary conditions at the (spatial) edge of the domain of integration, the generating integral is a constant of the motion, thanks to the equation of continuity, (A.20). Just as in mechanics, Noether's theorem relates infinitesimal invariant transformations to conserved quantities.

For what follows it is important to note that Eq. (A.19) does not determine the generating density C^ν uniquely. Choosing an arbitrary field of six components $E^{\rho\sigma} = -E^{\sigma\rho}$, one can construct an alternative generating density D^ν,

$$D^\nu = C^\nu + E^{\nu\sigma}{}_{,\sigma}\,, \qquad D^\nu{}_{,\nu} \equiv C^\nu{}_{,\nu}\,, \qquad \oint E^{\nu\sigma}{}_{,\sigma} \, d^3\Sigma_\nu \equiv 0\,, \tag{A.21}$$

which will serve as well as the original density C^ν.

If the formulation (A.19), (A.20) of Noether's theorem is to be applied to invariance groups, a distinction must be made between groups whose elements can be identified by a finite number of parameters, which are known as *Lie groups*, and those groups whose elements are identified by arbitrary functions of the four space-time coordinates. The latter are known as *function groups*, or more commonly, in current physicists' usage, as *gauge groups*. The Lorentz and Poincaré groups are examples of Lie groups, whereas the gauge group of electrodynamics and the group of curvilinear coordinate transformations of general relativity are gauge groups.

If a transformation belongs to a Lie group, and hence is determined throughout space-time once a finite number of parameters have been fixed, then that transformation will in general be nonidentity both in the interior and on the boundary of the chosen domain of integration D. If the transformation forms part of a gauge group, i.e., if it involves an arbitrary function, it is possible to make it the identity on the boundary of D though it is nonidentity in the interior. This distinction between Lie groups and function groups has significant implications for the structure of the generators.

Consider once more the identity (A.19), but on the assumption that the invariant transformation belongs to a gauge group, and that it has been chosen so that it represents the identity transformation on the bounding surface, including a sufficient (finite) number of derivatives out of the surface. As the generating density C^ν, or D^ν of Eq. (A.21), depends only on the properties of the transformation on the surface, a possible choice of the generating density is to make it vanish on that surface. Hence the linear combination of field equations implied by the first term of the identity (A.19) will have an identically vanishing volume integral if δy_A and its off-surface derivatives vanish on the bounding surface. Thus the existence of identities between the field equations and the vanishing of the generators of invariant transformations that form function groups are two results having a common root.

Einstein's vacuum field equations will serve to illustrate both. The action integral is given by Eq. (12.43), p. 190, and its variation under arbitrary variations of the metric by Eq. (12.54). In terms of the variation of the covariant metric (which is slightly more convenient for what follows), this expression may be turned into:

$$\delta \int \sqrt{-g}\, R\, d^4\xi \equiv -\int d^4\xi \sqrt{-g}\, G^{\mu\nu} \delta g_{\mu\nu} + \oint Q^\nu\, d^3\Sigma_\nu . \quad (A.22)$$

The last term on the right contains the remainder of integrations by part. If we wish to examine the change in action that is brought about by an infinitesimal coordinate transformation, the change in the metric as a function of the coordinates will be:

$$\delta_f g_{\mu\nu} = -(g_{\mu\rho} f^\rho{}_{,\nu} + g_{\nu\rho} f^\rho{}_{,\mu} + g_{\mu\nu,\rho} f^\rho) \equiv -(f_{\mu;\nu} + f_{\nu;\mu}) . \quad (A.23)$$

The symbol f^ρ stands for the infinitesimal change in the coordinate values of a given world point

$$f^\rho \equiv \delta\xi^\rho . \quad (A.24)$$

$f^\rho(\xi)$ forms an (arbitrary) vector field. The left-hand side of Eq. (A.22) is an integral over a scalar density. Its value will change only because of the (infinitesimal) shift in the boundary of the domain of integration, and this change can be represented as an integral over that bounding surface. With these substitutions made, Eq. (A.22) takes the form:

$$\int d^4\xi\sqrt{-g}\, G^{\mu\nu}(f_{\mu;\nu} + f_{\nu;\mu}) + \oint (Q^\nu + \sqrt{-g}\, Rf^\nu)\, d^3\Sigma_\nu \equiv 0. \qquad (A.25)$$

This identity has now precisely the form (A.19). It remains to establish the precise form of the integrand of the surface integral, that is to say, the generating density. One may trace in detail the form of Q^ν resulting from the several integrations by part. But it is more convenient to start directly from the first term of Eq. (A.25), using the contracted Bianchi identities. We have:

$$\begin{aligned} &\int d^4\xi\sqrt{-g}\, G^{\mu\nu}(f_{\mu;\nu} + f_{\nu;\mu}) \\ &\qquad \equiv 2\int d^4\xi\sqrt{-g}\, G^{\mu\nu}f_{\mu;\nu} \equiv 2\int d^4\xi(\sqrt{-g}\, G^\nu{}_\mu f^\mu)_{;\nu} \qquad (A.26) \\ &\qquad \equiv 2\oint \sqrt{-g}\, G^\nu{}_\mu f^\mu\, d^3\Sigma_\nu . \end{aligned}$$

Hence the generating density of an infinitesimal coordinate transformation is:

$$C^\nu = -2\sqrt{-g}\, G^\nu{}_\mu f^\mu . \qquad (A.27)$$

This vector field does indeed satisfy an equation of continuity when the vacuum field equations are satisfied, in spite of the appearance of the arbitrary functions f^μ. It accomplishes this feat by vanishing altogether, a somewhat disappointing result.

The expression (A.27) as a generating density suffers from the further drawback that it involves second-order derivatives of the field variables, the components of the metric tensor. Both disadvantages can be eliminated by the addition of a term of the type (A.21). However, the resulting generating density D^ν is no longer a vector density, like C^ν. The appropriate choice of $E^{\nu\rho}$ is[5]:

$$E^{\nu\rho} = U^{\nu\rho}{}_\tau f^\tau, \qquad U^{\nu\rho}{}_\tau = \mathfrak{g}_{\tau\beta}(\mathfrak{g}^{\nu\alpha}\mathfrak{g}^{\rho\beta} - \mathfrak{g}^{\nu\beta}\mathfrak{g}^{\rho\alpha})_{,\alpha} . \qquad (A.28)$$

---

[5] Ph. Freud, *Ann. Math.*, **40**, 417 (1939). L. Landau and E. Lifshitz, *The Classical Theory of Fields* (English translation), p. 317 f. Addison-Wesley, Reading, Mass., 1951.

The symbols $\mathfrak{g}^{\mu\nu}$, $\mathfrak{g}_{\mu\nu}$ are short for

$$\mathfrak{g}^{\mu\nu} \equiv \sqrt{-g}\, g^{\mu\nu}, \qquad \mathfrak{g}_{\mu\nu} = (-g)^{-1/2} g_{\mu\nu}. \tag{A.29}$$

The new generating density D^ν, Eq. (*A*.21), is quadratic in first-order derivatives of the metric, free of second-order derivatives, and non-zero. It satisfies an equation of continuity (i.e., its divergence vanishes) wherever the vacuum field equations hold.

The surface integral theorems. For general relativity the generating density D^ν in vacuum has the value

$$D^\nu = (U^{\nu\rho}{}_\tau f^\tau)_{,\rho}, \tag{A.30}$$

in view of the fact that C^ν vanishes. If the vacuum field equations hold in a four-dimensional domain, then the integral of D^ν over the bounding three-dimensional hypersurface must vanish,

$$\oint D^\nu\, d^3\Sigma_\nu = 0. \tag{A.31}$$

It follows in particular that if two three-dimensional domains have their two-dimensional boundaries in common, and if one of them can be carried over into the other by continuous deformation, sweeping out in the process a four-dimensional domain throughout which the vacuum equations are satisfied, then the two three-dimensional integrals must have the same value:

$$\int_{\mathrm{I}} D^\nu\, d^3\Sigma_\nu = \int_{\mathrm{II}} D^\nu\, d^3\Sigma_\nu. \tag{A.32}$$

Hence this value is determined by the common boundary; indeed, in view of Eq. (*A*.30), and applying the generalized Stokes' theorem, we must have:

$$\int D^\nu\, d^3\Sigma_\nu = \tfrac{1}{2} \oint U^{\nu\rho}{}_\tau f^\tau\, d^2\Sigma_{\nu\rho}. \tag{A.33}$$

This relation would be trivial if we were to use for D^ν the expression (*A*.30), but it is not if D^ν stands for (*A*.21), with C^ν given by Eq. (*A*.27) and $E^{\nu\rho}$ by Eq. (*A*.28). The integrand on the right of Eq. (*A*.33) is linear in the first-order derivatives of the metric, whereas the integrand on the left is quadratic in first-order derivatives, and free of second-order derivatives; the relationship holds only if the vacuum field equations are satisfied everywhere on the three-dimensional domain of integration of the left-hand side.

It is to be noted that theorems of the kind (*A*.33) can be obtained for any physical theory whose dynamical laws are derivable from an action principle that is form-invariant with respect to curvilinear coordinate transformations. For instance, if Einstein's action principle is augmented by a Maxwell-Lorentz term, so as to couple the gravitational field to the electromagnetic field, then the "vacuum" field equations will hold away from ponderable and electrically charged matter (but in the presence of electromagnetic as well as gravitational fields). Hence there will be an integral theorem of the general form (A.33), with the integrands on both sides containing additional terms involving the electromagnetic field. On the left, the field intensities will enter quadratically, on the right only linearly, but multiplied by undifferentiated potentials.[6]

Ponderomotive laws. Consider a number of discrete sources giving rise to a gravitational field. That is to say, each of the sources at any given time can be enclosed in a separate sphere-like two-dimensional surface. In space-time the enclosure of each source will be a three-dimensional tube, extending indefinitely in a time-like direction.

One such tube may now be surrounded by a three-dimensional hypersurface that lies everywhere in vacuum. Of course, this

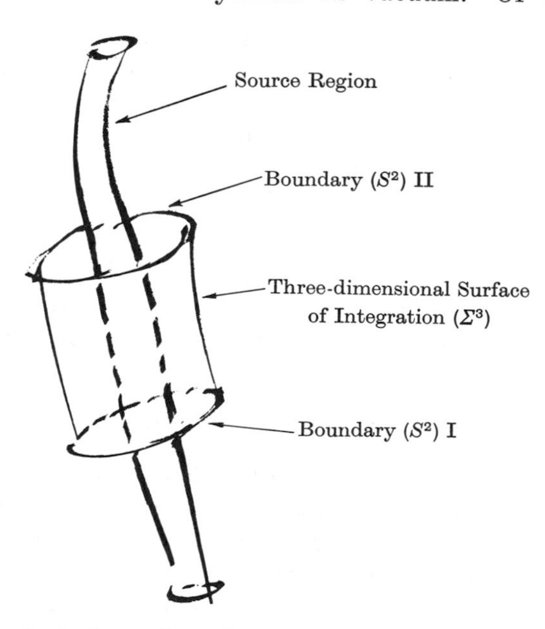

Fig. 10. The slender long tube encloses a source of the field. It is surrounded by a three-dimensional surface that intersects the source region nowhere and which is bounded by two disjoint two-dimensional domains.

---

[6] Expressions involving both electromagnetic fields and the scalar field, that has been conjectured, e.g., by R. H. Dicke, are to be found in P. G. Bergmann, *Intl. J. Theoret. Physics*, **1**, 25 (1968).

hypersurface Σ^3 (Fig. 10) is not the boundary of a four-dimensional domain. On the contrary, it has a two-dimensional boundary of its own, which consists of two disjoint pieces, each of which resembles the surface of an ordinary sphere (S^2 in standard topological notation). The equality ($A.33$) may be properly applied to Σ^3 and its boundary (II − I), as the vacuum conditions hold everywhere on it. It may be written in the form

$$\tfrac{1}{2}\oint_{\mathrm{II}} E^{\nu\rho}\, d^2\Sigma_{\nu\rho} - \tfrac{1}{2}\oint_{\mathrm{I}} E^{\nu\rho}\, d^2\Sigma_{\nu\rho} = \int_{\Sigma^3} D^{\nu}\, d^3\Sigma_{\nu}\,. \qquad (A.34)$$

This is to say, the two-dimensional integrals taken over II and I, respectively (both being surfaces enclosing the same source) differ by an amount that is determined by the total flux passing through that enclosing surface between the two "instants" I and II. Hence an integral taken over one enclosing two-dimensional surface, I or II, measures something that is conserved, in the sense that in ordinary field theories energy density, linear momentum density, or angular momentum density are conserved.

If the two-dimensional surfaces of integration in the relationship ($A.34$) are chosen to be coordinate hypersurfaces, e.g., ξ^0 = constant, and if they are infinitesimally close to each other, then one can infer the differential relation:

$$\frac{\partial}{\partial\xi^0}\oint E^n\, dS_n = \oint D^n\, dS_n\,, \qquad dS_n \equiv \tfrac{1}{2}\delta_{nkl}\, d^2\Sigma^{kl}, \qquad \text{etc.} \qquad (A.35)$$

The "time" derivative of one surface integral equals another surface integral.

Both integral equalities ($A.34$) and ($A.35$) involve fields of the four arbitrary functions f^{τ}. In principle, then, each of these relations is equivalent to an infinity of relations, which differ by the choice of f^{τ}. To interpret them is a challenging task, which has not been entirely completed. The discussion that follows will be concerned principally with the formulation ($A.34$).

Equations ($A.34$) are equivalent to the claim that

$$\int_{\Sigma^3} \sqrt{-g}\, G^{\nu}{}_{\tau} f^{\tau}\, d^3\Sigma_{\nu} = 0\,. \qquad (A.36)$$

This formulation facilitates the transition from one coordinate system to another. Assuming that the domain of integration is fixed independently of the choice of coordinates, the arbitrary functions f^{τ} must be transformed as a contravariant vector field in order to obtain in a new coordinate system the statement equivalent to the original assertion. Of course, only the totality of statements of the form (A.36), with a complete set of fields f^{τ} on the domain of integration, leaving out no linearly independent vector field, exhausts the simple assertion that on the domain Σ^3 the vacuum field equations are satisfied everywhere. With any particular choice of the field f^{τ}, Eq. (A.36) merely states that an integral with a certain weighting of the vacuum field equations vanishes. At any rate, the import of any chosen set of equations (A.36) in one coordinate frame can be matched by an equivalent set of statements in any other coordinate system. The domain of integration must be the same in both frames; it cannot be "dragged along" by the coordinate transformation.

The next natural step in the interpretation of the integral formulas would seem to involve a search for the physical significance of the two-dimensional surface integrals appearing on the left of Eq. (A.34). But their integrand lacks any intrinsic meaning. Superficially, that integrand resembles that of the charge integral in electrodynamics,

$$Q = \frac{1}{4\pi} \oint \mathbf{D} \cdot d\mathbf{S}, \qquad (A.37)$$

in that it involves the first derivatives of the (gravitational) potentials linearly. But whereas the electric displacement $\mathbf{D}$ forms part of a contravariant skew-symmetric tensor density (which moreover is gauge-invariant), so that $\mathbf{D} \cdot d\mathbf{S}$ is a scalar, the expression $E^{\mu\nu}$, Eq. (A.28), behaves like a tensor density only with respect to linear coordinate transformations; these have no special significance in general relativity. Indeed, by a suitable choice of coordinates one can make the components of $E^{\mu\nu}$ vanish at any one world point, simply by reducing the Christoffel symbols (and with them the first derivatives of $g_{\mu\nu}$) there to zero. Of course, if the curvature tensor does not vanish at that point, the second-order derivatives of the metric will be non-zero.

A few surface integrals $\frac{1}{2} \oint E^{\mu\nu} \, d^2\Sigma_{\mu\nu}$ are invariant in the linearized weak-field approximation, which was discussed in Chapter XII, on pp. 180ff. First it is to be remarked that the expression D^{ν}, Eq. (A.21), with C^{ν} substituted from Eq. (A.27) and $E^{\mu\nu}$ from Eq. (A.28), consists of terms in which f^{τ} occurs undifferentiated and terms in which f^{τ} is differentiated once. The first set of terms contains the Christoffel symbols quadratically; the second is linear in the Christoffel symbols. If one restricts oneself to f^{τ} that are constant in the chosen coordinate system, then only the terms of the first kind are non-zero, and these are proportional to the so-called canonical energy-momentum complex originally proposed by Einstein, the expression $\sqrt{-g}\, t^{\mu}{}_{\tau}$ introduced in Eq. (12.68). A three-dimensional integral of the form $\int \sqrt{-g}\, t^{\mu}{}_{\tau}\, d^3\Sigma_{\mu}$ generates an infinitesimal rigid displacement of the coordinate system along the ξ^{τ}-axis.

With the restriction to constant f^{τ} in mind, we now go over to the linearized approximation. In this approximation the energy-momentum complex vanishes (because it is homogeneous-quadratic in the Christoffel symbols). Outside the sources of the gravitational field it equals a divergence, Eq. (A.30), hence the transformation law of $E^{\mu\nu}$ must preserve its divergence-free character. Indeed, in the linearized approximation we find:

$$\begin{aligned} E^{\mu\nu *} &= E^{\mu\nu} + (v^{\mu\nu} f^{\rho} + v^{\nu\rho} f^{\mu} + v^{\rho\mu} f^{\nu})_{,\rho}\,, \\ v^{\mu\nu} &\equiv v^{\mu}{}_{,\sigma}\, \varepsilon^{\sigma\nu} - v^{\nu}{}_{,\sigma}\, \varepsilon^{\sigma\mu}\,, \end{aligned} \tag{A.38}$$

always provided f^{τ} is constant and unchanged by the coordinate transformation. The integral of such an expression taken over a closed two-dimensional domain vanishes identically, on account of Stokes' theorem (in the form appropriate to four-dimensional geometry). Hence the integral of $E^{\mu\nu}$ itself is invariant with respect to transformations of the type (12.21).

To accord this integral some physical significance, we shall return once more to its being equal to a three-dimensional volume integral extended over the volume enclosed by the surface S^2. In Fig. 10 this would be a space-like domain cutting across the tube containing the source, not the time-like surface Σ^3. As this latter integral is the generator of a rigid infinitesimal coordinate displacement, it is tempting to adopt the interpretation common to Newtonian and Lorentz-covariant physical theories, to the effect that the generators of rigid displacements are the total energy and the total linear momentum. This is indeed what is done. The choice $f^{\tau} = \delta^{\tau}{}_{0}$ results in the expression for the (relativistic) mass/energy. To adapt the domain of integration to a slice $\xi^0 =$ constant, one must give the two superscripts μ, ν the values of 0 and n ($n = 1, 2, 3$), respectively. The integrand E^{0n} then becomes:

$$E^{0n} = \gamma_{00,n} - \gamma_{ns,s}\,. \tag{A.39}$$

One can always construct a coordinate frame in which the second term vanishes over the whole domain of integration, and one is left with an integral of the first term, which is the gradient of the Newtonian gravitational potential. This integral represents the gravitating mass in the interior of the surface of integration, S^2. Thus, in spite of the approximation used, this formalism retains the principle of equivalence of gravitating mass and energy characteristic of the general theory of relativity.

These results look encouraging, but they cannot be extended beyond the realm of the linearized approximation. No corresponding invariant surface integrals have been discovered in the full theory, and this is no accident. When nonlinear terms are taken into consideration, the whole gravitational field inside a two-dimensional surface contributes to the energy, and it is well known that this field energy has no invariant meaning.

In the linearized theory translations and Lorentz transformations play a special role; they are the transformations that preserve the form of the zeroth-order metric (also referred to as the background metric). They lose this special significance in the full theory and are submerged in the group of curvilinear transformations. One may reasonably expect that likewise the generators of translations and Lorentz transformations have no outstanding significance among the generators of arbitrary coordinate transformations. In physical situations in which the metric is asymptotically flat at infinity there is the class of coordinate systems adapted to these boundary conditions, and with it the group of coordinate transformations preserving them. Their generators (i.e., total energy, total linear momentum, etc.) have intrinsic meaning, but not their local generating densities. Hence these properties cannot be attributed to individual particles in the presence of additional sources of the gravitational field.

If it is difficult to define the mass of a particle in the presence of an incident gravitational field in a satisfactory manner, this difficulty is compounded when it comes to such parameters as the dipole moment, the angular momentum (i.e., the spin), the mass quadrupole moment, and so forth. But from the analogy to Lorentz-covariant theories of physics we know that all of these parameters affect the motion of the particle. If we include at least terms quadratic in the Christoffel symbols in a weak-field approximation expansion applied to Eq. (A.35), then there will be nontrivial terms on the right-hand side of that dynamical equation. Suppose we expand, at any stage of the weak-field approximation, the Christoffel symbols in powers of r, the (Cartesian) distance from the particle's center, then we shall have both negative and positive powers. Somewhat rhetorically, the terms with negative powers of r might be said to belong to the "self-field" of the particle, those with positive powers to the "incident field." Those products of Christoffel symbols will be independent of the "radius" of the sphere of integration in which one factor will have a positive power n, the other a negative power $-(n + 2)$. For instance, the gradient of the Newtonian potential $-M/r$ will drop off with the second negative power of r, and it will be multiplied by the r^0 term of the ambient field, its value at the location of the particle itself.

In like fashion, the self-field associated with the particle's spin will be multiplied by the first derivative of the incident gravitational field strength, and so forth. All these terms will contribute to the rate of change of the particle's energy or linear momentum on the left-hand side of Eq. (A.35). At the higher-order stages of the weak-field approximation neither the left nor the right of Eq. (A.35) are invariant, or covariant, so no logical inconsistency arises. But there remains a paradox, which continues to beset the foundations of the theory of motion.

In spite of the remaining conceptual difficulties, the theory of motion has been applied with complete success to the so-called post-Newtonian effects in celestial mechanics, and also to the evaluation of the radiative reaction associated with accelerated ponderable bodies. On balance, general relativity has contributed significantly to our understanding of the relationship between field equations and the motions of the field's sources.

APPENDIX B
Supplementary Notes

Page 77

Transformations of the type (5.103), with the coefficients $\gamma^{\kappa}{}_{\iota}$ satisfying the conditions (5.106), are currently referred to as *Poincaré transformations*. If all four constants $x^{*\kappa}{}_0$ vanish, they continue to be called *Lorentz transformations*.

Page 99

A function of the canonical variables that is required to vanish is nowadays usually called a *constraint*. The role of constraints in physical theories has been investigated by P. A. M. Dirac and others. For their occurrence in general relativity cf., e.g., Dirac's paper in *Roy. Soc. London Proc.*, *A* **246**, 333 (1958) and *Phys. Rev.* **114**, 924 (1959).

Page 101

Only if the Lagrangian is homogeneous in the first degree with respect to the derivatives of the configuration variables with respect to the parameter τ or θ will the action integral be form-invariant with respect to the choice of that parameter. Because of Euler's law of homogeneous functions, a Hamiltonian formed routinely in accordance with Eq. (6.40), or its four-dimensional analog, will vanish identically. But as the momenta, being homogeneous in the zeroth degree, cannot all be algebraically independent of each other, the canonical variables will satisfy a constraint (cf. note for p. 99 above), which will serve as a nontrivial Hamiltonian.

Page 133

An examination by R. S. Shankland and co-workers, *Rev. Modern Phys.*, **27**, 167 (1955), indicates that Miller's results may have been caused by inadequate protection of his apparatus against thermal expansion when the laboratory was exposed to direct sunlight.

Page 189

H. Bondi, F. A. E. Pirani, and I. Robinson, *Roy. Soc. London Proc.*, *A* **251**, 519 (1959), have published solutions of Einstein's field equations that are plane waves. Cf. also note for p. 211 below.

Page 203

M. D. Kruskal, *Phys. Rev.*, **119**, 1743 (1960), has demonstrated the regularity of the Schwarzschild metric at the spherical surface $r = 2\kappa m$ by performing a coordinate transformation to an appropriate coordinate system ("Kruskal coordinates"). Only $r = 0$ is a singular region. If a concentration of matter were to retire to within the region $r \leqslant 2\kappa m$, the resulting structure would be a "black hole" or a "white hole." It is widely suspected, but not yet definitely known, that such stellar objects do exist.

Page 211

In recent years the experimental tests of the general theory of relativity described in this chapter have been performed with new techniques, and new tests have been added. The gravitational red shift has been confirmed through laboratory experiments by T. E. Cranshaw and collaborators, *Phys. Rev. Letters*, **4**, 163 (1960), by R. V. Pound and G. A. Rebka, Jr., *Phys. Rev. Letters*, **4**, 337 (1960), and by R. V. Pound and J. L. Snider, *Phys. Rev. Letters*, **13**, 539 (1964).

A review article by I. I. Shapiro, *N.Y. Acad. Sci. Ann.*, **224**, 31 (1973), discusses a number of experiments involving radio-astronomical and interplanetary-radar techniques. Some of these tests are concerned with the deflection of electromagnetic waves near the limb of the sun, others with the time delay in the propagation of electromagnetic signals passing close to the sun.

Finally, mention should be made of the detection of gravitational waves by J. Weber, which however has not been confirmed to date by other experimentalists and is as yet considered controversial. Review articles by V. Trimble and J. Weber, by J. A. Tyson, and the transcript of an "open discussion" at the Sixth Texas Symposium on Relativistic Astrophysics (December, 1972, New York City) are to be found at *N.Y. Acad. Sci. Ann.*, **224**, pp. 93, 74, and 101, respectively, with numerous references to the earlier literature.

Page 272

A different generalization of Kaluza's theory arises if one retains the A-cylindricity (now usually called the *isometry group*), but permits the norm of the A-vector to become a fifteenth field variable (a scalar in four-dimensional language), in addition to the ten gravitational and the four electromagnetic potentials. Aside from early attempts in this direction by A. Einstein and P. G. Bergmann, by P. Jordan, and by Y. R. Thiry, R. H. Dicke and his co-workers made considerable efforts in the 1960's to construct what they have called a scalar-tensor theory of gravitation, and to devise and perform experiments designed to test their theory against Einstein's ("orthodox") general theory of relativity.

Discussions of these theories, from different points of view, will be found in M.-A. Tonnelat, *Les Théories Unitaires de l'Électromagnétisme et de la Gravitation*, Gauthier-Villars, Paris, 1965, and S. Weinberg, *Gravitation and Cosmology: Principles and Applications of the General Theory of Relativity*, Wiley, New York, 1972. Both books contain numerous references to the original literature.

Index

N

O

P

Q

R

S

A CATALOG OF SELECTED
DOVER BOOKS
IN SCIENCE AND MATHEMATICS

NUMERICAL METHODS FOR SCIENTISTS AND ENGINEERS, Richard Hamming. Classic text stresses frequency approach in coverage of algorithms, polynomial approximation, Fourier approximation, exponential approximation, other topics. Revised and enlarged 2nd edition. 721pp. 5⅜ × 8½.
65241-6 Pa. $14.95

THEORETICAL SOLID STATE PHYSICS, Vol. I: Perfect Lattices in Equilibrium; Vol. II: Non-Equilibrium and Disorder, William Jones and Norman H. March. Monumental reference work covers fundamental theory of equilibrium properties of perfect crystalline solids, non-equilibrium properties, defects and disordered systems. Appendices. Problems. Preface. Diagrams. Index. Bibliography. Total of 1,301pp. 5⅜ × 8½. Two volumes. Vol. I 65015-4 Pa. $12.95
Vol. II 65016-2 Pa. $12.95

OPTIMIZATION THEORY WITH APPLICATIONS, Donald A. Pierre. Broad-spectrum approach to important topic. Classical theory of minima and maxima, calculus of variations, simplex technique and linear programming, more. Many problems, examples. 640pp. 5⅜ × 8½. 65205-X Pa. $13.95

THE MODERN THEORY OF SOLIDS, Frederick Seitz. First inexpensive edition of classic work on theory of ionic crystals, free-electron theory of metals and semiconductors, molecular binding, much more. 736pp. 5⅜ × 8½.
65482-6 Pa. $15.95

ESSAYS ON THE THEORY OF NUMBERS, Richard Dedekind. Two classic essays by great German mathematician: on the theory of irrational numbers; and on transfinite numbers and properties of natural numbers. 115pp. 5⅜ × 8½.
21010-3 Pa. $4.95

THE FUNCTIONS OF MATHEMATICAL PHYSICS, Harry Hochstadt. Comprehensive treatment of orthogonal polynomials, hypergeometric functions, Hill's equation, much more. Bibliography. Index. 322pp. 5⅜ × 8½. 65214-9 Pa. $9.95

NUMBER THEORY AND ITS HISTORY, Oystein Ore. Unusually clear, accessible introduction covers counting, properties of numbers, prime numbers, much more. Bibliography. 380pp. 5⅜ × 8½. 65620-9 Pa. $8.95

THE VARIATIONAL PRINCIPLES OF MECHANICS, Cornelius Lanczos. Graduate level coverage of calculus of variations, equations of motion, relativistic mechanics, more. First inexpensive paperbound edition of classic treatise. Index. Bibliography. 418pp. 5⅜ × 8½. 65067-7 Pa. $10.95

MATHEMATICAL TABLES AND FORMULAS, Robert D. Carmichael and Edwin R. Smith. Logarithms, sines, tangents, trig functions, powers, roots, reciprocals, exponential and hyperbolic functions, formulas and theorems. 269pp. 5⅜ × 8½. 60111-0 Pa. $5.95

THEORETICAL PHYSICS, Georg Joos, with Ira M. Freeman. Classic overview covers essential math, mechanics, electromagnetic theory, thermodynamics, quantum mechanics, nuclear physics, other topics. First paperback edition. xxiii + 885pp. 5⅜ × 8½. 65227-0 Pa. $18.95

CATALOG OF DOVER BOOKS

DE RE METALLICA, Georgius Agricola. The famous Hoover translation of greatest treatise on technological chemistry, engineering, geology, mining of early modern times (1556). All 289 original woodcuts. 638pp. 6¾ × 11.
60006-8 Pa. $17.95

SOME THEORY OF SAMPLING, William Edwards Deming. Analysis of the problems, theory and design of sampling techniques for social scientists, industrial managers and others who find statistics increasingly important in their work. 61 tables. 90 figures. xvii + 602pp. 5⅜ × 8½. 64684-X Pa. $15.95

THE VARIOUS AND INGENIOUS MACHINES OF AGOSTINO RAMELLI: A Classic Sixteenth-Century Illustrated Treatise on Technology, Agostino Ramelli. One of the most widely known and copied works on machinery in the 16th century. 194 detailed plates of water pumps, grain mills, cranes, more. 608pp. 9 × 12. (EBE)
25497-6 Clothbd. $34.95

LINEAR PROGRAMMING AND ECONOMIC ANALYSIS, Robert Dorfman, Paul A. Samuelson and Robert M. Solow. First comprehensive treatment of linear programming in standard economic analysis. Game theory, modern welfare economics, Leontief input-output, more. 525pp. 5⅜ × 8½. 65491-5 Pa. $13.95

ELEMENTARY DECISION THEORY, Herman Chernoff and Lincoln E. Moses. Clear introduction to statistics and statistical theory covers data processing, probability and random variables, testing hypotheses, much more. Exercises. 364pp. 5⅜ × 8½. 65218-1 Pa. $9.95

THE COMPLEAT STRATEGYST: Being a Primer on the Theory of Games of Strategy, J.D. Williams. Highly entertaining classic describes, with many illustrated examples, how to select best strategies in conflict situations. Prefaces. Appendices. 268pp. 5⅜ × 8½. 25101-2 Pa. $6.95

MATHEMATICAL METHODS OF OPERATIONS RESEARCH, Thomas L. Saaty. Classic graduate-level text covers historical background, classical methods of forming models, optimization, game theory, probability, queueing theory, much more. Exercises. Bibliography. 448pp. 5⅜ × 8¼. 65703-5 Pa. $12.95

CONSTRUCTIONS AND COMBINATORIAL PROBLEMS IN DESIGN OF EXPERIMENTS, Damaraju Raghavarao. In-depth reference work examines orthogonal Latin squares, incomplete block designs, tactical configuration, partial geometry, much more. Abundant explanations, examples. 416pp. 5⅜ × 8¼.
65685-3 Pa. $10.95

THE ABSOLUTE DIFFERENTIAL CALCULUS (CALCULUS OF TENSORS), Tullio Levi-Civita. Great 20th-century mathematician's classic work on material necessary for mathematical grasp of theory of relativity. 452pp. 5⅜ × 8½.
63401-9 Pa. $9.95

VECTOR AND TENSOR ANALYSIS WITH APPLICATIONS, A.I. Borisenko and I.E. Tarapov. Concise introduction. Worked-out problems, solutions, exercises. 257pp. 5⅜ × 8¼. 63833-2 Pa. $6.95